The Long-Term Perspective of Human Impact on Landscape for Environmental Change and Sustainability

The Long-Term Perspective of Human Impact on Landscape for Environmental Change and Sustainability

Special Issue Editors

Anna Maria Mercuri
Assunta Florenzano

MDPI • Basel • Beijing • Wuhan • Barcelona • Belgrade

MDPI

Special Issue Editors
Anna Maria Mercuri
Università di Modena e Reggio Emilia
Italy

Assunta Florenzano
Università di Modena e Reggio Emilia
Italy

Editorial Office
MDPI
St. Alban-Anlage 66
4052 Basel, Switzerland

This is a reprint of articles from the Special Issue published online in the open access journal *Sustainability* (ISSN 2071-1050) from 2018 to 2019 (available at: https://www.mdpi.com/journal/sustainability/special_issues/Human_Impact_on_Landscape)

For citation purposes, cite each article independently as indicated on the article page online and as indicated below:

LastName, A.A.; LastName, B.B.; LastName, C.C. Article Title. *Journal Name* **Year**, *Article Number*, Page Range.

ISBN 978-3-03921-796-0 (Pbk)
ISBN 978-3-03921-797-7 (PDF)

Contents

About the Special Issue Editors

Anna Maria Mercuri, Associate Professor—AMM is a biologist and professor of botany and palynology at the University of Modena and Reggio Emilia, Italy. She is the coordinator of the Group of Palynology and Palaeobotany of the Italian Society of Botany. Since the early 1990s, she has studied environmental transformations and archaeobotany in the Sahara. Her research interests are in long-term human impact and global change in the Mediterranean and North Africa, from the Late Glacial to recent times. She directed European and national projects on past/present vegetation and cultural landscape development, resource exploitation, and integration between palaeoecology and archaeology.

Assunta Florenzano, Adjunct Professor—AF received a BSc degree in Sciences of Cultural Heritage, an MSc degree in Sciences of Restoration and Conservation of Archaeological Heritage, and a Ph.D. degree in Earth System Sciences from the University of Modena and Reggio Emilia (Italy). In the same university, she teaches Botany to students of Natural Sciences. Her research interests are quaternary palynology and paleoecology, pollen analysis from archaeological deposits with an emphasis on non-pollen palynomorphs (e.g., coprophilous fungi as indicators of pastoral activities), anthropogenic impact on vegetation, and archaeopalynological evidence of cultural landscapes in the Mediterranean basin.

sustainability

MDPI

Editorial

The Long-Term Perspective of Human Impact on Landscape for Environmental Change (LoTEC) and Sustainability: From Botany to the Interdisciplinary Approach

Anna Maria Mercuri * and Assunta Florenzano

Laboratorio di Palinologia e Paleobotanica, Dipartimento Scienze della Vita, Università degli Studi di Modena e Reggio Emilia, 41100 Modena, Italy; assunta.florenzano@unimore.it
* Correspondence: annamaria.mercuri@unimore.it; Tel.: +39-059-205-8275

Received: 8 January 2019; Accepted: 9 January 2019; Published: 15 January 2019

This is not the first time the Earth has to experience dramatic environmental and climate changes but this seems to be the first time that a living species—humanity—is able to understand that great changes are taking place rapidly and that probably natural and anthropogenic forces are involved in the process that is under way. Interdisciplinary research is central to successfully promote sustainable development. The understanding of the Long-Term Environmental Changes—LoTEC—is urgent to facilitate sustainable development and it is based on the knowledge and description of environments at subsequent phases and degrees of human impact. Within the domains of Sustainability [1], this is a special matter for environmental and ecological studies. Many different perspectives are involved which study botany and have the goal of improving sustainability through the understanding of the role of people in shaping current landforms and landscapes, the importance of social and policy choices in the loss of species and equilibrium of old or new ecosystems, the correct application of scientific discoveries for renewable agriculture (e.g., [2]) or biodiversity conservation [3,4], new ideas for education and training of new generations working on/with the environment [5,6]. Not only crop, garden and decoration plant species but also wild plants are recognized as having an invaluable importance for new proposals for future agriculture [7] and additional ecosystem and cultural services exploitation [8]. Human culture is largely based on plants, which have always been at the center of past and present food, medicine, shelter, fuel, dress, feed, forage, and art, belief, feasts and traditions. Botanical studies have the largest potentiality to study the LoTEC with details on the different chronological phases, and regional features. Several studies have been, and are, carried out through paleoecology [9], geoarchaeology [10], agroecology [11], ethnobotany [12], archaeology [13], and demography [14]. Palynology is among the best tools to study high-resolution sequences formed under natural and anthropic (cultural) forces [15,16], and LoTEC can be studied by characterizing human action on tree crops and synanthropic wild plants, which prefer to grow in rural and urban environments (e.g., in the Mediterranean area: references [17–19]). Many studies outline the relationships between the rise and fall of past cultures [20–22] connecting the environmental changes with potential crisis of past societies (Terramare in N Italy: [23]; Negev in Israel: [24]; Beyşehir Occupation phase in southwest Turkey [25]; Khabur collapse in north-eastern Syria [26]).

Sustainable agriculture and land management are among the key themes that must be deepened in a long-term perspective. Palaeoenvironmental research has demonstrated that people have adopted a diffused pattern of land use involving a combination of diverse activities, using trees-crops-domesticate animals with a combination of wood exploitation, field cultivation and animal breeding (e.g., since the Neolithic in the Mediterranean area). The multifunctional land-use, adopted for millennia [27], seems to be the best way to develop our economy in the future. Moreover, the knowledge of the past is the most reliable basis to identify the vocation of different lands for future developments.

Archaeobotany can give fundamental information about the presence of people and various land uses along past millennia, in prehistoric and historic centuries [28–32]. For several reasons, however, the integration between results from past and present contexts is negligible, and it is more common to find papers written by archaeobotanists to botanists [33], by historians to palaeoecologists [34], and by palaeoecologists to ecologists [35], rather than by ecologists towards palaeoecologists [36].

This Special Issue seeks to engage an interdisciplinary dialogue on the dynamic interactions between nature and society, focusing on long-term environmental data as an essential tool to better-informed landscape management decisions that develop equilibrium between conservation and sustainable resources exploitation. In particular, this SI intends to provide papers from very different approaches studying the long term human impacts, possibly among the rare cases of diverse expertise and consideration on what 'human impact' means and what 'long-term' means (Figure 1).

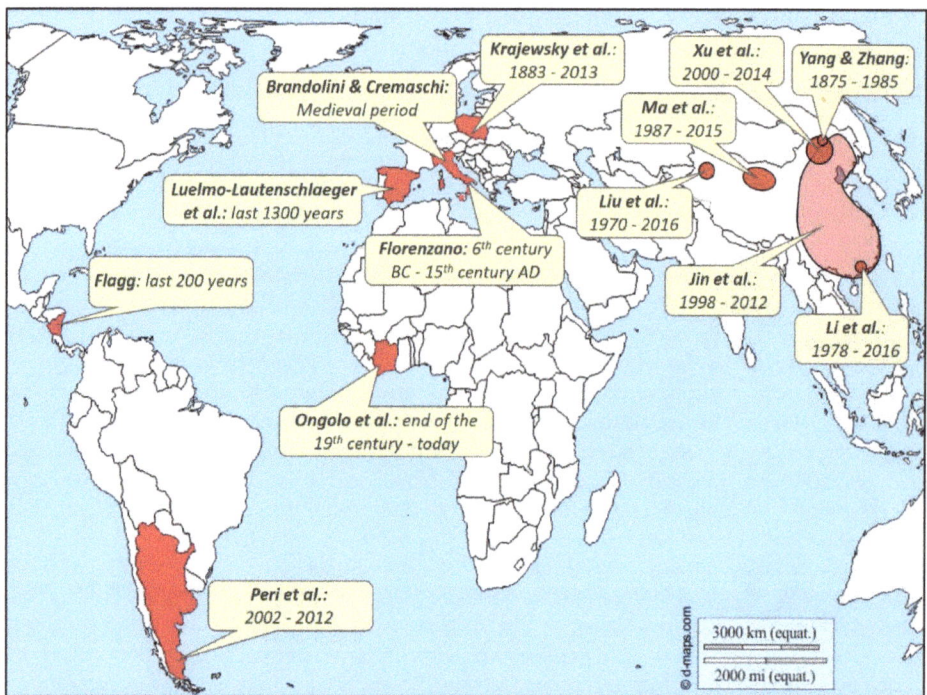

Figure 1. Location map of the study areas (in red) of the 13 research articles included in this Special Issue. For each country, a label showing the first Author names and the time frame of the research has been added (map modified from d-maps.com: free maps).

The Special Issue guest editors are archaeobotanists/palynologists and their 'long term' is on the range of millennia. Three papers—by Luelmo-Lautenschlaeger and colleagues, by Florenzano, and by Brandolini and Cremaschi—introduce the palaeoecological perspective focusing on the last millennia of southern Europe. Luelmo-Lautenschlaeger et al. present a palynological study of the 'mid-mountains' ecosystem dynamics driven by the climate and human actions in the Toledo Mountains, central Spain. This research, based on the multi-proxy study of the Bermú palaeoenvironmental record, is essential to understanding the long-term complex interactions between environment and highland people that led to the current high-value cultural landscape in that area. The same approach has been employed by Florenzano to detail past land-uses and pastoral activities in Basilicata, a region of southern Italy where animal breeding and pastoralism have a long tradition. Her study is intended to improve the awareness about biodiversity and human impact shaping the modern landscape of the region, and the Mediterranean landscapes. Brandolini and Cremaschi analyze the fluvial landscape development in the Central Po Plain (North Italy) by integrating geomorphological, archaeological and historical data. The analysis of the relationships between fluvial palaeoenvironments and past human activities is crucial for future sustainable management of these complex systems in order to maintain both their evolutionary processes and the fragile dynamic equilibrium of floodplains.

Besides this palaeoecological/palaeoenvironmental research, several studies on most recent landscape transformations are part of the collection of papers in the Special Issue. They offer a broad sweep across many of the current environmental issues by following different approaches and methodologies.

Krajewsky et al. detect the driving forces behind landscape changes of Ślęża Landscape Park, Poland. Their preliminary study on natural and socio-economic factors of changes introduces a new tool for assessing the level of historical landscape transformation (Landscape Change Index—LCI) and provides the first insights into the forest area transformation during the last 140 years.

The research by Jin et al. focuses on the study of the main factors (spatio-temporal variations of climate factors and urbanization impact) which govern the urban ecosystem, and explore their correlations. The authors utilize data from 71 large cities of eastern China using a 15-year window, estimating the impact of urbanization on vegetation change in urban and suburban areas. The main outcomes indicate that vegetation cover in cities has been negatively impacted by urbanization, whose effects were more severe in urban areas than in suburban spaces.

Yang and Zhang present an historical reconstruction of arable land based on the distribution of settlements and the farming radius in Zhenlai County, northeast China. This research contributes to the currently available arable land reconstruction studies applying a methodological approach combining the digitized land cover/use data from topographic maps, and the settlement information obtained from toponymy or gazetteer records within the research region. This "cultivation-settlement ratio-based" method is useful to reconstruct historical changes in arable land which could reflect human activities on environmental changes.

The paper by Ma et al. deals with an evaluation model for land-use intensity and its temporal-spatial variation in Shule River Basin, Gansu Province of central China. Land-use intensity change can be a useful indicator of the human impact on land-cover and its variation has important influence on biodiversity and ecosystem service function. These authors evaluate both the comprehensive land-use and the use intensity of a single type of land (artificial and semi artificial land-use) over a period of about 30 years and in doing so reveal some changes in the land-use structure in the study region.

Xu et al. estimate the human-induced vegetation change in arid and semi-arid ecosystems, which are extremely sensitive to external interference (especially over-grazing, over-farming and deforestation) and therefore need sustainable land use policies and programs. The research focuses on the discrimination of significant human-induced changes from 2000 to 2014 in two study areas with different levels of human influence in the Horqin Sandy Land, Northeast China. Climate and human actions have been responsible for the vegetation decrease in both regions, and human factors driving the changes (management of grassland, over-grazing, farming) varied in the two study areas.

Grazing pressure and land management are also recognized as the main factors influencing rangeland ecosystem functioning. Soil Organic Carbon (SOC) content in rangeland is essential for the capacity of the land to sustain plant and animal productivity, and mainly depends on environmental factors and land-use types. In their study, Peri et al. use climatic, topographic and vegetation variables to develop a model of SOC stocks in the Santa Cruz province (South Patagonia), useful tool to assess the sustainability of land management at local scale.

Liu et al. evaluate the expansion dynamics of agricultural oasis and its impact on oasis landscape patterns in Qira oasis, in the southern margin of Tarim basin, Northwest China, by using multi-source satellite images. The understanding of relations between agricultural oasis change and its landscape pattern is crucial to ensure that agricultural oasis sustainability persists in this arid region.

Multi-source remote sensing images are also used by Li et al. to monitor the evolution of dike-ponds in Shunde District of Southeast China. The authors examine both the spatial evolution and dynamics of change of dike-ponds to gain insights on the reduction/conversion of inland freshwater ecosystem due to the nowadays invasion of other land-use types.

The last two papers deal with the understanding of contemporary environmental issues through an analysis of long-term, historical, social, environmental, and political data. Flagg presents the case study on Costa Rica's 2007 mitigation pledge within a nearly 200-year long history of actions undertaken by the Costa Rica state. The human-environment interactions are investigated through an analysis of interview data, archival research, and secondary data. Ongolo et al. examine the sustainability of forest ecosystems in Côte d'Ivoire through an historic contextualisation of forestland use policies in Sub-Saharan Africa. The authors use a twofold approach that combines analysis of the qualitative data (from scientific literature and through historiographic revision of landscape transformations) together with a descriptive statistical analysis of the quantitative data (e.g., from scientific empirical research and FAO data). Both articles highlight the importance of using long-term historical data to understand environmental issues, and stress the prominent role of political elites and policy-makers in the present-day biodiversity and sustainability.

The research studies included in this Special Issue point to the fundamental contribution of the knowledge of past environmental history to conscious and efficient environment conservation and management. Therefore, the long-term perspective of the dynamics which govern the human-climate ecosystem is becoming one of the main focuses and a paramount interest in biological and earth system sciences. Modern biodiversity is the result of the long-term shaping that humans and climate made on vegetation, soils and landforms. Climate change and human impact are predicted to become significant risks to global biodiversity. Multidisciplinary bio-geo-archaeo investigations on the underlying processes of human impact on landscape are crucial to allow us to envisage possible future scenarios of biosphere responses to global warming and biodiversity losses.

Author Contributions: A.M.M. and A.F. contributed equally to this Editorial. A.M.M. and A.F. co-edited the Special Issue on "The Long-Term Perspective of Human Impact on Landscape For Environmental Change and Sustainability" (https://www.mdpi.com/journal/sustainability/special_issues/Human_Impact_on_Landscape).

Acknowledgments: We are very grateful to the different authors for their valuable contribution to this special issue on LoTEC and Sustainability. We warmly thank all the reviewers who dedicated time and assisted in improving the manuscripts.

Conflicts of Interest: The authors declare no conflict of interest.

References

1. Passet, R. *L'Économique et le vivant*; Payot: Paris, France, 1979; ISBN 9782717831047.
2. Beckford, C.L. Sustainable agriculture and innovation adoption in a tropical small-scale food production system: The case of Yam Minisetts in Jamaica. *Sustainability* **2009**, *1*, 81–96. [CrossRef]
3. Rogstad, S.H.; Pelikan, S. Plant species restoration: effects of different founding patterns on sustaining future population size and genetic diversity. *Sustainability* **2013**, *5*, 1304–1316. [CrossRef]
4. Fernández-Llamazares, Á.; Belmonte, J.; Boada, M.; Fraixedas, S. Airborne pollen records and their potential applications to the conservation of biodiversity. *Aerobiologia* **2014**, *30*, 111–122. [CrossRef]
5. Galaz, C.M.; Weil, C.G. University teachers' conceptions about science and science learning, and how they address the promotion of scientific skills in biology teacher-training. *Enseñanza de las Ciencias* **2014**, *32*, 51–81. [CrossRef]
6. Acevedo-Díaz, J.A.; García-Carmona, A. "Something old, something new, something borrowed". Trends on the nature of science in science education. *Revista Eureka sobre Enseñanza y Divulgación de las Ciencias* **2016**, *12*, 3–19. [CrossRef]

7. Mercuri, A.M.; Fornaciari, R.; Gallinaro, M.; Vanin, S.; di Lernia, S. Plant behaviour from human imprints and the cultivation of wild cereals in Holocene Sahara. *Nature Plants* **2018**, *4*, 71–81. [CrossRef] [PubMed]

8. Khoury, C.K.; Amariles, D.; Soto, J.S.; Diaz, M.V.; Sotelo, S.; Sosa, C.C.; Ramírez-Villegas, J.; Achicanoy, H.A.; Velásquez-Tibatá, J.; Guarino, L.; et al. Comprehensiveness of conservation of useful wild plants: An operational indicator for biodiversity and sustainable development targets. *Ecol. Indic.* **2019**, *98*, 420–429. [CrossRef]

9. Davies, A.L.; Bunting, M.J. Applications of paleoecology in conservation. *Open Ecol. J.* **2010**, *3*, 54–67. [CrossRef]

10. Cremaschi, M.; Zerboni, A.; Mercuri, A.M.; Olmi, L.; Biagetti, S.; di Lernia, S. Takarkori rock shelter (SW Libya): An archive of Holocene climate and environmental changes in the central Sahara. *Quat. Sci. Rev.* **2014**, *101*, 36–60. [CrossRef]

11. Wezel, A.; Bellon, S. Mapping Agroecology in Europe. New Developments and Applications. *Sustainability* **2018**, *10*, 2751. [CrossRef]

12. Sheng-Ji, P. Ethnobotanical approaches of traditional medicine studies: Some experiences from Asia. *Pharm. Biol.* **2001**, *39*, 74–79. [CrossRef] [PubMed]

13. Day, J. Botany meets archaeology: People and plants in the past. *J. Exp. Bot.* **2013**, *64*, 5805–5816. [CrossRef] [PubMed]

14. Dean, J.S.; Euler, R.C.; Gumerman, G.J.; Plog, F.; Hevly, R.H.; Karlstrom, T.N.V. Human Behavior, Demography, and Paleoenvironment on the Colorado Plateaus. *Am. Antiq.* **1985**, *50*, 537–554. [CrossRef]

15. Mercuri, A.M. Genesis and evolution of the cultural landscape in central Mediterranean: The 'where, when and how' through the palynological approach. *Landscape Ecol.* **2014**, *29*, 1799–1810. [CrossRef]

16. Edwards, K.J.; Fyfe, R.M.; Jackson, S.T. The first 100 years of pollen analysis. *Nature Plants* **2017**, *3*, 17001. [CrossRef]

17. Kouli, K.; Masi, A.; Mercuri, A.M.; Florenzano, A.; Sadori, L. Regional Vegetation Histories: An Overview of the Pollen Evidence from the Central Mediterranean. *Late Antiq. Archaeol.* **2018**, *11*, 69–82. [CrossRef]

18. Mercuri, A.M.; Bandini Mazzanti, M.; Florenzano, A.; Montecchi, M.C.; Rattighieri, E.; Torri, P. Anthropogenic Pollen Indicators (API) from archaeological sites as local evidence of human-induced environments in the Italian peninsula. *Ann. Bot.* **2013**, *3*, 143–153. [CrossRef]

19. Mercuri, A.M.; Bandini Mazzanti, M.; Florenzano, A.; Montecchi, M.C.; Rattighieri, E. *Olea*, *Juglans* and *Castanea*: The OJC group as pollen evidence of the development of human-induced environments in the Italian peninsula. *Quatern. Int.* **2013**, *303*, 24–42. [CrossRef]

20. Mercuri, A.M.; Sadori, L.; Uzquiano, P. Mediterranean and North-African cultural adaptations to mid-Holocene environmental and climatic changes. *Holocene* **2011**, *21*, 189–206. [CrossRef]

21. Zanchetta, G.; Bini, M.; Di Vito, M.A.; Sulpizio, R.; Sadori, L. Tephrostratigraphy of paleoclimatic archives in central Mediterranean during the Bronze Age. *Quatern. Int.* **2018**. [CrossRef]

22. Storozum, M.J.; Zhen, Q.; Xiaolin, R.; Haiming, L.; Yifu, C.; Kui, F.; Haiwang, L. The collapse of the North Song dynasty and the AD 1048–1128 Yellow River floods: Geoarchaeological evidence from northern Henan Province, China. *Holocene* **2018**, *28*, 1759–1770. [CrossRef]

23. Cremaschi, M.; Mercuri, A.M.; Torri, P.; Florenzano, A.; Pizzi, C.; Marchesini, M.; Zerboni, A. Climate change versus land management in the Po Plain (Northern Italy) during the Bronze Age: New insights from the VP/VG sequence of the Terramara Santa Rosa di Poviglio. *Quat. Sci. Rev.* **2016**, *136*, 153–172. [CrossRef]

24. Stavi, I.; Rozenberg, T.; Al-Ashhab, A.; Argaman, E.; Groner, E. Failure and collapse of ancient agricultural stone terraces: On-Site effects on soil and vegetation. *Water* **2018**, *10*, 1400. [CrossRef]

25. Roberts, N.; Esatwood, W.J.; Lamb, H.; Tibby, J.C. The age and causes of mid-late Holocene environmental change in Southwestern Turkey. In *Third Millennium BC Climate Change and Old World Collapse (NATO ASI Series I: Global Environmental Change)*; Nüzhet Dalles, H., Kukla, G., Weiss, H., Eds.; Springer: Berlin, Germany, 1997; Volume 49, pp. 409–429. ISBN 9783642644764.

26. Koliński, R. On grain, bones, and Khabur collapse. In *Folia Praehistorica Posnaniensia*; Instytut Praehistorii, Uam Poznań: Poznań, Poland, 2011; Volume XVI, pp. 201–217. ISBN 9788323224358.

27. Mercuri, A.M.; Florenzano, A.; Burjachs, F.; Giardini, M.; Kouli, K.; Masi, A.; Picornell-Gelabert, L.; Revelles, J.; Sadori, L.; Servera-Vives, G.; et al. From influence to impact: The multifunctional land-use in Mediterranean prehistory emerging from palynology of archaeological sites (8.0–2.8 ka BP). *Holocene* **2019**, in press.

28. Willcox, G. Evidence for plant exploitation and vegetation history from three Early Neolithic pre-pottery sites on the Euphrates (Syria). *Veget. Hist. Archaeobot.* **1996**, *5*, 143–152. [CrossRef]
29. Mercuri, A.M.; Sadori, L.; Blasi, C. Archaeobotany for cultural landscape and human impact reconstructions. *Plant Biosyst.* **2010**, *144*, 860–864. [CrossRef]
30. Miller, N.F. An archaeobotanical perspective on environment, plant use, agriculture, and interregional contact in South and Western Iran. *IJAS* **2011**, *1*, 1–8. [CrossRef]
31. Hunt, C.O.; Rabett, R.J. Holocene landscape intervention and plant food production strategies in island and mainland Southeast Asia. *J. Archaeol. Sci.* **2014**, *51*, 22–33. [CrossRef]
32. Pérez-Jordà, G.; Peña-Chocarro, L.; Picornell-Gelabert, L.; Carrión Marco, Y. Agriculture between the third and first millennium BC in the Balearic Islands: The archaeobotanical data. *Veget. Hist. Archaeobot.* **2018**, *27*, 253–265. [CrossRef]
33. Mercuri, A.M.; Marignani, M.; Sadori, L. 2013 Palynology: The bridge between paleoecology and ecology for the understanding of human-induced global changes in the Mediterranean Area. *Ann. Bot.* **2013**, *3*, 107–113. [CrossRef]
34. Izdebski, A.; Holmgren, K.; Weiberg, E.; Stocker, S.R.; Büntgen, U.; Florenzano, A.; Gogou, A.; Leroy, S.A.; Luterbacher, J.; Martrat, B.; et al. Realising consilience: How better communication between archaeologists, historians and natural scientists can transform the study of past climate change in the Mediterranean. *Quat. Sci. Rev.* **2016**, *136*, 5–22. [CrossRef]
35. Hjelle, K.L.; Kaland, S.; Kvamme, M.; Lødøen, T.K.; Natlandsmyr, B. Ecology and long-term land-use, paleoecology and archaeology—The usefulness of interdisciplinary studies for knowledge-based conservation and management of cultural landscapes. *Int. J. Biodivers. Sci. Ecosyst. Serv. Manag.* **2012**, *8*, 321–337. [CrossRef]
36. Marignani, M.; Chiarucci, A.; Sadori, L.; Mercuri, A.M. Natural and human impact in Mediterranean landscapes: An intriguing puzzle or only a question of time? *Plant Biosyst.* **2017**, *151*, 900–905. [CrossRef]

sustainability

MDPI

Article

Vegetation History in the Toledo Mountains (Central Iberia): Human Impact during the Last 1300 Years

Reyes Luelmo-Lautenschlaeger [1,2,*], Sebastián Pérez-Díaz [1], Francisca Alba-Sánchez [3], Daniel Abel-Schaad [3] and José Antonio López-Sáez [1]

[1] G.I. Arqueobiología, Instituto de Historia (CCHS), CSIC, Albasanz 26-28, 28037 Madrid, Spain; sebas.perezdiaz@gmail.com (S.P.-D.); joseantonio.lopez@cchs.csic.es (J.A.L.-S.)

[2] Departamento de Geografía, Facultad de Filosofía y Letras, Universidad Autónoma de Madrid, 28049 Madrid, Spain

[3] Departamento de Botánica, Facultad de Ciencias, Universidad de Granada, 18071 Granada, Spain; falba@ugr.es (F.A.-S.); dabel222@hotmail.com (D.A.-S.)

* Correspondence: reyes.luelmo@gmail.com

Received: 22 June 2018; Accepted: 18 July 2018; Published: 23 July 2018

Abstract: Mid-mountain ecosystems provide a broad diversity of resources, heterogeneous relief, and a mild climate, which are all very useful for human necessities. These features enable different strategies such as the terracing of the slopes as well as wide crop diversification. Their relations lead to a parallel co-evolution between the environment and human societies, where fire and grazing become the most effective landscape management tools. This paper presents the results obtained from a multi-proxy study of the Bermú paleoenvironmental record, which is a minerotrophic mire located in the Quintos de Mora National Hunting Reserve (Toledo Mountains, central Spain). The bottom of this core has been dated in the Islamic period (ca. 711–1100 cal AD), and the study shows how the landscape that was built over time in the Toledo Mountains up to the present day is narrowly linked to human development. This study shows the increasing human pressure on the landscape, as well as the subsequent strategies followed by the plant and human communities as they faced diverse environmental changes. Thus, it is possible to attest the main role played by the humans in the Toledo Mountains, not only as a simple user, but also as a builder of their own reflexion in the environment.

Keywords: mid-mountains; paleoecology; Late Holocene; central Spain

---

1. Introduction

Mountain ecosystems are usually quite isolated spaces that are difficult to access, with particular features and dynamics according to their own nature, topography, and climatic conditions. Despite this, they show a high biodiversity and attractive wide resource richness that meet human requirements [1–4]. This interaction unleashes continuous changes in mountain landscapes, which generate adaptive strategies, leading to a parallel co-evolution with human societies [5].

This shared evolution has been usually determined by human demography. This has varied depending on climate change, natural disturbances, and human history. Therefore, it has not been a linear, continuous, or uniform process, but it has suffered shrinking and regressions. Human activities began to draw an impact in the environment in the Western Mediterranean Region ca. 5500 cal BC. The sharpest influence ever seen was reached during the "industrial period" [3,6,7]. Since the beginning, fire has been the most effective management tool, and became essential in human history since it has had unequal effects on different forests depending on their nature, height, or age [8,9]. It generated open areas that were used for human habitat, crop cultivation, or grazing. These activities had deep consequences on the forest cover, such as deforestation or soil erosion. Humans have used woodlands for survival, getting building materials and fuel (firewood, charcoal, stubble . . .), but also for hunting or apiculture. Along with climatic events over centuries, they have drawn a high-value cultural landscape [4,10,11].

Paleoecological research studies, along with modern pollen rain studies [12–15] are one of the best tools to understand this complex process. Quite a few paleoecological studies have been carried out in the high-mountain spaces of the Iberian Peninsula, from the Pyrenees (e.g., [16]) to the Cantabrian mountains (e.g., [17,18]) including the Spanish Central System (e.g., [19,20]) among others, composing a very complete knowledge about these mountain ecosystems. However, despite many studies carried out in some European mid-mountain ranges (e.g., [21,22]), these ecosystems have been less studied, as Sancho-Reinoso [23] recognized. Although the term "mid-mountain" has been widely used, the characterization of this environment has been always been built in contrast with lowlands and high-mountain spaces [24]. The main criteria that has been used to define these spaces has been the kind of relief and the altitude, forgetting about climatic, biological, and land-use features [24,25]. Mid-mountain ecosystems show a high diversity of resources, based on a heterogeneous relief and a mild climate that allows a number of strategies such as terracing the slopes or a huge crop variety. Therefore, it's so easy to find human traces: not only deforestation and soil erosion, but also activities made over the slopes with a terrace−based systemand other elements performed in order to reduce the effort and make it easier to work in a mean environment [23,25,26].

This dynamic is perfectly traceable in the Toledo Mountains, which is one of the best spots for these kinds of studies. Besides, they are nearly crossed by many main cattle roads such as the Cañada Real Leonesa and the Cañada Real Segoviana, as well as other important paths that are now placed in the current road network [27,28]. In spite of this interesting position and the rich anthropogenic evidence in this area, paleoecological studies are scarce and unlinked in this area, although they highlight the diversity, richness, and hazards present in the territory [29–31]. The current vegetation shows important ecological, natural, socio-economical, and cultural unarguable values that make the Toledo Mountains' landscape a complex and changing reality that needs understanding and protection in order to assure its identity.

According to the above, this work presents the results obtained from the study of Bermú mire and compares them with other pollen records in the Toledo Mountains. The main aim of this study is to understand the landscape evolution and the role played by humans and climatic processes during the late Holocene. This study delves into the vegetation dynamics and other natural processes that happened in this area, as well as the relationship between human and plant communities focusing on the important natural and cultural heritage currently present there.

2. Study Area

Bermú mire (783 m a.s.l., 39°26′2.71″ N, 4°8′45.67″ W) (Figure 1) is located in the southern slope of the Torneros range (Los Yébenes, Toledo, Spain), inside Los Quintos de Mora. Bermú is a minerotrophic mire placed in a property hunting reserve (since 1942), and a privileged witness of the consequences of anthropic management. In addition, the mire lies at the oriental border of the Toledo Mountains, so it presents an intense Mediterranean influence.

The Toledo Mountains are located in the center of Iberian Peninsula, on the southern plateau, separating the Tagus and Guadiana basins. Geologically, this area is part of the western half of the Iberian Hercynian Massif, which was formed during the Ordovician period, and whose main reliefs were originated during the Hesperian Orogeny and later uplifted by the Alpine Orogeny [32]. The landscape is of an Appalachian type, with low ranges (Rocigalgo is the highest massif with 1448 m a.s.l.; medium altitude, 800–1100 m a.s.l.) and alluvial valleys. The space between these mountainous reliefs, the alluvial layer called "rañas", is where the less resistant materials such as slates or quartzite sandstones were deposited. Soils are acid and not very fertile [32,33]. The study site is located close to one of those alluvial layers (raña) in the Torneros range. The climate of Bermú mire typically has dry and warm summers, and an oceanic influence on the western side of the massif. The average temperature is 17 °C, and the annual precipitation range is 600 mm to 800 mm [31,33].

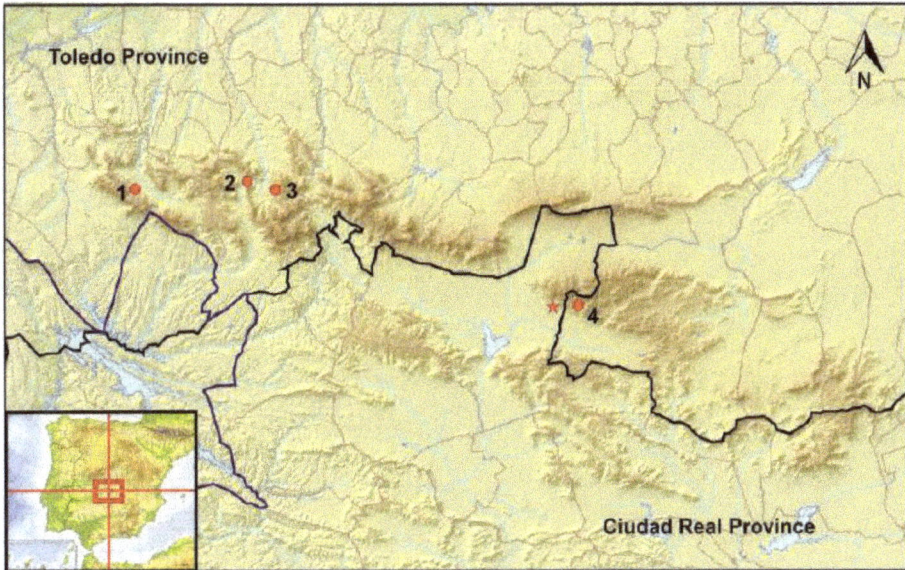

Figure 1. Map of Bermú mire location in the Toledo Mountains (red star) and other mires mentioned in the text (red points): 1. Las Lanchas; 2. Patateros; 3. La Botija; 4. Valdeyernos.

The current vegetation is mainly composed by holm oak (*Quercus ilex*) and cork oak (*Q. suber*) forests in the meso-Mediterranean foothills. The first forest type includes mesothermophilous taxa such as *Arbutus unedo*, *Phillyrea angustifolia*, *Pistacia terebinthus*, and *Pyrus bourgaeana*. The second ones are associated with deciduous trees such as *Quercus faginea* subsp. *broteroi*, *Q. pyrenaica*, and *Acer monspessulanum*. The supra-Mediterranean belt (>900 m a.s.l.) is composed of deciduous oak forests of *Quercus pyrenaica* [31,33,34]. We can also find some relict species in this area such as *Corylus avellana*, *Betula pendula*, or *B. pubescens* [35–37]. In the mire surroundings, we can find *Erica tetralix*, *Dactylorhiza elata* subsp. *sesquipedalis*, *Drosera rotundifolia*, *Lobelia urens*, and *Sphagnum capillifolium*.

3. Material and Methods

3.1. Sampling and Chronology

A 72-cm deep core was collected using a Russian peat corer (50-cm long and 5 cm in diameter). Peat sections were placed in plastic tubes, protected in plastic guttering, and stored under cold conditions (4 °C) prior to laboratory sub-sampling and analysis. The core was sectioned into continuous 1-cm thick portions. The core stratigraphy consists of five parts that are mainly composed of peat and little gravels (Figure 2).

Eight bulk organic sediment samples were ^{14}C dated using the Accelerator Mass Spectrometry technique (onwards AMS). The AMS dating was conducted at Centro Nacional de Aceleradores (Seville, Spain) and the Poznań Radiocarbon Laboratory (Poznán, Poland). The dates were calibrated using CALIB 7.1 with the IntCal13 curve [38], except for the most recent sample, which was calibrated using the CALIBomb program with the calibration dataset NH zone 1 [39] (Table 1). An age–depth model (Figure 2) was produced using Clam 2.2 software [40]. The best fit was obtained applying a smoothing spline to the available radiocarbon dates. Confidence intervals of the calibrations and the age–depth model were calculated at 95% (2σ) with 1000 iterations. According to the model, the base of the 72-cm core corresponds to the Islamic period (711 cal AD), and the sequence extends up to the present day without any recorded sedimentary hiatus.

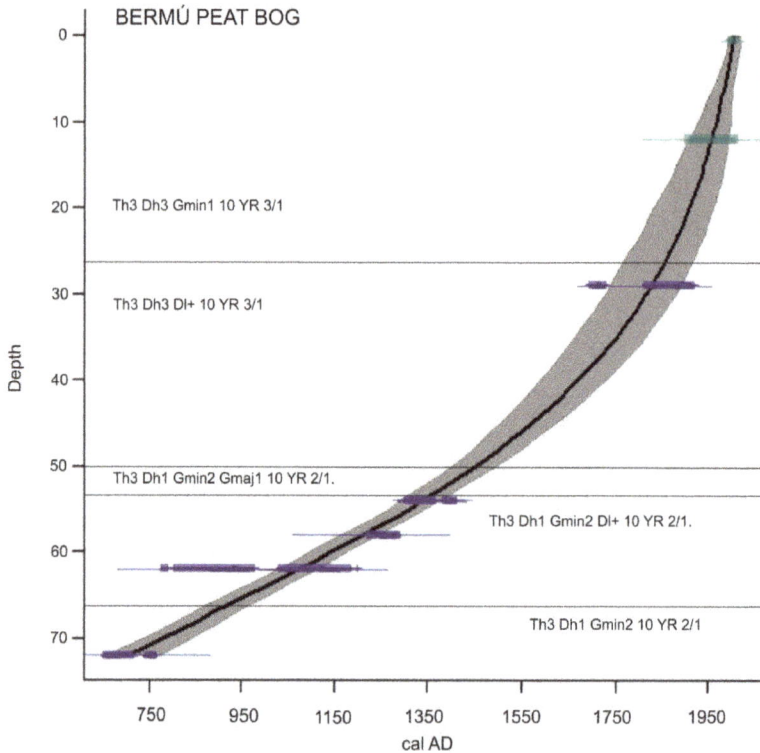

Figure 2. Age–depth model of Bermú mire. Lithostratigraphic description according to Aaby and Berglund [41].

Table 1. AMS radiocarbon data with 2σ range of calibration from Bermú mire. (Before Present Age timescale, years before 1950)

Depth (cm)	Laboratory Code	BP Age	Calibrated AD Age	Median Probability Cal AD
12	CNA-723	114.47 ± 0.38	1958–1994	1992
29	CNA-724	75 ± 30	1692–1920	1847
44	CNA-725	335 ± 30	1691–1923	1560
54	CNA-726	558 ± 30	1474–1640	1346
58	CNA-2150	745 ± 35	1299–1415	1265
62	CNA-727	915 ± 35	1219–1293	1105
67	CNA-2151	1140 ± 35	1029–1204	904
72	CNA-094	1327 ± 32	649–767	711

3.2. Pollen Analysis

Pollen analysis was carried out on 72 sub-samples of 1 cm^3 along the core length. All of the samples were treated chemically (HCl, KOH, HF) following the standard procedures described by Moore et al. [42], using Thoulet solution for the densimetric extraction of pollen and non-pollen microfossils [43]. Acetolysis was not carried out to allow the identification of any contamination by modern pollen. Macrofossils were not discerned throughout the core. Palynomorphs were identified by the use of the reference collection of the Institute of History of the Consejo Superior Iinvestigaciones Científicas (onwards CSIC) (Madrid), and diverse identification keys and photo atlases [42,44–47]. *Erica* pollen was discriminated following Mateus [48], while *Betula* and *Corylus* pollen types were identified according to Blackmore et al. [49]. Anthropogenic Pollen Indicators were designed following Mercuri et al. [50] and López-Sáez et al. [13]. Pollen concentration (grains cm^{-3}) was estimated by adding a *Lycopodium* tablet to each sample [51]. Pollen counts of up to 500 grains of total land pollen per sample were identified and counted. Pollen of aquatic or wetland plants as well as spores, non-pollen palynomorphs (NPPs), and *Erica tetralix* were excluded from the pollen sum. The pollen and summary diagrams have been plotted against age using TGview [52]. To establish the zonation of the pollen sequence, we tested several divisive and agglomerative methods with the program IBM SPSS Statistics 21. Based on the ecological meaning of the obtained zones, eight local pollen assemblage zones (LPAZ-1 to LPAZ-8) were constructed on the basis of agglomerative constrained cluster analysis of an incremental sum of squares (Coniss) with square root transformed percentage data [53]. The number of statistically significant zones was determined using the broken-stick model [54].

3.3. Charcoal Analysis

Charcoal particles were counted under a microscope along with the identification of pollen. The findings were classified into >125 μm and <125 μm, indicating local and regional fires, respectively [55,56]. Both are shown in the figures in spite of their strong correlation [57]. Charcoal accumulation rate (CHAR) was calculated by sedimentation rate (cm year^{-1}), and is expressed in particles cm^{-2} year^{-1} [58]. It is noted that the treatment process can split original large charcoal remains into smaller pieces. Nevertheless, as all of the samples were treated with the same procedure, we assume that this systematic bias was uniform and did not influence the variation of charcoal abundance. Many studies comparing the sieving charcoal series versus the pollen-slide charcoal series showed that both of them display a similar pattern [59].

4. Results and Discussion

Thirty-four samples were studied identifying pollen, spores, and non-pollen palynomorphs (NPPs onwards). The results are presented on diagrams (Figures 3–5), where tree and shrub taxa are grouped according to their modern distribution. Herbs and NPPs were classified based on their ecological affinities. The percentage pollen diagrams can be divided into two pollen zones: (i) Bermú-1 (onwards BM), from the bottom of the record at 72 cm depth to 24 cm depth, and (ii) BM-2, from the 24 cm depth to the top of the record. The first one is also divisible into four different pollen sub-zones: BM-1.1. (72–62 cm; ca. 711–1100 cal AD); BM 1.2. (62–50 cm; ca. 1100–1450 cal AD); BM. 1.3. (50–38 cm; ca. 1450–1703 cal AD); and BM 1.4. (38–24 cm; ca. 1703–1876 cal AD).

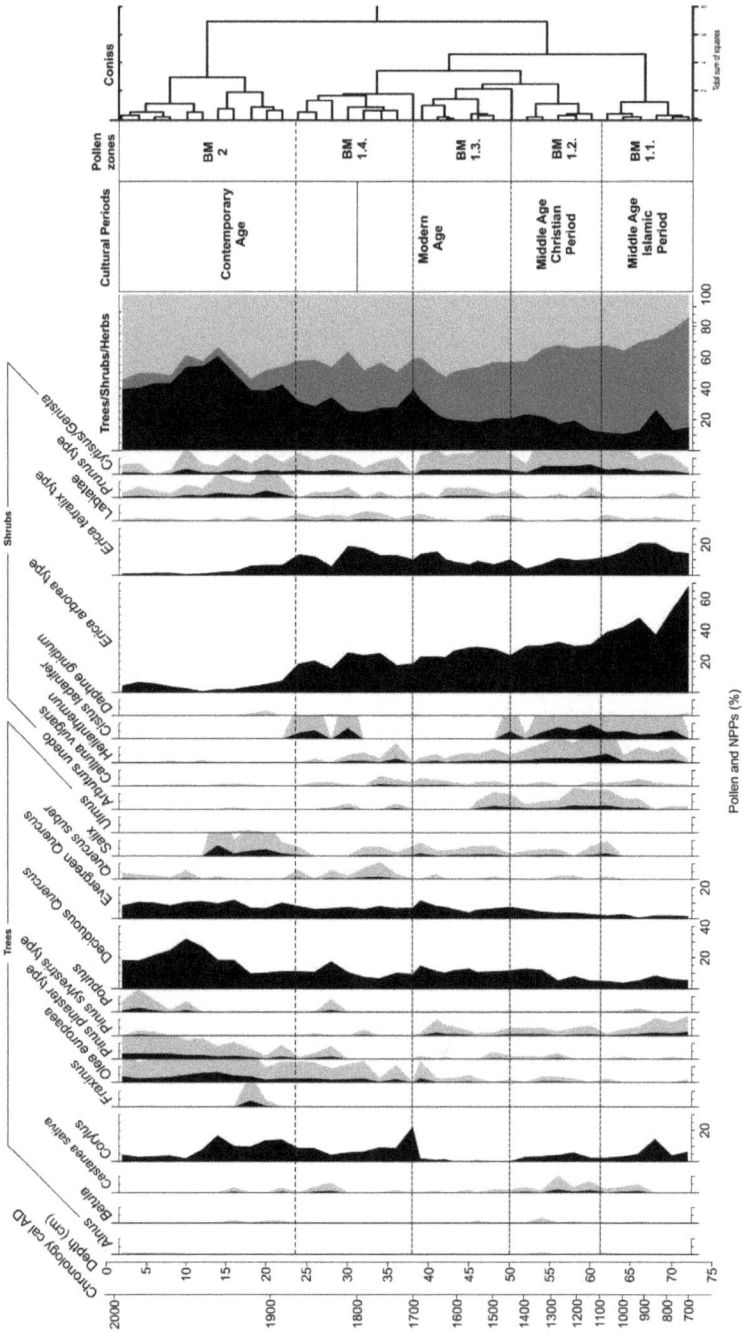

Figure 3. Bermú mire trees and shrubs pollen diagram plotted against depth. The black silhouettes show the percentage curves of the taxa; the grey silhouettes show the ×5 exaggeration curves.

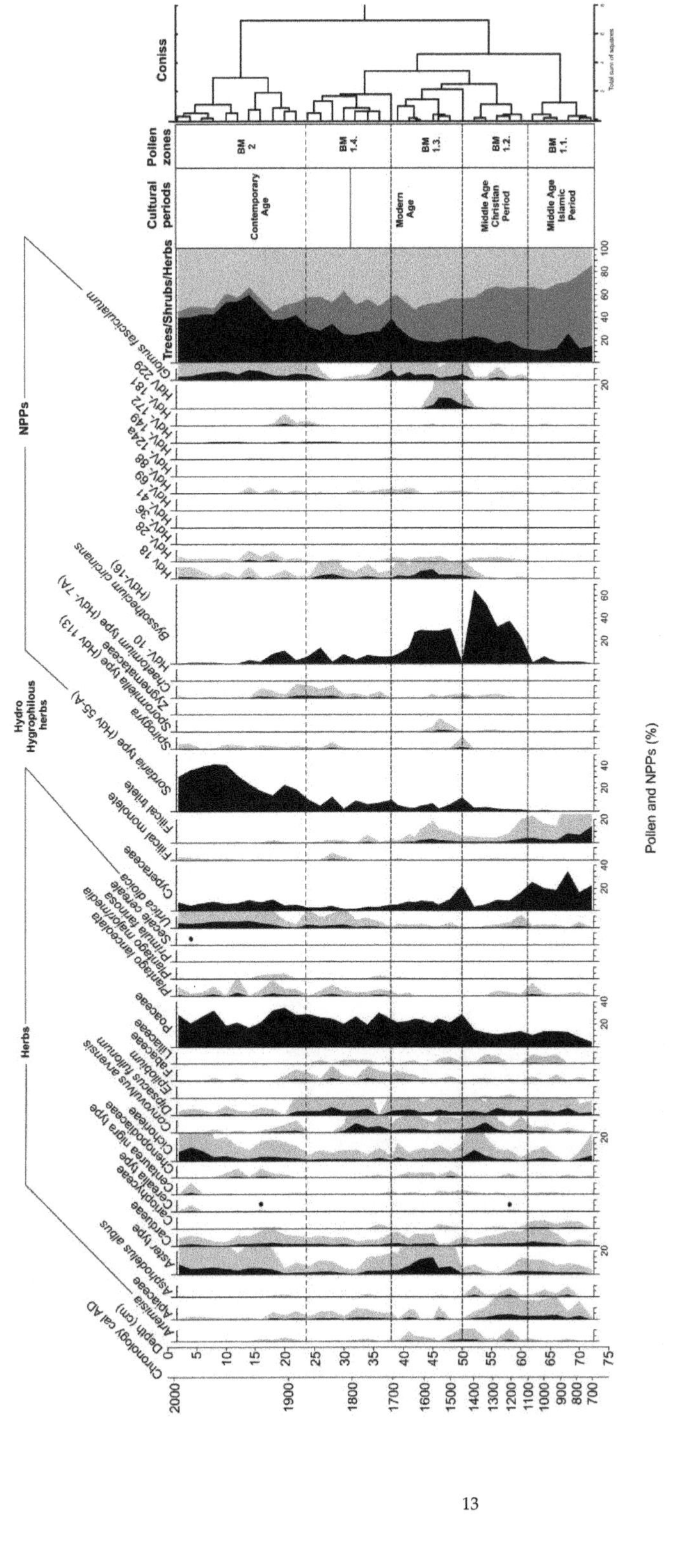

Figure 4. Bermú herbs and non-pollen palynomorphs (NPPs) diagram plotted against depth. The black silhouettes show the percentage curves of the taxa, the grey silhouettes show the ×5 exaggeration curves. Dots represent percentages below 0.5%.

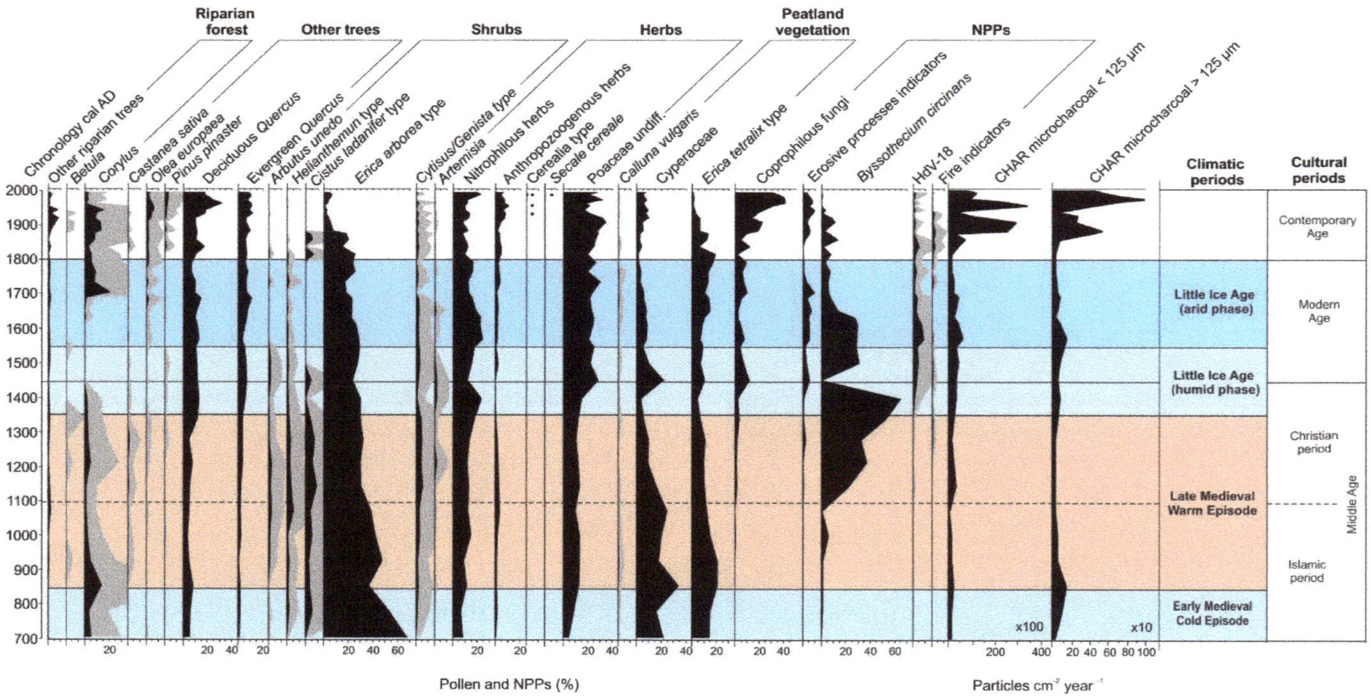

Figure 5. Synthetic Bermú mire pollen diagram plotted against age. The black silhouettes show the percentage curves of the taxa; the grey silhouettes show the ×5 exaggeration curves. Dots represent percentages below 0.5%. Other riparian trees: *Alnus, Fraxinus, Populus, Salix, Ulmus*. Anthropogenic-nitrophilous herbs: *Aster, Asphodelus albus, Carduncus, Centaurea nigra, Cichorieae, Convovultus arvensis, Dipsacus fullonum*. Anthropozoogenous herbs: Chenopodiaceae, *Plantago lanceolata, Urtica dioica. Coprophilous fungi: Sordaria* (HdV-55), *Sporormiella. Erosive processes indicators: Glomus, Pseudoschizaea. Fire indicators: Chaetomium* (HdV-7A).

4.1. Islamic Period: Survival in a Warfare Land (ca. 711–1100 cal AD)

The Bermú mire bottom is dated in the Islamic period (ca. 711–1100 cal AD), at the Early Medieval Age. By this moment (BM 1.1 pollen zone), Bermú shows an open landscape with a scarce forest cover. Riparian species such as *Alnus*, *Fraxinus*, *Populus*, *Salix*, or *Ulmus* among others are also extended around Bermú mire. Despite the regular tendency shown by trees, at middle 19th century cal AD, there was a decrease of deciduous oaks and a reduction of *Corylus avellana* percentages, highlighting the end of a cold and humid phase, also indicated by the descending levels of Cyperaceae. At the same time, *Castanea sativa* appears with low percentages (1.1%). This taxon is not widely spread along the Toledo Mountains, but it was also found in Patateros record at this time [29] (Figures 1 and 6). This landscape stems from the human communities development in the area, and endures over the whole Islamic period.

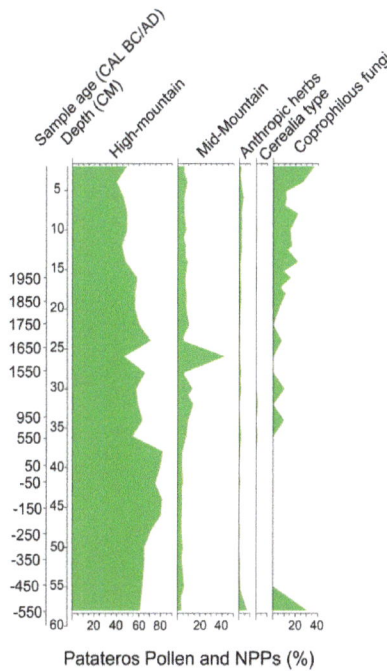

Figure 6. Patateros mire synthetic diagram. High-mountain species: *Pinus pinaster*, *Pinus sylvestris*, *Calluna vulgaris*, *Erica* spp. *Cytisus* type. Mid-mountain species: Deciduous *Quercus*, evergreen *Quercus*, *Castanea sativa*, *Olea europaea*, *Cistus ladanifer*. Anthropic herbs: Aster, *Asphodelus albus*, Cardueae, *Centaurea nigra*, Cichorieae, *Convolvulus arvensis*, *Dipsacus fullonum*. Coprophilous fungi: *Sordaria* (HdV-55), *Sporormiella*.

This scarce forest cover, which is a consequence of a disturbed landscape, led a wide brushwood line. *Erica* is the most representative element in this landscape, according to the results showed in every single place studied until this moment. *E. tetralix* (20.5%) is the main species associated to the mire environment. By this reason, it was excluded from the pollen sum, and it is present in all of the studied places [29,30,59,60]. *Cistus ladanifer* (6.3%), *Cytisus*/*Genista* type (3.7%), and *Helianthemum* type (5.6%), as well as *Calluna vulgaris* (0.9%) or *Arbutus unedo* (2.5%), cover the wide open space under the canopy, as they also do in places such as the nearby Valdeyernos or La Botija [29,30,61] (Figure 1, Figure 7, and Figure 8).

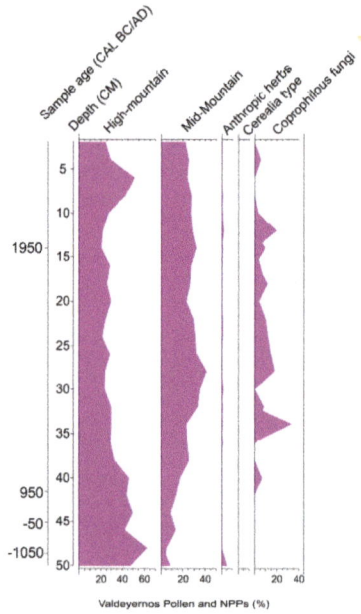

Figure 7. Valdeyernos mire synthetic diagram. High-mountain species: *Pinus pinaster*, *Pinus sylvestris*, *Calluna vulgaris*, *Erica* spp. *Cytisus* type. Mid-mountain species: Deciduous *Quercus*, evergreen *Quercus*, *Castanea sativa*, *Olea europaea*, *Cistus ladanifer*.

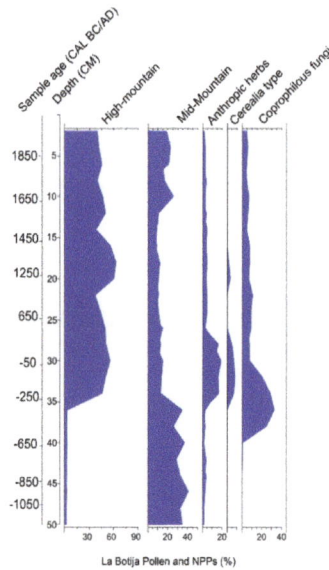

Figure 8. La Botija mire synthetic diagram. High-mountain species: *Pinus pinaster*, *Pinus sylvestris*, *Calluna vulgaris*, *Erica* spp. *Cytisus* type. Mid-mountain species: Deciduous *Quercus*, evergreen *Quercus*, *Castanea sativa*, *Olea europaea*, *Cistus ladanifer*.

The herbaceous layer is mainly composed by Poaceae (13.4%) and anthropogenic nitrophilous herbs pollen taxa (*Aster*, Cardueae, Cichorieae, *Centaurea nigra*, *Asphodelus albus*, *Convolvulus arvensis* or *Dipsacus fullonum*). These are indicators of human activities next to the mire, and are probably related to the use of many routes crossing close to the study site, although no cereal, legume pollen, nor anthropozoogenic taxa have been found in this pollen zone. A local fire occurred during the eigth century cal AD, when CHAR microcharcoal >125 μm registered a peak (138 particles cm^{-2} year^{-1}), but the regional fire activity is very low, and there was not an increase of human presence in the mire surroundings.

By this time, the Al−Andalus northern border facing the Christian territory reached the Central System. The Toledo Mountains are part of the central Islamic district called *Marca Media*. During this period, the war was one of the main features of this frontier territory, which was threatened by the sporadic but constant Christian raids, more frequently as the Middle Age went on. This insecure situation is also linked to many internal troubles of the Islamic kingdom, making a safe development difficult [62,63]. The population living there was scarce; it was composed of basically *mozarabs*—Christians living in Islamic territory—because, as Ladero-Quesada [64] and Boloix-Gallardo [63] remarked, the Arabic and *bereber* immigration was scarce till these northern domains. So, in the end, we cannot assure a very intense exploitation of the mire in the Islamic period.

The economy was characterized by agriculture in the lowlands, and livestock prospered in the mountainous lands [65–68]. The mid-mountain areas were more used for grazing pasturelands than for agricultural practices, as the absence of cereal pollen grains and the presence of pastureland indicators mark, although they did not focus around Bermú (Figure 5). Islamic society divided the rural space in *alquerías*, or territories belonging to many small farms. These little farms managed the wide space that was available, drawing a scattered population pattern, and made grazing as its economic code, which fit perfectly into the warfare and a not very populated context [69,70]. Transhumance was practiced covering short distances, and it is interesting to remark on the main role played by the secondary products such as leather and wood. Muslims introduced the Merino race, which will become essential on Castilian economy, but which did not play a main role over the landscape by this time [69–71]. The *Coprophilous fungi* (*Sordaria* sp.) found at this pollen zone were very scarce (0.5%), and the levels of grazing indicators were also too low to assess human exploitation of the Bermú surroundings, as also happens at Valdeyernos (Figure 7). On the contrary, La Botija and Patateros (Figures 6 and 8) show higher coprophilous fungi values, which are likely related to an increasing livestock pressure, especially at the end of the period, when cereal type pollen is recorded in La Botija (Figure 8) [29,30,61].

Cyperaceae (24.9%) and Filicales trilete spores (14.2%) highlight the climatic conditions—mainly cold and humid—which define the Early Medieval Cold Episode (ca. 450–950 cal AD) [20].

4.2. Christian Period: Building a New Order (ca. 1110–1450 cal AD)

The pollen diagram of Bermú still shows a scant forest cover (23.7%), although some transformations are noticeable (bottom of BM 1.2 pollen zone), such as the increase of both deciduous and evergreen oaks (12.7% and 7.3%%, respectively) since the eighth century cal AD (Figures 3–5), except for a scant descent in the second half of the eighth century (1278 cal AD). The same dynamic is found in the other studied sites, such as Patateros and Valdeyernos. It is also interesting to note the significant *Castanea sativa* pollen percentages (2.1%), in contrast to Patateros, which shows a soft decreasing trend at the end of this period. Human management could be tracked in these growths, but they could have been also favored by the more humid conditions present during the Late Medieval Warm Episode [72,73]. *Pinus* type pollen is lower than in previous steps (1.2%), and *Corylus avellana* keeps the average percentages, despite a tiny descending trend. The opposite situation is found in the nearby Valdeyernos or La Botija, but is similar to Patateros mire (Figure 8).

Brush layer (53.4%) is mainly composed by *Erica* in Bermú, which is similar to in the other analyzed mires. In this study site, total brush pollen values diminished (top of BM 1.2 pollen zone), especially in the end of 14th century (1339–1397 cal AD, 42.8–34%). This reduction fits with the increasing trend shown by the trees layer. This reduction is also noticeable in western placements such as La Botija, but not in Patateros mire, where the end of this period includes an increase of *Erica* type. The other taxa remained low, with a reduction in their pollen concentration since the 15th century cal AD. This tendency is especially clear in *Cistus ladanifer* (9.7–0.7%). In Valdeyernos, the total shrub layer tends to increase by the end of the period.

The Poaceae pastureland increased, while the above-mentioned high percentages of coprophilous fungi and the absence of cereal pollen point out the presence of livestock in the mire's surroundings (Figures 5 and 7).

Furthermore, there is also an increase of erosion indicators (e.g., *Glomus* (5.3%) in Bermú and in Valdeyernos pollen records (Figures 5–8), which may indicate the mentioned loss of tree cover. It is interesting to highlight that all those processes came accompanied by a soft increase in CHAR trends, especially by the end of the period. CHAR microcharcoal >125 µm reached 71.94 particles cm^{-2} year^{-1}, and regional fire indicators were also higher than at the beginning of the Christian conquer (CHAR microcharcoal <125 µm 3735.97 particles cm^{-2} year^{-1}).

The Christian period is a warfare time as well, especially in the first centuries [74,75]. Repopulation, as a mechanism to stabilize the Castilian presence, became a royal task, whose pace and advance was directly linked to internal al-Andalus conflicts [76]. Thus, repopulation was not well established until the 14th century, as reflected by the increase of anthropogenic indicators and the tree cover drop (Figures 5 and 8). Repopulation implies land ownership change, triggering the property concentration mainly by Toledo's Cathedral and the Crown. An interesting episode involving the study site is the sale of Toledo Mountains land, including the castles and fortifications, made by Fernando III to the Toledo Council in 1246 AD. However, there is not a very noticeable impact over the surrounding Bermú landscape after that. The territory was disperse and composed by not very numerous little towns depending on the city; this structure guaranteed Toledo's supply, obtaining among others wood, hunting, and two very highly estimated products: honey and wax [75–77].

Transhumance movements, which were present in a low scale due to the Islamic tradition, were favored after Toledo surrender in 1085, but especially after the eighth century, when the Christian border was finally fixed south enough to keep these lands safe. The new inhabitants could connect the Cuenca range and western Tagus river lands as summer and winter grazing pasturelands, respectively, where open woodland landscape was prepared once again for the cattle supply [76,78]. Grazing practices had a parallel development with the repopulation movements as shown in the diagrams (Figures 3–8). Livestock in this area begins with local flocks associated with the neighborhoods and reached its maximum when the La Mesta Council was founded in the eighth century [67,79]. New pastureland was necessary, which was a facility along with the tax exemptions that constituted a lure for new inhabitants [78]. Since this moment, the La Mesta Council extended, protected, and regulated livestock movements. Many paths or *Cañadas* crossed close to these mountains, such as *Cañada Real Leonesa* and *Cañada Real Segoviana*, connecting with the northern grazing lands of the Central System, or *Cañada Real Manchega*, which were bound for Murcia and Andalucía. The Toledo Mountains were one of those extreme lands used for grazing during the summertime.

The Christian period is climatically defined by the Late Medieval Warm Episode (ca. 900–1300 cal AD) [72,73], with higher temperatures and drier conditions in the northwest hemisphere. By the beginning of this moment, the amount of humid indicators is scarce: (i) HdV-18 percentages are almost absent (0.2–1.4%), and (ii) Cyperaceae or fern spores display a descending trend. However, this temperature rising joins the increasing human environment exploitation, which has a direct consequence on Bermú nature: the mire suffers an eutrophication process reflected in the great growth of *Byssothecium circinans* (HdV-16) (64.6%). The Little Ice Age (ca. 1350–1850 cal AD) starts at the end of Middle Age [20]. Temperatures descend [80,81] and humid indicators rise not only in Bermú (3.3%), but also in Valdeyernos or La Botija mire [30,61], while the presence of HdV-16 is lower, in part due to the increasing rainfall in the first Little Ice Age humid phase (ca. 1300–1570 cal AD) [81].

4.3. Modern Age Period: Grazing Shaping the Landscape (ca. 1450–1800 cal AD)

Along the Modern Age period (BM 1.3 and bottom of BM 1.4 pollen zones), the forest canopy shows a modest recovery trend (39.7%). *Olea europaea* reappears (2.7%), and there is no evidence of *Castanea sativa* during this period. The main genus is still *Quercus*, especially deciduous oaks (14.6%) such as in Valdeyernos, while *Pinus* pollen maintains very low percentages (2%). In the western border of the Toledo Mountains the trend is opposite, except for Las Lanchas mire, where the amount of evergreen and deciduous oaks is similar, and there is also a remarkable *Quercus suber* population. This duality between east and west is also perceptible in relation to *Corylus avellana* development. There is an abrupt reappearance of this species in Bermú at the end of the 17th century cal AD (1686–1703 cal AD, from 1.8% to 22%; top of BM 1.3 pollen zone), reaching its highest peak in the sequence, as it also happened in Valdeyernos, with an opposite trend at the western side of the Toledo Mountains, as shown in La Botija, and by its absence in Las Lanchas and Patateros [29,30,59,61]. However, the *Corylus* population diminished and stabilized in Bermú at the end of this period (6.1%) when the oaks recovered.

This modest tree recovery triggered a soft brushwood reduction (36.2–26.6%), specifically of *Erica*, while many taxa such as *Cistus ladanifer*, *Arbutus unedo*, and *Prunus* disappeared. Poaceae is the main herbaceous type at this period, maintaining the previous period levels with a little increasing trend. Nevertheless, it is necessary to mention the great amount of anthropogenic indicators: (i) nitrophilous herbs (*Aster*, *Asphodelus albus*, Cardueae, *Centaurea nigra*, Cichorieae, *Convovulvus arvensis*, *Dipsacus fullonum*) (33.1%); (ii) anthropozoogenous herbs (Chenopodiaceae, Plantago lanceolata, *Urtica dioica*) (3.8%), especially after the 18th century cal AD; and (iii) high coprophilous fungi values (9.6%), highlighting the impact of grazing activities (Figures 3–5). Due to the lack of pasturelands and crop fields, the Toledo Council allowed the neighbors to use forest resources, generating new open spaces. This spread is indicated in the Bermú pollen diagram with the pastureland increasing [73,82]. The big amount of erosion indicators (9.2%) confirms this new forest exploitation. However, fire is not the most used tool; CHAR show a descending trend: CHAR microcharcoal >125 µm decreased to 4.69 particles cm^{-2} $year^{-1}$, and CHAR microcharcoal <125 µm decreased to ca. 364.83 particles cm^{-2} $year^{-1}$.

A change in the land use began in the 17th century. Due to the subsistence crisis happening during those years, many royal properties were sold, changing their traditional use. Forest lands were specially affected, despite the forestall conservation trials applied [83]. In fact, woodland use with livestock purposes was at this moment so intense that the Toledo Council prohibited the felling and promoted the first forest repopulations; they were also focused on trying to avoid fire episodes, forbidding the flocks from getting into burned pastureland [75,82,83]. Fire is not just a tool anymore, but rather a threat for forest resources. The Modern Age is the period when important conservation policies are promoted, especially during the 14th and 15th centuries [83]. Lots of these rules were focused on fire, such as those establishing fire calendars excluding the warmest months or the obligation to intervene when someone detected a fire. Since then, stockbreeder interests were protected, and wood or charcoal use was regulated. All of those rules were actualized at the beginning of the 16th century, when deforestation became more evident in the whole kingdom [75].

During this period, it is necessary to focus on the La Mesta Council's movements. By the Modern Age, this organization is powerful enough to dominate the main flocks in the whole country. The owners built a large and solid net to manage these livestock movements along the cattle roads [27]. This has a noteworthy impact over the landscape configuration, as it points out the high presence of oaks and Poaceae pastureland in many studied sites, building an open woodland landscape that meets human necessities.

During the Modern Age, the climate is determined by the Little Ice Age (ca. 1300–1850 cal AD) [84]. In the first years, the humid phase maintained rainy and cold conditions, but it was mainly the arid phase (ca. 1570–1800 cal AD) that marked this cultural time [72,81]. In Bermú record, it is possible to appreciate the growth of HdV-18 at the beginning of this period, but at the end, it descends alongside Cyperaceae, which falls during the whole phase. Many drought and floods episodes periodically affected the area during the 17th century [84]. These episodes are very noticeable in the little amount of humid indicators and the second eutrophization process that suffered by the studied mire, which was pointed out by the great percentages of HdV-16 (30.8%) (Figure 5).

4.4. Contemporary Age Period: a Changing Tendency (ca. 1800 cal AD-Present)

By this time (top of BM 1.4 and BM 2 pollen zones), the trend in Bermú is a progressive forest recovery (arboreal pollen percentages increased 25.8–60.7%, see Figure 3), while in the nearby Valdeyernos, there is a better equilibrium between trees, shrubs, and herbs. Westwards, only Las Lanchas was characterized by a higher tree cover, while the other two study sites, marked by a livestock economy, presented an open landscape [61]. In Bermú, the main forest elements remained deciduous oaks (10.4–31.8%) followed by evergreen oaks (6.9–11.7%), composing a wide mixed forest that replaced the shrub layer, although Poaceae pastureland was still dominant.

The riparian forest reappeared in the second half of the century, and it was more present at the river banks. *Pinus* species, and more specifically *Pinus pinaster* (3.6%), were widely used in the repopulations that developed during the 20th century. These additions are perfectly recognizable in Bermú record and also in Patateros, where it is the main tree layer component. Diagrams show that the recent landscape has been shaped by the laws and initiatives that were trying to recover the forest throughout the whole country. They were especially intense from 1940 to 1970, when many afforestation programmes were carried out on public and private lands. In the Toledo Mountains, these plans respected the original species such as Pyrenean oak and holm oak, as the pollen diagram displays. The main introduced species in this area were pines, which quickly invaded the abandoned agricultural fields, generating second-generation forests [9,85,86].

Corylus avellana also had a great role in Bermú, where it was present since the first half of this period. Among the cultivated species, it is interesting to highlight the importance gained by *Olea europaea* at this time (1.9–6.6%), getting prominence in the area alongside the *Castanea sativa* slight reintroduction (1.2%), which disappeared at the beginning of the 20th century (BM 2 pollen zone). *Olea europaea* is not very relevant in Valdeyernos. It is possible that the pollen recorded there came from the Bermú surrounds. At the western side, only La Botija recorded a great amount of *Olea europaea* but very low percentages of *Castanea sativa*, which is also present in Patateros or Las Lanchas, but absent in Valdeyernos [29,30,60,61].

On the other hand, a decreasing trend is found in the brushwood layer, which was 38.3% at the beginning of the period at 1800 cal AD, and 7.4% at the end of 19th century (top of BM 1.4 pollen zone), and only 5.9% at the top of the pollen record (BM 2 pollen zone). This change is especially noticeable in Ericaceace, except in the nearby Valdeyernos, where a big increase is shown. *Cistus ladanifer* disappeared as well as *Erica* type at the beginning of the 19th century, while *Prunus* began to increase (3%), similar to in Valdeyernos. On the contrary, at the western edge, *Erica* type was the main species composing the brush layer. In Las Lanchas, it tended to grow despite the soft decreasing tendency found in many places such as Patateros [29,30,60,61].

Herbs (36–54.2%) were mainly still composed by Poaceae and nitrophilous herbs, but with a major presence of anthropozoogenic herbs (6.0%), especially since the beginning of the 20th century cal AD. Human presence at Bermú is very clear, which is also indicated by the appearance at this moment of cereal pollen in the record. There is not enough pollen to confirm its cultivation in the mire edge (only 0.2–1.2%), but it reveals its cultivation in the area, particularly during the last decades of the 20th century cal AD. On the contrary, in Las Lanchas, where cereal pollen reached a great amount during the Modern Age, this type is absent during the Contemporary Age (Figures 5 and 9). On the western side, in Patateros, and in Valdeyernos, the cereal pollen concentration was also very low (Figures 6 and 7).

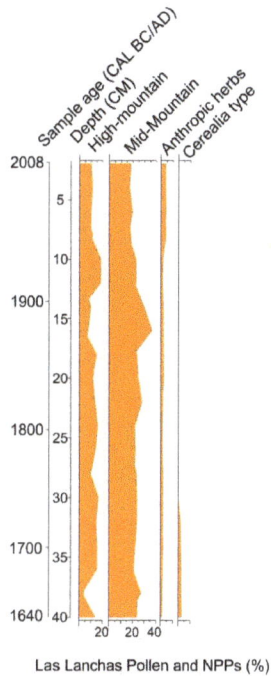

Figure 9. Las Lanchas mire synthetic diagram. High-mountain species: *Pinus pinaster, Pinus sylvestris,* and *Calluna vulgaris, Erica* spp. *Cytisus* type. Mid-mountain species: Deciduous *Quercus*, evergreen *Quercus, Castanea sativa, Olea europaea,* and *Cistus ladanifer*.

There is an evident increase in erosion indicators (2.4%) despite the forest recovery, which points to a major land use by humans and higher pressure on it. The CHAR trend in Bermú showed two peaks: 1883 cal AD presents values ca. $25,077.77$ cm^{-2} year^{-1}, which increased to $28,789.99$ cm^{-2} year^{-1} in 1908 cal AD, showing a noticeable regional fire activity. On the other hand, the local fires also had an intense impact over the landscape close to Bermú mire, despite the descending tendency of CHAR microcharcoal <125 µm (1883 cal AD values ca. 520.79 cm^{-2} year^{-1} and 1908 cal AD values ca. 230.07 cm^{-2} year^{-1}). These charcoal peaks came along with evidence of fire indicators (2.4%), although they were scarce and almost absent at the end of the period. During the 20th century cal AD, CHAR presented their maximum values (CHAR microcharcoal <125 µm $33,074.97$ cm^{-2} year^{-1} and CHAR microcharcoal >125 µm 968.93 cm^{-2} year^{-1}).

The first local fire episode could be associated with livestock because it occured simultaneously with an increase of coprophilous fungi and anthropozoogenic herbs. It's very interesting to highlight here the abrupt stop of fire indicators at the same time that the *Pinus* species are reintroduced, following the reforestation policies. This new fire management is not recorded in the studied mires on the western side, so we could associate the second peak to an extra local fire or a very located event not affecting vegetation dynamics around Bermú. *Byssothecium circinans* (HdV-16) doesn't show great values (its maximum is 13.8%), but rather an irregular trend that could also point out the use made by the human communities of this space.

During this period, the population behaves according to two different trends, which produce two tendencies over the landscape as well. During the 19th century, the population kept increasing in the area, whereas in the 20th century, there was a great movement from the rural areas to the cities, leaving the former with a very low population density. Therefore, during the first part of the period, the resources demand was higher, and agriculture spread its limits throughout these mountains, using low–quality and distant lands [26]. During the second half of this period, due to the lack of workforce, previously forced lands were abandoned and recovered back to forest [9,26,87].

The most important action performed in the Toledo Mountains during the Contemporary Age, affecting the whole country, were the Confiscation Laws. These laws were accomplished during the 19th century and triggered a land owning change with evident consequences for the forest all over the country [83]. The first disentailment event was performed in 1837, and especially affected the Church properties [83,85,88,89]. Madoz's disentailment was ordered in 1855. In this case, it tried to exclude mountainous lands from the sale with a relative success. In 1868, the Treasury also sold the State Mountains if they didn't have a particular ecological value [88,89]. In the Toledo Mountains, the public lands were divided up between the neighbors, ending the monopoly held by the Toledo Council.

Agriculture was the main argument in these interventions for two reasons: (i) the population rising and (ii) Spain becoming a grain exporter during the first half of the 19th century. This last circumstance was tied with an increase the agricultural surface area and a diversification of the products, which now included fruit trees and oil. This change can be seen in the diagram by the increase of *Olea europaea* percentages (Figures 3–5). Thus, 40% the Toledo Mountain's surface was catalogued as able to be ploughed and sold. Even today, the consequences of this sale are present, as 60% of the Toledo Mountain territory is catalogued as a "Mountainous Agricultural Zone" [83,85,86,89].

In the Toledo Mountains, where agricultural practices were not very worthwhile because of land properties, husbandry became necessary, as it has been pointed out before. However, the La Mesta Council disappeared in 1836 when transhumance dynamics were reduced. The main reasons were the drop in the number of transhumant livestock, the appearance of new grazing models, such as the stabled cattle, and the lost of importance of the Spanish wood in the international markets [83,87,90]. These changes are perceptible on the Bermú pollen diagram (Figures 3–5). Trees increased and shrubs suffered an evident drop, while the coprophilous fungi concentration grew as a result of the bigger exploitation of the mire. Alongside livestock, hunting was the main economic activity performed in the Toledo Mountains, which means that Bermú itself was placed within an area designated for this activity. Deer became the more widespread species in the whole shire. Also, many vegetal taxa were affected by the uses that these animals made of them for feeding or cleaning procedures, such as the horn scrub against the trunk or branches. In fact, many studies have been conducted in order to determine the damage that the heath species suffered. One of them made precisely inside the Los Quintos de Mora property [91] claimed that the preference shown by this ungulate for the arbutus (*Arbutus unedo*), rockrose (*Cistus ladanifer*), or false olive (*Phillyrea angustifolia*) was perceptible in the Bermú pollen diagram (Figure 3) [91,92].

Along with the increasingly important role played by hunting, it is possible to trace the traditional activities supporting population, which are now shown with more intensity. Among them, one of the most important resources was wood charcoal, which people used as the main fuel until the mineral charcoal appeared [93]. Heather charcoal was much appreciated, alongside holm oak and other shrubs. The drop of *Erica* sp. and the fluctuating trend of evergreen oaks are good reflections of this trend. This activity was fiercely regulated in the Toledo Mountains, but it was not always made under license or legal conditions [7,75]. Nevertheless, this activity decreased by the second half of the 20th century when the population migrated to the cities, reducing considerably its demand on the territory and the use of these fuels.

This period seems to be arid, as shown by the absence of fern spores and the low levels of HdV-18 (4.6% maximum); similar Figures were also recorded in the other study sites except in Valdeyernos, where there was an intense increase of this NPP as well as in Las Lanchas, which was likely linked to their humid natural conditions.

5. Conclusions

The Toledo Mountains' landscape is the consequence of the effect of the climatic dynamic and the human land use through time. As a mid-mountain ecosystem, the possibilities offered to the human communities were so wide due to the wide variety in resources and mild climate. These particular circumstances allowed the inhabitants to develop different activities in order to make the most of the means. Bermú mire interdisciplinary paleoecological study is a long-term pollen-based research. These kinds of studies provide a long-term perspective of all those climatic and anthropic synchronous changes unleashing landscape transformations [13,14,94]. In this case, it shows an environment shaped by the human communities through time with a dynamic that is very similar to the other study sites near this land [29,30,60,61]. By the Islamic period (711–1100 cal AD), the area was dominated by an open landscape, although the trending change was produced in the Late Iron Age, as shown in other study sites.

Despite the low agricultural quality, the Toledo Mountains' landscape was intensely used in many ways, especially after the Christian conquer and repopulation movements. Grazing, hunting, and forest exploitation were the most important activities performed in these lands, which generated open landscapes and pastureland for the cattle support in a way that is still present even today.

The cultural landscape that has been built over time in the Toledo Mountains is narrowly linked to human development. Political settings and different socio-economic trends forced human communities to either put higher pressure on the landscape and sometimes promote forest restoration; both of these trends reflect the main role played by the wood business during the Middle and Modern Age. The demographic trends sometimes generated a higher pressure as when the repopulation movements succeeded in the Late Middle Age or moved the tendency toward a softer interaction, as happened in the Contemporary Age, when people abandoned the rural areas and moved toward the cities. The different protection movements or reforestation attempts made on these mountains are another example of how much human beings influence their environment. Thus, it's possible to attest the main role played by humans in the Toledo Mountains, not only as a simple user but also as a builder of their own reflexion in the environment.

Author Contributions: Conceptualization, R.L.-L., S.P.-D. and J.A.L.-S.; Data curation, S.P.-D. and J.A.L.-S.; Formal analysis, R.L.-L., S.P.-D. and F.A.-S.; Funding acquisition, J.A.L.-S.; Investigation, R.L.-L., S.P.-D. and J.A.L.-S.; Methodology, J.A.L.-S.; Project administration, J.A.L.-S.; Resources, S.P.-D.; Supervision, S.P.-D. and J.A.L.-S.; Validation, S.P.-D. and J.A.L.-S.; Writing—original draft, R.L.-L.; Writing—review and editing, S.P.-D., F.A.-S., D.A.-S. and J.A.L.-S.

Funding: This research was funded the project REDISCO-HAR2017-88035-P (Plan Nacional I+D+I, Spanish Ministry of Economy and Competitiveness). Reyes Luelmo is funded by a FPU grant (Spanish Ministry of Education, Culture and Sports).

Acknowledgments: This publication was funded by the *Laboratorio di Palinologia e Paleobotanica—CEA 2018 Award* for the oral presentation titled 'A mid-mountain landscape shaped during fourteen centuries in the heart of Toledo Mountains (central Iberia): the Bermú peat bog record' by Luelmo-Lautenschlaeger Reyes, López-Sáez José-Antonio and Pérez-Díaz Sebastián presented at XIVth Confererence of Environmental Archaeology.

Conflicts of Interest: The authors declare no conflicts of interest.

References

1. Galop, D. *La Forêt, L'homme et le Troupeau Dans les Pyrénées. 6000 ans D'histoire de L'environnement Entre Garonne et Méditerranée*; Geode: Toulouse, France, 1998; 285p, ISBN 9782912025012.
2. Lozny, L.R. *Continuity and Change in Cultural Adaptation to Mountain Environments: From Prehistory to Comtemporary Threats*; Springer: New York, NY, USA, 2013; 410p, ISBN 978146145702.
3. Pérez-Díaz, S.; Núñez de la Fuente, S.; Frochoso Sánchez, M.; González Pellejero, R.; Allende Álvarez, F.; López-Sáez, J.A. Seis mil años de gestión y dinámica antrópica en el entorno del Parque Natural de los Collados del Asón (Cordillera Cantábrica Oriental). *Cuaternario y Geomorfología* **2016**, *30*, 49–74. [CrossRef]

4. Robles-López, S.; Luelmo-Lautenschlaeger, R.; Pérez-Díaz, S.; Abel-Schaad, D.; Alba-Sánchez, F.; Ruiz-Alonso, M.; López-Sáez, J.A. Vulnerabilidad y resiliencia de los pinares de alta montaña de la Sierra de Gredos (Ávila, Sistema Central): Dos mil años de dinámica socioecológica. *Cuaternario y Geomorfología* **2017**, *31*, 51–72. [CrossRef]

5. Mirás, Y.; Ejarque, A.; Orengo, H.; Mora, S.R.; Palet, J.M.; Poiraud, A. Prehistoric impact on landscape and vegetation at high altitudes: An integrated palaeoecological and archaeological approach in the Eastern Pyrenees (Perafita valley, Andorra). *Plant Biosyst.* **2010**, *144*, 924–939. [CrossRef]

6. Pérez-Obiol, R.; Jalut, G.; Julià, R.; Pèlachs, A.; Iriarte, M.J.; Otto, T.; Hernández-Beloqui, B. Mid-Holocene vegetation and climatic history of the Iberian Peninsula. *Holocene* **2011**, *21*, 75–93. [CrossRef]

7. Vanwalleghem, T.; Gómez, J.A.; Infante-Amate, J.; González-de Molina, M.; Vanderlinden, K.; Guzmán, G.; Laguna, A.; Giráldez, J.V. Impact of historical land use and soil management change on soil erosion and agricultural sustainability during the Antropocene. *Anthropocene* **2017**, *17*, 13–29. [CrossRef]

8. Pausas, J.G.; Keeley, J.E. A burning story: The role of fire in the history of life. *BioScience* **2009**, *59*, 593–601. [CrossRef]

9. Bebi, P.; Seidl, R.; Motta, R.; Fuhr, M.; Firm, D.; Krumm, F.; Conedera, M.; Ginzler, C.; Wohlgemuth, T.; Kulakowski, D. Changes of forest cover and disturbance regimes in the mountain forests of the Alps. *For. Ecol. Manag.* **2017**, *388*, 43–56. [CrossRef] [PubMed]

10. Kohler, T.; Giger, M.; Hurni, H.; Ott, C.; Wiesmann, U.; Wyman-von Dach, S.; Maselli, D. Mountains and climate change: A global concern. *Mt. Res. Dev.* **2010**, *30*, 53–55. [CrossRef]

11. López-Sáez, J.A.; Serra-González, C.; Alba-Sánchez, F.; Robles-López, S.; Pérez-Díaz, S.; Abel-Schaad, D.; Glais, A. Exploring seven hundred years of transhumance, dynamic, fire and human activity through a historical mountain pass in central Spain. *J. Mt. Sci.* **2016**, *13*, 1139–1153. [CrossRef]

12. López-Sáez, J.A.; López-Merino, L.; Alba-Sánchez, F.; Pérez-Díaz, S. Modern pollen analysis: A reliable tool for discriminating *Quercus rotundifolia* communities in Central Spain. *Phytocoenologia* **2010**, *40*, 57–72. [CrossRef]

13. López-Sáez, J.A.; Sánchez-Mata, D.; Alba-Sánchez, F.; Abel-Schaad, D.; Gavilán, R.G.; Pérez-Díaz, S. Discrimination of Scots pine forests in the Iberian Central System (*Pinus sylvestris* var. *iberica*) by means of pollen analysis. Phytosociological considerations. *Lazaroa* **2013**, *34*, 191–208. [CrossRef]

14. López-Sáez, J.A.; Alba-Sánchez, F.; Sánchez-Mata, D.; Abel-Schaad, D.; Gavilán, R.G.; Pérez-Díaz, S. A palynological approach to the study of *Quercus pyrenaica* forest communities in the Spanish Central System. *Phytocoenologia* **2015**, *45*, 107–124. [CrossRef]

15. Mercuri, A.M. Applied palynology as a trans-disciplinary science: The contribution of aerobiology data to forensic and palaeoenvironmental issues. *Aerobiología* **2015**, *31*, 323–339. [CrossRef]

16. Pérez-Sanz, A.; González-Sampériz, P.; Moreno, A.; Valero-Garcés, B.; Gil-Romerá, G.; Rieradevall, M.; Tarrats, P.; Lasheras-Álvarez, L.; Morellón, M.; Belmonte, A.; Sancho, C. Holocene climate variability, vegetation dynamics and fire regime in the central Pyrenees: The Basa de la Mora sequence (NE Spain). *Quat. Sci. Rev.* **2013**, *73*, 149–169. [CrossRef]

17. Valero-Garcés, B.L.; Moreno, A.; González-Sampériz, P.; Morellón, M.; Rico, M.; Corella, J.P.; Jiménez-Sánchez, M.; Domínguez-Cuesta, M.J.; Farias, P.; Stoll, H.; et al. Evolución climática y ambiental del Parque Nacional de Picos de Europa desde el último máximo glaciar. In *Proyectos de Investigación en Parques Nacionales: 2006–2009*; Organismo Autónomo de Parques Nacionales (España): Madrid, Spain, 2010; pp. 55–71. ISBN 9788480147224.

18. Ruiz-Fernández, J.; Oliva, M.; Cruces, A.; López, V.; da Conceiçao-Freitas, M.; Andrade, C.; García Hernández, C.; López-Sáez, J.A.; Geraldes, M. Environmental evolution in the Picos de Europa (Cantabrian Mountains, SW Europe) since the Last Glaciation. *Quat. Sci. Rev.* **2016**, *138*, 87–104. [CrossRef]

19. Franco-Múgica, F.; García-Antón, M.; Sainz-Ollero, H. Impacto antrópico y dinámica de la vegetación durante los últimos 2000 años BP en la vertiente septentrional de la Sierra de Gredos: Navarredonda (Ávila, España). *Revue de Paleobiología* **1997**, *16*, 29–45.

20. López-Sáez, J.A.; Abel-Schaad, D.; Pérez-Díaz, S.; Blanco-González, A.; Alba-Sánchez, F.; Dorado, M.; Ruiz-Zapata, B.; Gil-García, M.J.; Gómez-González, C.; Franco-Múgica, F. Vegetation history, climate and human impact in the Spanish Central System over the last 9000 years. *Quat. Int.* **2014**, *353*, 98–122. [CrossRef]

21. Treml, V.; Jankovská, V.; Petr, L. Holocene timberline fluctuations in the mid mountains of Central Europe. *Fennia* **2006**, *184*, 107–119.

22. Doyen, E.; Vannière, V.; Bichet, V.; Gauthier, E.; Richard, H.; Petit, C. Vegetation history and landscape management from 6500 to 1500 cal. BP at Lac d´Antre, Gallo-Roman sanctuary of Villards d´Heria, Jura, France. *Veg. Hist. Archaeobot.* **2011**, *22*, 83–97. [CrossRef]

23. Sancho-Reinoso, A. Fighting for survival. Planning and development issues in two European rural border mid-mountain regions. *J. Geogr.* **2010**, *5*, 67–81.

24. Diry, J.P. Moyennes montagnes d´Europe occidentale et dynamiques rurales. *Rev. Geogr. Alp.* **1955**, *83*, 15–26. [CrossRef]

25. Bettinger, L.; Ormaux, S. La moyene montagne européene, approche d´un concept.problème à partir de l´exemple français. *Insaniyat* **2011**, *53*, 17–39. [CrossRef]

26. Lasanta-Martínez-Martínez, T. La transformación del paisaje en montaña media por la actividad agrícola en relación con las condiciones ambientales. In *Acción Humana y Desertificación en Ambientes Mediterráneos*; García-Ruiz, J.M., López-García, P., Eds.; Instituto Pirenaico de Ecología: Zaragoza, Spain, 1997; pp. 145–172. ISBN 9788492184224.

27. Klein, J. *La Mesta: Estudio de la Historia Económica Española, 1273-1836*; Alianza Editorial: Madrid, Spain, 1990; 480p, ISBN 9788420622378.

28. Molenat, J.P. *Campagnes et Monts de Tolede du XIIe au XVe Siècle*; Casa de Velázquez: Madrid, Spain, 1997; 724p, ISBN 9788486839789.

29. Dorado-Valiño, M.; López-Sáez, J.A.; García-Gómez, E. Patateros, Toledo Mountains (central Spain). *Grana* **2014**, *53*, 171–173. [CrossRef]

30. Dorado-Valiño, M.; López-Sáez, J.A.; García-Gómez, E. Valdeyernos, Toledo Mountains (central Spain). *Grana* **2014**, *53*, 315–317. [CrossRef]

31. López-Sáez, J.A.; García-Río, R.; Alba-Sánchez, F.; García-Gómez, E.; Pérez-Díaz, S. Peatlands in the Toledo Mountains (central Spain): Characterisation and conservation status. *Mires Peat* **2014**, *15*, 1–23.

32. Martín-Serrano, A.; Molina, E.; Nozal, F.; Carral, M.P. Itinerario A2. Transversal en los Montes de Toledo. In *Itinerarios Geomorfológicos por Castilla-La Mancha: Libro de las Excursiones Desarrolladas Durante la VIII Reunión Nacional de Geomorfología, Celebrada en Toledo, 22–25 de Septiembre de 2004*; Benito, G., Díez Herrero, A., Eds.; Sociedad Española de Geomorfología-CSIC Centro de Ciencias Medioambientales: Madrid, Spain, 2004; pp. 51–82. ISBN 849219586X.

33. San Miguel, A.; Rodríguez-Vigal, C.; Perea García-Calvo, R. Los Quintos de Mora. Gestión integral del monte mediterráneo. In *Pastos, Paisajes Culturales Entre Tradición y Nuevos Paradigmas del Siglo XXI. Visitas de Campo*; López-Carrasco, C., Rodríguez, M.P., San Miguel, A., Fernández, F., Roig, S., Eds.; Sociedad Española para el Estudio de los Pastos: Madrid, Spain, 2011; 704p, ISBN 9788461487134.

34. Perea, D.F.; Perea, R. *Vegetación y Flora de los Montes de Toledo*; Ediciones Covarrubias: Toledo, Spain, 2008; 296p, ISBN 9788493603519.

35. Punt, W.; Marcks, A.; Hoen, P.P. Myricaceae. *Rev. Paleaeobot. Palynol.* **2002**, *123*, 99–105. [CrossRef]

36. Sánchez-del Álamo, C.; Sardinero, S.; Bouso, V.; Hernández-Palacios, G.; Pérez-Badía, R.; Fernández-González, F. Los abedulares del Parque Nacional de Cabañeros: Sistemática, demografía, biología reproductiva y estrategias de conservación. In *Proyectos de Investigación en Parques Nacionales: 2006–2009*; Organismo Autónomo Parques Nacionale: Madrid, Spain, 2010; pp. 275–310.

37. Luengo-Nicolau, E.; Sánchez-Mata, D. A hazel tree relict community (*Corylus avellana* L., Betulaceae) from the Guadiana River Middle Basin (Ciudad Real, Spain). *Lanzaroa* **2015**, *36*, 133–137. [CrossRef]

38. Reimer, P.J.; Bard, E.; Bayliss, A.; Beck, J.W.; Blackwell, P.G.; Bronk Ramsey, C.; Buck, C.E.; Cheng, H.; Edwards, R.L.; Friedrich, M.; et al. Intcal13 and marine13 radiocarbon age calibration curves 0–50,000 years cal BP. *Radiocarbon* **2013**, *55*, 1869–1887. [CrossRef]

39. Hua, Q.; Barbetti, M. Review of tropospheric bomb ^{14}C data for carbon cycle modelling and age calibration purposes. *Radiocarbon* **2004**, *46*, 1273–1298. [CrossRef]

40. Blaauw, M. Methods and code for classical age-modelling of radiocarbon sequences. Quaternary. *Geochronology* **2010**, *5*, 512–518. [CrossRef]

41. Aaby, B.; Berglund, B.E. Characterization of peat and lake deposits. In *Handbook of Holocene Palaeoecology and Palaeohydrology*; Berglund, B.E., Ed.; John Wiley and Sons Ltd.: Chichester, UK, 1986; pp. 231–246. ISBN 9781930665804.

42. Moore, P.D.; Webb, J.A.; Collinson, M.E. *Pollen Analysis*; Blackwell: London, UK, 1991; 216p, ISBN 9780632021765.

43. Goeury, C.; de Beaulieu, J.L. À propos de la concentration du pollen à l'aide de la liqueur de Thoulet dans les sédiments minéraux. *Pollen Spores* **1979**, *21*, 239–251.

44. Reille, M. *Pollen et spores d'Europe et d'Afrique du Nord*, 2nd ed.; Laboratoire de Botanique Historique et Palynologie: Marseille, France, 1999; 543p, ISBN 2950717500.

45. Van Geel, B. Non-pollen palynomorphs. In *Tracking Environmental Change Using Lake Sediments, Vol. 3, Terrestrial, algal, and Siliceous Indicators*; Smol, J.P., Birks, H.J.B., Last, W.M., Eds.; Kluwer: Dordrecht, The Netherlands, 2001; pp. 99–119. ISBN 9781402006814.

46. Beug, H.J. *Leitfaden der Pollenbestimmung für Mittleleuropa und Angrenzende Gebeite*; Gustav Fisher Verlag: Stuttgart, Germany, 2004; ISBN 9783899370430.

47. Cugny, C.; Mazier, F.; Galop, D. Modern and fossil non-pollen palynomorphs from the Basque mountains (western Pyrenees, France): The use of coprophilous fungi to reconstruct pastoral activity. *Veg. Hist. Archaeobot.* **2010**, *19*, 391–408. [CrossRef]

48. Mateus, J.E. Pollen Morphography of Portuguese Ericales. *Revista Biología* **1989**, *14*, 135–208.

49. Blackmore, S.; Steinmann, J.A.J.; Hoen, P.P.; Punt, W. Betulaceae and Corylaceae. *Rev. Palaeobot. Palynol.* **2003**, *123*, 71–98. [CrossRef]

50. Mercuri, A.M.; Bandini Mazzanti, M.; Florenzano, A.; Montecchi, M.C.; Rattighieri, E.; Torri, P. Anthropogenic Pollen Indicators (API) from archaeological sites as local evidence of human-induced environments in the italian península. *Ann. Bot. Coenol. Plant Ecol.* **2013**, *3*, 143–153. [CrossRef]

51. Stockmarr, J. Tablets with spores used in absolute pollen analysis. *Pollen Spores* **1971**, *13*, 614–621.

52. Grimm, E.D. *TGView*; Illinois State Museum, Research and Collection Center: Springfield, MA, USA, 2004.

53. Grimm, E.C. Coniss: A Fortran 77 program for stratigraphically constrained cluster analysis by the method of incremental sum of squares. *Comput. Geosci.* **1987**, *13*, 13–35. [CrossRef]

54. Bennett, K.D. Determination of the number of zones in a biostratigraphical sequence. *New. Phytol.* **1996**, *132*, 155–170. [CrossRef]

55. Whitlock, C.; Larsen, C. Charcoal as a fire proxy. In *Tracking Environmental Change Using Lake Sediments: Volume 3. Terrestrial, Algal, and Siliceous Indicators*; Smol, J.P., Birks, H.J.B., Last, W.M., Eds.; Kluwer: Dordrecht, The Netherlands, 2001; pp. 75–97. ISBN 9780306476686.

56. Vannière, B.; Colombaroli, D.; Chapron, E.; Leroux, A.; Tinner, W.; Magny, M. Climate versus human-driven fire regimes in Mediterranean landscapes: The Holocene record of Lago dell'Accesa (Tuscany, Italy). *Quat. Sci. Rev.* **2008**, *27*, 1181–1196. [CrossRef]

57. Tinner, W.; Hu, F.S. Size parameters, size-class distribution and area-number relationship of microscopic charcoal: Relevance for fire reconstruction. *Holocene* **2003**, *13*, 499–505. [CrossRef]

58. Long, C.J.; Whitlock, C. Fire and vegetation history from the coastal rain forest of the Western Oregon Coast Range. *Quat. Res.* **2002**, *58*, 215–225. [CrossRef]

59. Carcaillet, C.; Bouvier, M.; Fréchette, B.; Larouche, A.C.; Richard, P.J.H. Comparison of pollen-slice and sieving methods in lacustrine charcoal analyses for local and regional fire history. *Holocene* **2001**, *11*, 467–476. [CrossRef]

60. Luelmo-Lautenschlaeger, R.; López-Sáez, J.A.; Pérez-Díaz, S. Las Lanchas, Toledo Mountains (central Spain). *Grana* **2018**, *57*, 246–248. [CrossRef]

61. Luelmo-Lautenschlaeger, R.; López-Sáez, J.A.; Pérez-Díaz, S. Botija, Toledo Mountains (central Spain). *Grana* **2018**, *57*, 322–324. [CrossRef]

62. Herrera-Casado, A. La Marca media de Al-Andalus en tierras de Guadalajara. *Wad-al-Hayara* **1985**, *12*, 9–26.

63. Boloix-Gallardo, B. La taifa de Toledo en el Siglo XI. Aproximación a sus límites y extensión territorial. *Tulaytula* **2001**, *8*, 23–57.

64. Ladero-Quesada, M.A. *Toledo en Época de la Frontera*; Anales de la Universidad de Alicante. Historia Medieval; Universidad de Alicante: Alicante, Spain, 1984; Volume 3, pp. 71–88.

65. Blanco-González, A.; López-Sáez, J.A.; López-Merino, L. Ocupación y uso del territorio en el sector centromeridional de la cuenca del Duero entre la Antigüedad y la Alta Edad Media (siglos I-XI d.C.). *Archivo Español de Arqueología* **2009**, *82*, 275–300. [CrossRef]

66. Blanco-González, A.; López-Sáez, J.A.; Alba, F.; Abel-Schaad, D.; Pérez-Díaz, S. Medieval landscapes in the Spanish Central System (450–1350): A palaeoenvironmental and historical perspective. *J. Mediev. Iber. Stud.* **2015**, *7*, 1–17. [CrossRef]

67. Carrobles, J.; Morín, J.; Rodríguez, S. La génesis de un paisaje medieval II: Los espacios ganaderos bajomedievales. In *Alquerías, Cigarrales y Palacios: La Quinta de Mirabel*; Carrobles, J., Morín, J., Eds.; AUDEMA S.A.: Toledo, Spain, 2016; pp. 115–134. ISBN 9788416450145.

68. López-Sáez, J.A.; Abel-Schaad, D.; Robles-López, S.; Pérez-Díaz, S.; Alba-Sánchez, F.; Nieto-Lugilde, D. Landscape dynamics and human impact on high-mountain woodlands in the western Spanish Central System during the last three millennia. *J. Archaeol. Sci. Rep.* **2016**, *9*, 203–218. [CrossRef]

69. Pastor-de Togneri, R. *Del Islam al Cristianismo. En las Fronteras de dos Formaciones Económico-Sociales*; Toledo Siglos XI-XIII; Península: Barcelona, Spain, 1985; 186p, ISBN 9788429711479.

70. Izquierdo-Benito, R. *Vascos: La Vida Cotidiana en una Ciudad Fronteriza de al-Andalus*; Junta de Comunidades de Castilla-La Mancha: Toledo, Spain, 1999; p. 175. ISBN 8477882282.

71. Gerbet, M.C. *La Ganadería Medieval en la Península Ibérica*; Crítica: Barcelona, Spain, 2002; 281p, ISBN 9788484324164.

72. Bradley, R.S. Climate of the last millenium. In Proceedings of the Holocene Working Group Workshop, Hafslo, Norway, 27–29 August 2003; Bjerknes Centre for Climate Research: Bergen, Norway, 2003.

73. Sánchez-López, G.; Hernández, A.; Pla-Rabes, S.; Trigo, R.M.; Toro, M.; Granados, I.; Sáez, A.; Masqué, P.; Pueyo, J.J.; Rubio-Inglés, M.J.; Giralt, S. Climate reconstruction for the last two millenia in central Iberia: The role of East Atlantic (EA), North Atlantic Oscillation (NAO) and their interplay over the Iberian Peninsula. *Quat. Sci. Rev.* **2016**, *149*, 135–150. [CrossRef]

74. Izquierdo-Benito, R. *Reconquista y Repoblación de la Tierra Toledana*; Diputación provincial de Toledo-Instituto Provincial de Investigaciones y Estudios Toledanos: Toledo, Spain, 1983; 45p, ISBN 0211-4607.

75. Carrobles-Santos, J. El Cuidado del Monte. Evolución de las políticas de protección de la masa forestal en Toledo entre los siglos XIV y XVI. *Cuadernos de la SECF* **2009**, *30*, 143–151.

76. Izquierdo-Benito, R. *Monografías. Castilla la Mancha en la Edad Media*; Servicio de Publicaciones de la JUNTA de Comunidades de Castilla La Mancha: Toledo, Spain, 1985; 160p, ISBN 9788450510485.

77. Martín-Martín, J.L. El campesinado en los Montes de Toledo en los siglos XVIII y XIX. *Beresit* **2005**, *5*, 93–121.

78. Pastor-de Togneri, R. La lana en Castilla y León antes de la organización de la Mesta. In *Contribución a la Historia de la Trashumancia en España*; García Martín, P., Sánchez Benito, J.M., Eds.; Ministerio de Agricultura, Pesca y Alimentación, Secretaría General Técnica: Madrid, Spain, 1996; pp. 363–390. ISBN 9788474794960.

79. Izquierdo-Benito, R. *Privilegios Reales Otorgados a Toledo Durante la Edad Media: 1101–1494*; Instituto Provincial de Investigaciones y Estudios Toledanos: Toledo, Spain, 1990; 327p, ISBN 9788487103063.

80. Mann, M.E.; Bradley, R.S. Northern Hemisphere Temperatures during the past millenium: Inferences, uncertainties and limitations. *Geophys. Res. Lett.* **1999**, *26*, 759–762. [CrossRef]

81. Oliva, M.; Ruiz-Fernández, J.; Barriendos, M.; Benito, G.; Cuadrat, J.M.; Domínguez-Castro, F.; García-Ruiz, J.M.; Giralt, S.; Gómez-Ortiz, A.; Hernández, A.; et al. The Little Ice Age in Iberian mountains. *Earth-Sci. Rev.* **2018**, *177*, 175–208. [CrossRef]

82. Jiménez-de Gregorio, F. *La Comarca Histórica Toledana de los Montes de Toledo*; Instituto Provincial de Investigación y Estudios Toledanos-Diputación de Toledo: Toledo, Spain, 2008; 149p, ISBN 8487103943.

83. De Linares, V.G.G. Los bosques en España a lo largo de la Historia. In *Historia de Los Bosques. El Significado de la Madera en el Desarrollo de la Civilización*; Perlin, J., Ed.; Gaia Proyecto 2050: Madrid, Spain, 1999; pp. 429–480. ISBN 9788493023218.

84. Domínguez-Castro, F.; Santisteban, J.I.; Barriendos, M.; Mediavilla, R. Reconstruction of drought episodes for central Spain from rogation ceremonies recorded at the Toledo Cathedral from 1506 to 1900: A methodological approach. *Glob. Planet. Chang.* **2008**, *63*, 230–242. [CrossRef]

85. Martín-Lou, M.A.; López-Vizoso, J.M.; Martínez-Vega, J. Las actividades forestales en los Montes de Toledo. *Anales de Geografía de la Universidad Complutense* **1992**, *12*, 233–241.

86. Martínez-Vega, J. La actual situación de los bosques en la Comarca de los Montes de Toledo. *Estud. Geogr.* **1990**, *51*, 577–588.

87. Diago-Hernando, M. *Mesta y Trashumancia en Castilla. Siglos XIII-XIX*; Arco Libros: Madrid, Spain, 2002; ISBN 8476355181.

88. Bauer-Maanderscheid, E. *Los Montes de España en la Historia*; Ministerio de Agricultura: Madrid, Spain, 1980; 610p, ISBN 8474790840.

89. González-González De Linares, V.; García-Viñas, J.I.; Carrero Díez, L.; Cuevas Moreno, J.; González-Doncel, I.; Gil, L. Effects on vegetation of historical charcoal making in central spain: The "Montes de Toledo" case. In Proceedings of the International Conference of the European Rural History Organisation (EURHO), Bern, Switzerland, 19–22 August 2013.

90. Sánchez-Miguel, J.M. Las tenerías o tinterías en los Montes de Toledo. *Revista de Estudios Monteños* **1988**, *42*, 25–28.

91. Moreno-Gómez, A.; Rodríguez-Vigal, C.; Ferrandis-Gotor, P.; de las Heras-Ibáñez, J. Impacto del escodado del ciervo (*Cervus elaphus* L.) sobre la cornicabra (*Pistacia terebinthus* L.) en ≪Quintos de Mora≫ (Los Yébenes, Toledo). *Investig. Agrar. Sist. Recur. For.* **2001**, *10*, 81–93.

92. Orueta, J.F.; Aranda, Y.; García, F.J. Efecto del ramoneo del ciervo (*Cervus elaphus*) sobre dos especies del matorral mediterráneo en los Montes de Toledo (Centro de España). *Galemys SECEM* **1998**, *10*, 27–36.

93. Huerta-González, M. El carbón vegetal en los Montes de Toledo. *Revista de estudios Monteños* **2006**, *114*, 8–23.

94. Mercuri, A.M. Genesis and evolution of the cultural landscape in central Mediterranean: The ≪where, when and how≫ through the palinological approach. *Landsc. Ecol.* **2014**, *29*, 1799–1810. [CrossRef]

![sustainability logo]

sustainability

MDPI

Article

The History of Pastoral Activities in S Italy Inferred from Palynology: A Long-Term Perspective to Support Biodiversity Awareness

Assunta Florenzano

Laboratorio di Palinologia e Paleobotanica, Dip. Scienze Vita, Università degli Studi di Modena e Reggio Emilia, 41100 Modena, Italy; assunta.florenzano@unimore.it; Tel.: +39-059-205-8276

Received: 13 December 2018; Accepted: 11 January 2019; Published: 15 January 2019

Abstract: The present-day Mediterranean landscape is a result of the long-term human–environment–climate interactions that have driven the ecological dynamics throughout the Holocene. Pastoralism had (and still has) an important role in shaping this landscape, and contributes to maintaining the mosaic patterns of the Mediterranean habitats. Palaeoecological records provide significant multi-proxy data on environmental changes during the Holocene that are linked to human activities. In such research, the palynological approach is especially useful for detailing the complexity of anthropogenically-driven landscape transformations by discriminating past land uses and pastoral/breeding activities. This paper focuses on the palynological evidence for the impact of centuries of grazing on the vegetation of Basilicata, a region of southern Italy where animal breeding and pastoralism have a long tradition. A set of 121 pollen samples from eight archaeological sites (dated from the 6th century BC to the 15th century AD) and five modern surface soil samples were analyzed. The joint record of pollen pasture indicators and spores of coprophilous fungi suggests that continuous and intense pastoral activities have been practiced in the territory and have highly influenced its landscape. The palaeoecological results of this study provide us with better knowledge of the diachronical transformations of the habitats that were exposed to continuous grazing, with a shift toward more open vegetation and increase of sclerophyllous shrubs. The palynological approach gives insights into the vocation and environmental sustainability of this southern Italy region on a long-term basis.

Keywords: pollen; NPPs; pasture indicators; palaeoecology; archaeological sites; palaeoenvironmental reconstruction; southern Italy; Mediterranean

---

1. Introduction

Since prehistoric times, humans have directly (by activities) or indirectly (by presence) influenced ecosystems and landscapes around the world [1]. Human actions have resulted in several changes in the biosphere by influencing and driving the main variations in biodiversity and ecosystem processes [2]. From a long-term perspective, human influence and then—depending on the time and space scale—impact have generated notable patterns of landscape complexity [3–6]. This is especially evident in the Mediterranean area, where, since ancient times, the continuous succession of different civilizations has had direct and indirect effects on the environment that led to changes in the vegetation cover [7–9]. Grazing is one of the most important factors determining patterns of vegetation in Mediterranean ecosystems [10–12]. In the Mediterranean basin, long-term pastoral activities have had impressive consequences for the current biodiversity [13]. It is widely recognized that grazing prevents the growth of woody vegetation and reduces the species richness, and that intensive grazing profoundly affects the succession in such systems [14–17]. Moreover, it is well-known that grazing also affects the morphological and functional traits of plants at the global level [18].

Nevertheless, grazing should not be considered a negative factor that has a detrimental impact on plant biodiversity. It actually improves the quality of vegetation by removing old plants and stimulating grass production [19]. Recent multidisciplinary studies (integrating palaeoecological, pedological, phylogenetic, and palaeontological evidence) have shown the key role that wild herbivores and domestic livestock play in grassland–forest dynamics, contributing to the maintenance of the European mosaic-pattern vegetation that was created by megaherbivore disturbances at the Pleistocene–Holocene boundary [20,21].

Over the last few decades, several ecological studies dealing with grazing's effects on vegetation and the environment have been carried out in order to obtain information essential to improving landscape management and sustainable resources exploitation [22–25]. These studies are based on short-term ecological data that span only a few decades, and they do not provide potentially significant trends in these records, nor allow us to predict possible future scenarios of biosphere responses to global changes. Indeed, it is the study of long-term ecological records (greater than 50 years) that actually has great potential in conservation biology, as it provides us with a valuable historical perspective on the dynamics of contemporary ecological systems [26–30]. Palaeoecological records, which provide multi-proxy palaeoenvironmental data, are essential for understanding details of long-term human impacts on ecosystems [31,32]. Among the multi-proxy analyses, palynological investigations allow us to acquire a high-resolution data set on environmental changes during the Holocene that are linked to human activities (e.g., [4,33]). Microscopic biological remains (pollen and non-pollen palynomorphs (NPPs), such as fungal spores and algal elements) from sedimentary records are meaningful bio-indicators for palaeoecological reconstruction (e.g., [34–37]). In particular, the combined evidence of pollen and NPPs from archaeological sites is useful for tracing the impact of past human activities on the local ecosystem; these biomarkers can also be used to discriminate past land uses and pastoral/breeding practices (e.g., [38–42]).

A palynological approach to the study of Mediterranean landscapes is ideally suited for detecting the land-use history and ecosystem changes that gave rise to the present-day Mediterranean environment. Such challenging issues have been addressed, for example, by Jouffroy-Bapicot and colleagues in the framework of the multidisciplinary research project on the making of the Cretan Mountain Landscape, Greece [43]. Multi-proxy analyses of the Asi Gonia peat bog sediments have shown that, during the two last millennia, pastoralism had a predominant role in vegetation dynamics. Other studies dealing with palynological investigations in grazing areas have been conducted on modern analogues in the Pyrenees Mountains to characterize pollen and non-pollen palynomorphs (NPP) indicators of types of highland vegetation and grazing pressure [44–47]. In the same study area, previous palaeoenvironmental studies that were carried out in several peat basins have proved that these wetlands were grazed in the past [48,49].

Previous Studies and the Aim of This Paper

In Italy, an outstanding number of studies that include pollen or other plant remains from archaeological contexts (BRAIN database: *brainplants.successoterra.net*) [50] have testified to the widespread occurrence of breeding/pastoral activities on the peninsula over the last few millennia. This long-term practice can also be inferred from pollen spectra from the top cores (Holocene sediments) of biostratigraphical records [51–55]. However, only a few papers focused solely on past pastoral practices and their legacy on the current Italian landscape have been published thus far [56]. On this issue, palaeoenvironmental research in southern Italy—and especially research carried out in Basilicata, on which this paper is based—can serve as a most suitable example. Here, animal breeding and pastoralism have a long tradition and have certainly played an important role in shaping the landscape. The PhD project on the archaeo-environmental reconstruction of this region has provided a palynological dataset that may be of key relevance for understanding the pastoralism that has been practiced in the area over the last 2500 years [57]. Some aspects of this research have already been published, such as the environmental and economic settings of the rural sites of the

Greek colonial system on the coastal plain [58–61], and the agro-pastoral characterization of the inland indigenous sites [62–65]. Besides this, some specific issues that emerged from the research have been further explored (e.g., the value of Cichorieae as a pastoral indicator [66] and the *Olea* pollen representativeness in the modern local olive groves [67]).

This work focuses on the palynological evidence of pastoral/breeding activities in the studied archaeological records and explores the impact of centuries of grazing on the vegetation of Basilicata. The main aim of this study is to better understand the complexity of landscape transformations that have occurred in these lands, which have continuously been exploited as pastures, thus as to improve our awareness of the biodiversity in, and the long-term human impact on, the current landscape of a typical region of southern Italy. As part of the historically underdeveloped area of *Mezzogiorno*, in the last few decades, Basilicata has faced several challenges due to its geological instability, peripherality, and marked depopulation. To improve its economic development, the Italian government has implemented policy interventions and development strategies that are supported by European programmes, which threaten to transform the natural and cultural assets of the region. A long-term perspective on the development of its cultural landscape is fundamental to provide strategies to avoid the loss of environmental peculiarities and biodiversity that comply with the cultural identity and vocation of this territory.

2. Materials and Methods

2.1. Study Area

This research has been carried out on samples from the main archaeological sites of the Basilicata region, which are located in a transect from the southern Apennines (1 site) to the Ionian coast along the Bradano river (7 sites). The eight sites (Figure 1, Table 1) have been studied in collaboration with the University of Basilicata (sites 1–4) [68,69] and the University of Texas at Austin (sites 5–8) [70–72].

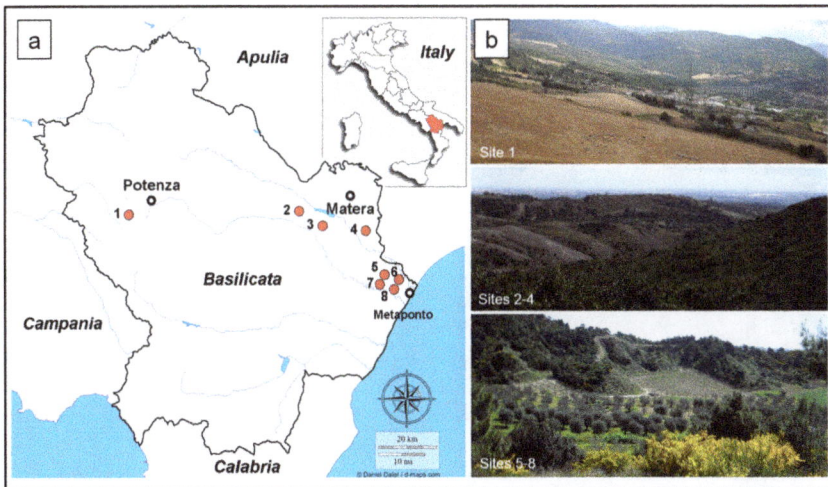

Figure 1. Study area: (**a**) Location map of the eight archaeological sites studied for pollen and non-pollen palynomorphs (NPPs) in Basilicata, southern Italy; (**b**) the current landscapes in which the sites are located.

Table 1. The studied sites in the Basilicata region, S Italy: Geographical coordinates, chronology, cultural phase, archaeological context, and the number of pollen samples.

Site	Geographical Coordinates	Chronology	Cultural Phase	Archaeological Context	Samples
1 - Torre di Satriano–TS	40°34′12.28″N; 15°38′15.26″E 930 m asl	6th cent. BC	Archaic	indigenous rural settlement	5
2 - Altojanni–ALJ	40°39′73″N; 16°20′57″E 375 m asl	3rd-5th and 12th-15th cent. AD	Roman; Medieval	fortified rural village; aggers	26
3 - Miglionico–MGL	40°34′03.42″N; 16°29′58.97″E 445 m asl	14th/15th cent. AD	Medieval	castle	10
4 - Difesa San Biagio–DSB	40°30′21.82″N; 16°40′51.03″E 138 m asl	5th-1st cent. BC	Hellenistic	indigenous rural settlement	24
5 - Fattoria Fabrizio–FF	40°24′46.58″N; 16°44′28.33″E 57 m asl	6th-4th cent. BC	Archaic-Hellenistic	Greek farmhouse	14
6 - Pizzica–PZZ	40°24′45.09″N; 16°47′28.14″E 36 m asl	5th/4th cent. BC	Archaic-Hellenistic	drainage channel in Greek necropolis	5
7 - Sant'Angelo Vecchio–SAV	40°23′39.96″N; 16°43′10.71″E 46 m asl	6th/5th-1st cent. BC	Archaic-Hellenistic	Greek rural settlement	29
8 - Pantanello–PNT	40°23′21.53″N; 16°47′11.89″E 8 m asl	2nd-1st cent. BC	Hellenistic	dump of a productive area	13

Torre di Satriano (site 1) is the only high-ground site (c. 900 m asl); although it is subject to very cold winter temperatures, it is located in a territory rich in natural resources, such as a vast supply of water, extensive hilly areas for pastures and cultivations, and wood. The archaeological excavations brought to light the remains of an outstanding dwelling, which has been interpreted as the center of power within the local community between 560/550 BC and 480 BC [73]. The sites of the low-hill inland area (2–4) lie along the Bradano river, which was an important waterway for the transport of goods. These sites are fortified rural settlements that span from the Archaic/Hellenistic (4) to the Roman periods and the Middle Ages/Renaissance (2,3), and their displacement on the top of the hills is due to both defense and control reasons [68,69]. In fact, the river has always represented the way for foreign populations (such as the Greek settlers) who were attracted to the lands for farming to access the territory. The sites of the coastal plain (5–8), which all belong to the *chora* (rural territory) of the Greek colonial city of Metaponto, are located mainly on the first marine terraces a few meters above the plain, which is often subject to floods. The network of small farmhouses and rural villages along the Bradano river had a crucial role in setting up the local farming system and regiment the water course and its floods [71].

The climate of the study area has a typical Mediterranean seasonality. The vegetation typologies that occur in the region depend on the high environmental diversity of the territory [74]. Most of the study area is characterized by the Mesotemperate Turkey oak vegetation series of *Lathyro digitati-Quercetum cerridis* [75]. In the Apennine sector, anthropogenic meadows (including grassland species such as *Trifolium nigrescens* Viv., *Medicago hispida* Gaertner, *Dactylis glomerata* L. subsp. *glomerata*, *Centaurea centaurium* L. and *Scorpiurus muricatus* L.), and mixed oakwoods (mainly composed of *Quercus cerris* L., *Q. frainetto* Ten., *Q. virgiliana* Ten., *Q. pubescens* Willd., usually with *Carpinus orientalis* Mill., *Fraxinus ornus* L., *Acer monspessolanum* L.) are the main vegetation [75]. The low-hill inland area is covered with shrubby Mediterranean plants (*Pistacia lentiscus* L., *Phillyrea latifolia* L., *Cistus creticus* L., *Helianthemum jonium* Lacaita and other elements related to *Pistacio Rhamnetalia alaterni* plant communities) that are associated with grasses and annual herbs included in the *Camphorosmo-monspeliacae-Lygeetum sparti* and *Medicago coronate-Hedysaretum glomerati* vegetation series. In the Metaponto area, the relatively recent *Pinus halepensis* reforestations influence the dune communities, which the *Salsolo kali-Cakiletum maritimae* and *Malcomietalia* series dominate [75,76]. Vegetables and fruits, especially cherry orchards, citrus groves, and vineyards, grow in the mild climate of the coastal plain [76].

2.2. Pollen Samples and Analysis

A set of 121 pollen samples were taken from archaeological layers—small trenches, rooms or floors of houses, and spot samples from the eight sites. Sometimes also fillings of pottery were analysed, but data were not used for palaeoenvironmental reconstructions. In addition, five pollen samples were collected from the current surface soils to obtain a reference of the modern composition of local pollen rain. According to the chronology of the archaeological sites, the temporal transect spans about 20 centuries—from the 6th century BC to the 15th century AD (Table 1).

Pollen samples, of about 8 g each, were prepared using tetra NaPyrophosphate, HCl 10%, acetolysis, separation with NaMetatungstate hydrate, HF 40%, and ethanol [77]. *Lycopodium* tablets were added to calculate concentrations (p/g = pollen per gram of sediment, and npp/g = npp per gram) [78]. Permanent pollen slides were mounted on glycerol jelly. Identification was made at 400× and 1000× magnification, with the help of keys, atlases [79–81], and the reference pollen collection of the Laboratory of Palynology and Palaeobotany of the University of Modena and Reggio Emilia. Non-pollen palynomorphs (NPPs, microfossils of a great variety of organisms—mainly fungi and algae—sensitive to various ecological parameters or to human presence) were identified according to the reference literature [82–84]. Among NPPs, particular attention has been paid to the coprophilous fungal spores (e.g., *Sordaria*, *Sporormiella*, *Delitschia*, *Podospora* types) as indices of the presence of dung, and therefore associated with pastoral/breeding activities [38,47].

Pollen and NPPs were counted in the same samples. On average, about 350 pollen grains and 200 NPPs per sample were counted. The percentage of pollen diagrams were basically calculated from pollen sums including all pollen counted. Five pollen groups which refer to the plant landscape and human activities indicators were calculated: (a) Mixed oakwood (including broadleaved *Quercus*-oaks, *Carpinus betulus*-hornbeam, *Ostrya carpinifolia*-hop hornbeam, *Fraxinus*-ash, *Tilia*-linden, *Ulmus*-elm, and *Corylus*-hazel); (b) Mediterranean plants (*Quercus ilex*-evergreen oak type, *Daphne* cf. *gnidium*-flax-leaved daphne, *Helianthemum*-rock rose, *Juniperus*-juniper type, *Erica*-heat, *Olea*-olive tree, *Phillyrea*-green olive tree, *Pistacia* cf. *lentiscus*-lentisk, *Rhamnus*-buckthorn type); (c) Hydro-hygrophilous plants (sum of plants from wet environments: *Alnus*-alder, Cyperaceae-sedges, *Nymphaea* cf. *alba*-white water lily, *Phragmites* cf. *australis*-common reed, *Populus*-poplar, *Ranunculus* cf. *macrophyllus*-large leaved buttercup, *Sagittaria*-arrowhead, *Typha/Sparganium*-cattail/bur-reed type); (d) cultivated/cultivable plants (the woody plants *Castanea*-chestnut, *Corylus*-hazelnut, *Juglans*-walnut, *Olea*, *Prunus*-plum, *Malus/Pyrus*-apple/pear, *Vitis*-grapevine, and cereals *Avena/Triticum*-oats/wheat group and *Hordeum*-barely group); (e) local pastoral pollen indicators (LPPI) [85], including plants strictly linked to local pastoral activities (*Cirsium*, *Centaurea nigra* type and other Asteroideae, Ranunculaceae, *Galium* type, *Potentilla* type, and Cichorieae—the tribe with fenestrate pollen within the Cichorioideae subfamily [66]).

The pollen diagrams were drawn with TGView [86]. On-site pollen data were chronologically ordered and grouped according to the main time phases of the sites. Following Mercuri [87,88], the chronological pollen samples groups were treated as a single 'regional site' to interpret the regional data set by comparing coeval samples and checking the main floristic differences among the chronological phases. Moreover, though their value of local palaeoecological indicators is well known, in the elaborations, NPPs were also treated as 'regional proxies' in order to compare the different sites.

3. Results and Discussion

The pollen dataset shows evidence of the diachronic evolution of the Basilicata cultural landscape over about 2500 years of (mainly pastoral) human activities. Below, after a description of the state of preservation and concentration of pollen and other palynomorphs, the main results on the vegetation cover and land-uses in the region are briefly reported; greater emphasis has been placed on the signs of pastoral/breeding practices in the landscape, which are treated in different subsections. Data are presented according to the main chronological/cultural phases at a regional scale; the principal palynological results site-by-site are summarised in Table 2.

Table 2. Summary of pollen and NPP data from each of the eight studied sites.

Site	Concentr.	AP	Mixed Oakwood	Mediterr. Plants	Hydro-hygroph. Plants	Cereals	Cultiv. Plants (tot.)	LPPI	Coproph. Fungi	Main Features
1 - TS	c. 2100 p/g; c. 7000 npp/g	6.6%	2.5%	1.5%	11.6%	6.4%	6.7%	38.1%	c. 3500 npp/g	Open environment, with spread wet environments; cereal fields and pastures widespread
2 - ALJ	c. 16,200 p/g; c. 62,200 npp/g	13.7%	6.4%	4.9%	4.1%	0.7%	2.8%	34.4%	c. 17,600 npp/g	Thin oakwood combined with Mediterranean vegetation; dry grasslands/pastures widespread
3 - MGL	c. 1640 p/g; <100 npp/g	11%	1.9%	1.6%	5.5%	1.6%	2.6%	35.2%	<50 npp/g	Reduced oakwood and Mediterranean habitats; land-use with cereal fields and pastures
4 - DSB	c. 29,200 p/g; c. 87,000 npp/g	7.5%	0.3%	4.5%	0.1%	0.2%	2.7%	39%	c. 9000 npp/g	Open vegetation, in which Mediterranean shrubs and dry grasslands/pastures predominate
5 - FF	c. 16,700 p/g; c. 5400 npp/g	19.4%	4.8%	9%	2.3%	1.7%	4.6%	23.9%	c. 900 npp/g	Thin oakwood and local growing of Mediterranean shrubs; economy based on cereals and pastures
6 - PZZ	c. 1900 p/g; c. 300 npp/g	11.6%	0.7%	10.1%	2.5%	-	0.1%	47.8%	<100 npp/g	Mediterranean habitats widespread; no evidence of local cultivations; dry grasslands/pastures widespread
7 - SAV	c. 5500 p/g; c. 2600 npp/g	12%	3.4%	5.7%	6.9%	0.7%	3.2%	34.6%	c. 550 npp/g	Mediterranean habitats combined with wet environments; land-use with pastures and reduced cereal fields
8 - PNT	c. 22,000 p/g; <100 npp/g	3%	1.5%	0.8%	1.3%	4.5%	4.9%	48.2%	<50 npp/g	Open vegetation predominates; dry-grasslands/pastures spread and land-use with cereal fields

3.1. Pollen Preservation and Concentration of Palynomorphs

Pollen was found in almost all samples in a generally good state of preservation, although many grains were crumpled, broken, or rearranged due to the physical–chemical and post-depositional disturbance processes that are fairly common in the archaeological deposits [89–92]. In addition, some pollen grains were thinned and pale, likely as a consequence of their passage through the digestive tract before their inclusion in the deposits. Several pollen clumps—small single, or mixed-type pollen clusters (Figure 2)—were observed at three sites (1, 5, and 7). These clumps could be interpreted as the remains of anthers (flowers standing in/transported to the sites) or as fecal pellets dropped by arthropods and herbivorous mammals [88,93–95]. Interestingly, no pollen clumps have been observed in the modern soil samples.

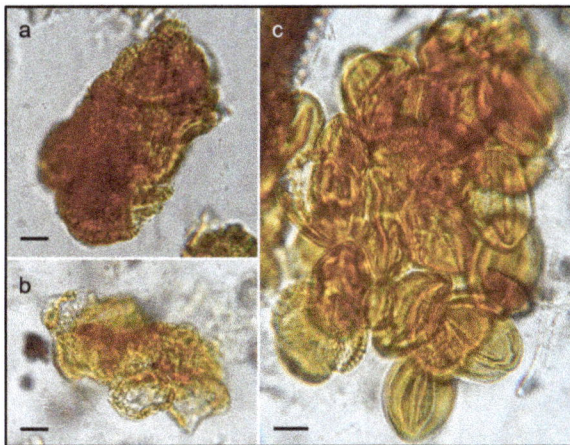

Figure 2. Pollen clumps from the archaeological layers of site 7: (**a,b**) Single-type pollen clumps; (**c**) mixed-type cluster, including more than one pollen type. The scale bar is 10 µm.

The pollen and the NPP concentrations were quite high in most of the archaeological samples (overall mean concentrations: about 11×10^3 p/g and 20×10^3 npp/g). The modern samples showed a notable pollen concentration (26×10^3 p/g), while NPPs were less recurring in the slides (7×10^3 npp/g).

Pollen spectra were characterized by a high number of taxa, mainly belonging to herbs (mostly Cichorieae and Poaceae wild grass group), with floristic lists that include between 31 and 152 taxa per sample. The good mean pollen concentration and the high taxa diversity suggest that any selective deterioration affected pollen grains in the deposits [62]. Herbs also prevailed in the modern samples (again, Cichorieae and Poaceae were the most recurring taxa); their floristic lists (from 52 to 71 taxa) suggest a less-diversified vegetation composition as compared to the past.

3.2. Vegetation Cover: Wood Composition, Mediterranean Habitats, and Wet Environments

Woody plants were poorly represented in the pollen spectra (the mean percentage ratio of woodland/herbaceous plants is 11/89), and the tree cover, in particular, was very low. The forest cover was <20% at all sites and over the whole considered period (from the 6th century BC to the present; Figure 3). The woods were represented by both Mediterranean trees and shrubs (4% on average in the archaeological samples) and mixed oakwoods (3%). Fagaceae (deciduous *Quercus*, 1.9%) and Pinaceae (1.8%) prevailed in the forest cover, followed by *Olea* (1.3%), *Daphne* (0.9%), and *Quercus ilex* type (0.6%), which were the main elements of the Mediterranean vegetation.

The striking trait of the vegetation cover corresponded to a Mediterranean shrubland that was widespread during the Hellenistic period (>5% on average; up to 10.2% in the 5th–4th century BC), while its minimum values (0.2%) were observed during the 3rd–5th century AD, when the lowest values of the mixed oakwoods (1.2%) were recorded (Figure 3). In particular, Mediterranean plants recurred in the pollen spectra from the Greek rural sites 5 (9%) and 7 (5.7%) (Table 2) and were also significant in the Medieval samples from site 2 (7.1%) (Figure 3). Although the spread of sclerophyllous shrubs is linked to dryer environmental conditions, there was no local evidence of a concurrent significant reduction in wet environments (see below). This suggests that this was mostly a degraded environment rather than natural vegetation. In fact, the Mediterranean plants (*Olea* may have been cultivated, and the shrubs developed under continued grazing pressure) may have been the result of several human activities that were practised in the area for a long time.

In the surface sediment samples, the oakwoods and Mediterranean vegetation were well-represented. The modern pollen spectra indicate a prevalence of pines and Mediterranean shrubs in the current wood composition. The increase of sclerophyllous shrubs in the modern samples compared to the post-Hellenistic ones may be reasonably explained by a less-diversified land-use rather than by a true environmental change. Likewise, the extension of pinewoods is due to the modern afforestation of the abandoned farmlands in the region.

In past pollen spectra, wet environments were represented by *Alnus*, *Populus*, *Salix* (0.2%) and to a lesser extent *Populus* (0.1%), among trees, and by Cyperaceae (2.7%), *Ranunculus* cf. *macrophyllus* (0.8%), *Typha* types (0.3%) and *Nymphaea alba* type (0.2%), accompanied by other plants of wetlands in traces (e.g., *Lemna*, *Potamogeton*, *Hydrocharis*, <0.1%). These pollen grains are ubiquitous, and their highest mean value dated to the first Hellenistic phases (11.9%). All Roman samples have pollen from wet environments that never exceeds 3%, and then rises up to 5.6% in the Medieval Ages (Figure 3). Aquatic and hygrophilous plants matched the presence of the HdV-181, Zygnemataceae and *Pseudoschizaea* algae, suggesting that there were some small springs or stagnant shallow water pools not far from the sites [83,96–98].

In modern samples, pollen from plants living near or in the water showed fairly similar percentages (4.5% on average), while there are few algae. This is evidence of the occurrence of a low number of wet environments in the area.

Overall, the past pollen spectra describe open areas with scanty woodlands, shrubby grasslands, and a very local presence of wet environments. These data agree with the assumption that the area was largely treeless, with few variations in the wood composition. The Mediterranean elements that recur in the spectra could reflect the impact of pastoralism on the landscape.

Today, as in the past, the thin forest cover has a strong Mediterranean imprint, and the wet environments are similar to those that were present in the area after the Greek period. The ensemble of this pollen evidence, together with that from human environments (the following paragraphs), indicates that the natural environment of the region has always had a strong 'contamination imprint' by a human presence.

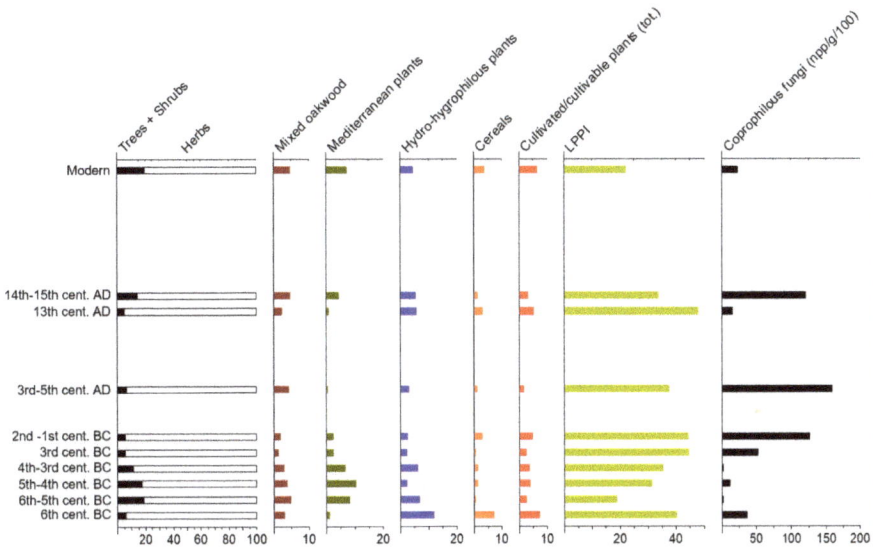

Figure 3. Percentage pollen diagram (average data from the eight sites, and modern soil samples): Main sums useful for palaeoenvironmental reconstruction, and concentration of coprophilous fungal spores (npp/g/100). The samples are grouped according to their chronology.

3.3. Crops and Fields

Cultivated plants were mainly represented by *Olea* (1.1%; up to 2.8% in the 5th–4th century BC samples from the sites 5, 6, and 7) and *Juglans* (0.2%) and traces of *Castanea* and *Vitis* (<0.1%) among woody plants. Unlike the other mentioned taxa, *Olea* was well-represented in all samples, and could reasonably have resulted from a process of cultivation instead of from the wild subspecies living in the area. In fact, *Olea* pollen is known to have been spreading since about 3600 cal. years BP in southern Italy, when a strong human impact was recorded (e.g., [99,100]). Moreover, olive tree cultivation has been well-attested to at the sites of Basilicata, where evidence of plant processing has also been found (e.g., olive presses at site 4) [62]. Other trees and shrubs that produce edible fruits that may be evidence of possibly cultivated plants were *Corylus* (0.3%), *Ficus* (0.1%), and *Prunus* (0.03%). Their low total amount (4% on average) suggests that they were grown/cultivated fairly far away from the sites.

Cereals were represented by the oat/wheat type (1.8%; up to 6.4% in the 6th century BC samples from site 1), barley type (0.3%), and traces of rye (0.04%). Therefore, wheat, rather than other cereals, were mainly cultivated in the fields in all phases. Weeds, such as Papaveraceae (0.1%), *Anagallis*, *Aphanes* type, *Cerastium*, *Convolvulus*, *Mercurialis*, *Polygonum*, and *Solanum* (total: <0.2%), are other indicators of crop practices [101]. Overall, the low amount of pollen from fields suggests that these cultivations were carried out quite far away from the sites (with the exception of site 1, where the cereal farming should have been very well-developed; Table 2). Other cultivated herbs may have been Fabaceae (3.4%, including *Astragalus*, *Lotus*, *Medicago*, and *Trifolium* type, legumes possibly cultivated for fodder) and some Apiaceae (0.6%, including aromatic species). Interestingly, *Cannabis* was attested to at site 7 (only in 2 samples dated 3rd century BC, 0.2% each), suggesting that some hemp was grown on the Metaponto plain during the Hellenistic period.

Agricultural activities were also well-attested to in the modern spectra (6.6%) by orchard crops, cereal fields, and olive groves. The cultivated woody plants (mainly olives) had values comparable to those obtained from the archaeological samples, while slightly higher percentages of cereals were observed in the current surface soils (3.9%). This suggests a continuity of the agricultural practices in the area, probably linked to and in line with the traditional land-uses of the territory, and despite the introduction of mechanized farming in recent years.

3.4. Grazing and Pasturelands

3.4.1. Pollen Pasture Indicators

All of the archaeological samples were characterized by plants and other evidence from grazing environments (Figure 4). Pollen spectra were dominated by Cichorieae (26.5%) and Poaceae wild grass group (18.7%), together with other Asteraceae (8.4%), Chenopodiaceae (5.5%), and Brassicaceae (4.5%). All of these taxa are indicative of dry pastures and are commonly included in the pollen assemblage that is employed for detecting the impact of pastoralism on the landscape (Figure 5) [44,45]. In particular, Cichorieae was recognized as one of the main pasture pollen indicators reflecting animal breeding and pasturelands [66]. The controversial significance of the overrepresentation of Cichorieae in pollen records from the Mediterranean context (as an effect of selective pollen deterioration in the deposits or due to the selection by animal browsing?) had been resolved in the studied sites through an integrated approach comparing the main results from palaeoecological and ecological studies that were carried out in the region. The study showed that Cichorieae prevail in secondary pastures and some types of primary open habitats; the recovery of high percentages of this pollen is, therefore, a good indicator of these habitats even in past environments [66] (p. 163).

Besides Cichorieae, the recovery of high percentages of Poaceae wild grass group—strongly correlated with open and grazed areas [44]—is further evidence of open habitats characterized by dry pastures. These xeric environments are, in fact, favored by the overexploitation of thinned plant resources, including overgrazing [62,102].

Cichorieae had prevalent values in the Roman (32%) and Medieval phases (31%; up to 41% in the 13th century samples). In the Hellenistic samples (23%), there was a steady increase in Cichorieae percentages from the 5th (9%) to the 2nd–1st century BC (34%) after a first peak (24%) in the 6th century BC (Figure 4). Similarly, the most significant values of Poaceae were found in the Roman (31%) and Medieval samples (18%), with high percentages also in the Greek phases (17%). These pollen data, along with the constant presence of Asteraceae (e.g., *Carduus*, *Centaurea nigra* type, *Cirsium*), Chenopodiaceae and Brassicaceae, indicate that dry grasslands/pastures might have been well-extended, probably with seasonal oscillations, in the past landscape.

Constantly recorded in the spectra were other herbs linked to grazing/breeding activities, such as *Plantago* (2.5%, recurring in trampled areas) [103,104] and Fabaceae (3.4%, palatable herbs present in meadow-pastures or used for fodder). Plantain had its maximum values in the Roman samples (5.6%) when legumes were less represented (0.8%). This probably means that animals grazed in open pastures without any additional feed, maybe due to the mild climate of this period. Fabaceae and *Plantago* instead had quite comparable values in almost all of the other studied phases (Figure 4).

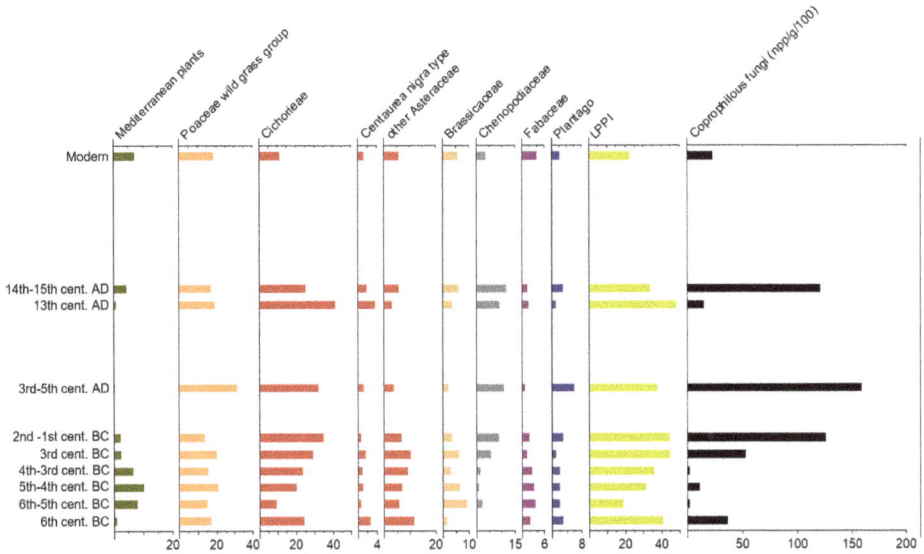

Figure 4. Percentage pollen diagram (average data from the eight sites, and modern soil samples): Mediterranean plants, pollen taxa indicators of (or linked to) pastures, and concentration of coprophilous fungal spores (npp/g/100). The samples are grouped according to their chronology.

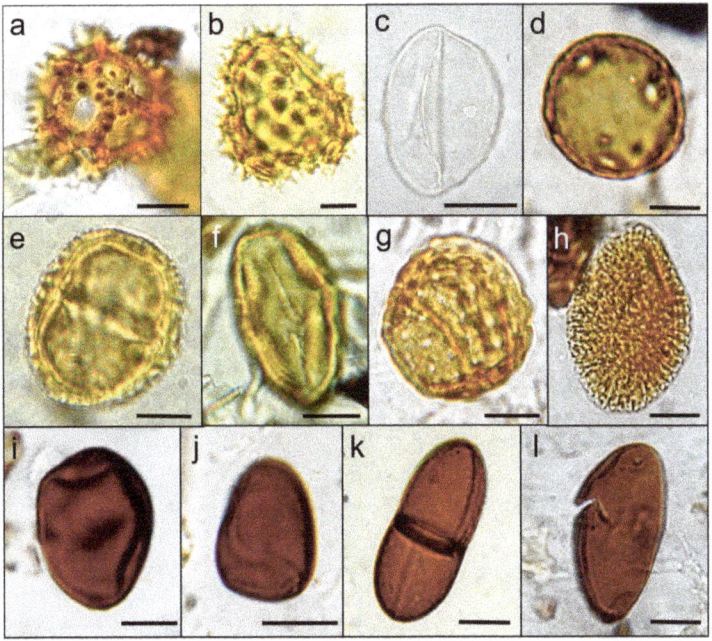

Figure 5. Main pollen and non-pollen palynomorphs indicators of grazing environments from the studied archaeological sites: (**a**) Cichorieae; (**b**) *Aster* type; (**c**) Poaceae wild grass group; (**d**) *Plantago*; (**e**) *Centaurea nigra* type; (**f**) *Trifolium* type; (**g**) Chenopodiaceae; (**h**) *Brassica* type; (**i**) *Sordaria* type; (**j**) *Sporormiella* type; (**k**) *Delitschia* type; (**l**) *Podospora* type. The scale bar is 10 µm.

According to Mazier [45,85], the LPPI—Local Pastoral Pollen Indicators is a sum that is helpful when investigating the response of plants to browsing and the local presence of animal breeding or pastoral activities in a territory. In the archaeological samples, the sum of LPPI had mostly high percentages (37%, up to 48% in the samples dated 13th century AD), suggesting that there were widespread breeding activities in the sites' surroundings. In particular, this sum reached its highest values in the samples from sites 8 and 6 (c. 48%; Table 2); due to their peculiar archaeological contexts (a dump and a drainage channel, respectively), these samples record and provide detailed information on the human activities in the area.

Pollen pasture indicators were similarly represented in the current pollen spectra. Poaceae was the most-recurring pollen taxa (18%), followed by Cichorieae (11%) and Brassicaceae (5.5%). This seems to be in line with the reduction in the grazed areas in the region at present. Conversely, the good values of Fabaceae (3.8%) suggest the use of such plants as fodder (cultivated and then grazed or stored).

The amount of LPPI plants was decidedly lower in the modern surface soils than in the archaeological samples (22% vs. 37%; Figure 4), indicating that there were more breeding activities in the past than today.

3.4.2. Non-Pollen Palynomorphs as Grazing Markers

The dung-loving fungi are powerful environmental proxies used to assess the presence of fauna in the past, in particular, herbivores [38,105–108]. In the life-cycle of saprobic fungi growing on animal dung, the spores of coprophilous ascomycetes are accidentally eaten by herbivores, survive passage through the intestinal tract, germinate, and then grow in the deposited dung. The fungi then release their spores, which are deposited on vegetation, that is then consumed by grazing animals to complete the cycle. Therefore, the coprophilous fungi are indices of the presence of dung and are associated with pastoral/breeding activities.

The main dung-loving fungi are *Sordaria* (HdV-55a), *Apiosordaria verruculosa* (169), *Sporormiella* (HdV-113), *Cercophora* (HdV-112), *Delitschia*, and *Podospora* (HdV-368) types (Figure 5). As shown by recent studies on modern analogues in grazed areas [44,47,105,106], *Sporormiella* (including obligate coprophilous species) and *Sordaria* (a non-strictly coprophilous genus) can be considered the most reliable indicators of the local presence of herbivores.

Altogether, these proxies of pastoral activities were found in the archaeological samples in variable concentrations (Figure 4). Coprophilous spores (mostly *Sordaria* and *Sporormiella* types) recurred in the Late Hellenistic, Roman, and Medieval samples (from c. 12,000 up to c. 16,000 npp/g) when the highest LPPI values were observed. The significant values of LPPI, together with the coprophilous fungal spores, strongly suggest that pastoral/breeding activities were widely practiced in the study area.

The largest amount of coprophilous fungi was observed at sites 2 and 4 (c. 17,600 npp/g and 9000 npp/g; Table 2) in samples that were mainly from layers within natural or artificial enclosures where animals were housed or moved to graze. Interestingly, in those samples, several spores of *Diporotheca rhizophila* have also been found (c. 2000 npp/g as the average of the two sites; up to c. 9000 npp/g in the 3rd century BC samples from the site 4). The genus *D. rhizophila* is a root parasite that grows in meso- to eutrophic conditions and wet habitats with extensive soil erosion due to the impact of agricultural activities and livestock trampling [109]. The spores of this fungus may generally be a valuable soil disturbance indicator for pastoral activities. In particular, in highly trampled or overgrazed areas, the soil is left barren and exposed to erosion with the dispersion of *Diporotheca* spores. The combined evidence of *D. rhizophila* and coprophilous fungal spores suggests that these enclosed areas (probably with the water sources needed to supply livestock) had been intensively grazed.

Sporormiella and *Cercophora* types reached their highest values at site 2 (c. 13,000 npp/g and 8300 npp/g), where *Sordaria* was also quite well-attested to (c. 3500 npp/g). *Delitschia* and *Podospora* were the lowest represented morphotypes in all the archaeological samples (always <200 npp/g).

Other microfossils that are related to the presence of animals at the sites are eggs of intestinal parasites, such as *Dicrocoelium*, which is a parasite of ruminants [77,110]. Parasitological remains were quite rare in the samples (few eggs of *Dicrocoelium* were found in late Hellenistic and Roman phases at the sites 2 and 7) but are further evidence that some animal excrements became incorporated into these deposits.

Overall, the NPP record evinces the continual presence of animals in the past landscapes, supporting the evidence of pastures and pastoral activities in the region.

In the modern samples, the NPP indicators of grazing pressure were less-recorded (Figure 4). The coprophilous fungal spore concentration was decidedly lower (c. 2500 npp/g) than that found in the archaeological layers. Nevertheless, the morphotypes observed were the same and with a proportionately similar presence: mostly *Sordaria*, Sordariaceae, and *Sporormiella* (c. 1600 npp/g in total), and few *Cercophora* (c. 150 npp/g), *Delitschia* and *Podospora* types (<100 npp/g each). No intestinal parasite eggs were found. Altogether, these microfossils data indicate that animals frequent the area less as compared to the past, probably as an effect of the reduction in pastoral practices in the region at present.

3.5. The Relevance of Past Pastoral Activities and Their Legacy on Today's Landscape

The integrated analyses of microscopic records from the studied archaeological sites indicate wide and continuous pastoral activities practiced in the region. In particular, the combined evidence from pollen pasture indicators and NPP markers of grazing (mainly coprophilous fungal spores) point out that pastures were the main type of land-use in the territory surrounding each of the eight study sites. As evidenced by the high values of the LPPI sums in the past pollen records, this region has long been a grazed area, with more intense pastoral activities documented from the end of the Hellenistic age to the Medieval and Renaissance periods; besides this, land used for animal feed was also attested to by fodder plants pollen.

The interpretation of the archaeological pollen data was actually based on a comparison of several contexts, chronologies, geomorphologies, and vegetation belts, offering different environmental contexts from the sites located on mountains and hills to those placed on the coastal plain. Interestingly, the pollen spectra suggest that similar agrarian systems exist along this transect, mainly with pastoral activities and cereal and olive cultivations. According to the palynological dataset, during ancient Greek times, crop and tree cultivation was the prevalent activity in the *chora* of Metaponto, and agriculture was performed close to the settlements along with pastoral/herding practices. The same land-use had continued throughout the late Hellenistic/early-Roman phases up to the Medieval period, when pastoral farming was the predominant form of land-use, especially in the inland areas. Here, the determining factor of such a type of farming was probably the conformation of the land, since ploughing is not practicable on hilly slopes. The geomorphology may have also influenced the types of livestock that were used in the territory, where—in accordance with the zooarchaeological record [64]—sheep and goats rather than cattle were reared and grazed. Moreover, the widespread shrubby vegetation that was documented in the pollen record is further evidence that the prevalent domestic animals may have been ovicaprines, as most of the Mediterranean shrub species contain secondary metabolites such as tannins, terpenes, and volatile oils that are non-edible for many other herbivores [111].

Conversely, in the modern spectra, pollen and NPP indicators of grazing are not as common as in previous times. The current landscape is characterized by intense agricultural activities, including orchard crops, cereal fields, and olive groves, and by shrubby grasslands and a well-developed Mediterranean *macchia*. The modern development of Mediterranean vegetation may also reasonably be a response to overgrazing in the past [12,13]. Grazing is regarded as one of the main actions responsible for deforestation as well as the major factor preventing the evolution of *macchia* to forests [14,16]. Overgrazing or continual grazing pressure by sheep and goats has changed most of the once-forested areas of the Mediterranean into the maquis shrublands that typically occur in the current landscapes.

Sustainability **2019**, *11*, 404

This environmental change has been documented in most of the reference biostratigraphical records from the central Mediterranean. For example, the Holocene pollen record from Lago Grande di Monticchio in Basilicata revealed a forest cover decrease from c. 3000 cal. years BP to the present, and a concurrent increase in the relative abundance of *Pistacia* and other Mediterranean plants, *Olea*, *Juglans*, *Castanea* (OJC group, evidence of human-induced environments) [6], Poaceae wild grass group, *Artemisia*, and *Plantago* (PAZ 1b-1a) [112]. Similarly, a sharp reduction in forested areas, and an increase in synanthropic plants and Mediterranean vegetation was systematically documented in the mid—late Holocene sections of off-site cores from southern Italy [52–55,99,100,113–116]. Overall, these biostratigraphical records suggest that a human impact on vegetation has occurred since about 3500 cal. years BP. Therefore, it is not surprising that the natural environment had been shaped by the human presence in the study area since ancient times, as shown by the very low tree cover that was expressed in the 6th century BC samples. In addition, the pollen sums indicate that intense pastoral activities and broad agrarian practices have highly influenced the landscape surrounding the sites. In the modern pollen spectra, the low forest cover is thereby the result of millennia of human presence and activities in the region. The prevailing Mediterranean vegetation can be considered to be a degraded environment related to the long-term impact of livestock grazing.

The current plant biodiversity of the region mainly depends on past human activities, closely intertwined with environmental and climatic factors. The microfossil records suggest that continuous livestock activity had greatly contributed to a shift in the floristic composition towards the current plant biodiversity and a patchwork of habitats, including open areas. Therefore, the conservation and sustainable use of this biodiversity can be achieved mainly by recognizing the important role of pastoral practices in regional cultures, traditions, and livelihoods.

This study can support biodiversity awareness by providing useful information on the vocation and environmental sustainability of the Basilicata region. For example, the current environmental resource management strategies intend to replace the abandoned grazing areas with woodlands, as well as introduce new cost-effective crops in a major renovation and development of the territory. These actions may carry significant risks of biodiversity alteration in the context of unsustainable development. Palynological data from the regional archaeological sites provide insights into anthropogenically-driven landscape dynamics and inform us about a landscape adaptation and biodiversity changes over time. In addition, the long-term perspective of this research provides a knowledge base on the effects of many different forms of farming (cultivations or grazing) on local biodiversity and also testifies to the impact of agricultural intensification or abandonment on the ecosystem, which the region has already experienced in the past. Understanding the responses of ecosystems to anthropogenic disturbance and the long-term scale of these processes are essential points to the development of sustainable management strategies that can help to reduce biodiversity loss.

4. Conclusions

The palynological dataset from eight archaeological sites in Basilicata (southern Italy) is relevant to understanding the farming activities that have been practised in the area over the last 2500 years. The data point to an open plant landscape that has been continuously dominated by pastures, cereal fields, and olive groves. Important evidence of pastoral farming has arisen from the joint record of pollen pasture indicators and NPP markers of grazing (mainly spores of coprophilous fungi). The combined evidence from these microfossils points out that the region's environmental setting has been greatly influenced by intense pastoral/breeding activities since ancient times.

Pastoral farming was the predominant activity from the late Hellenistic period to the Middle Ages/Renaissance, especially in the inland areas, where the geomorphological features may have favoured the breeding of ovicaprines. The high values of both LPPI sums and coprophilous fungal spores suggest that pastures were widespread in the sites' surroundings. In addition, the high percentages of the Cichorieae and Poaceae wild grass group indicate open habitats characterized by dry pastures. Besides these grazing lands, the good percentages of Fabaceae suggest a combination of pastures and land used for animal feed production. In addition, pollen from sclerophyllous plants (especially shrubs) recurs in all of the spectra. Mediterranean vegetation is commonly viewed as a degraded landscape and could be considered to be a long-term pasture marker. While pollen and NPP indicators of grazing are not common in modern spectra, Mediterranean plants are quite well-attested to; these data suggest a reduction in pastures in current land-uses and support the idea of the importance of past pastoral practices as a major agent of the regional landscape's transformation. To improve environmental sustainability, we cannot ignore both the landscape evolution and bio-cultural value of a territory. The palaeoecological results from the study sites are useful information, especially for the understanding of the various environmental conservation measures. By knowing the past biodiversity and ecosystem dynamics, future scenarios can be taken into account by stakeholders and decision makers, helping them figure out the consequences of different conservation strategies. In order to better understand the dynamics and processes that govern modern ecosystems and habitats, an integrated discussion of the main results obtained from palaeoecological and ecological studies carried out in a certain region should be pursued. In fact, the insights into long-term events greatly contribute to improving modern concepts of community organization, evaluating climate dynamics, and understanding disturbance processes that cannot be understood by analyzing modern systems alone [117]. In the same way, the study of modern analogues could provide further information on pollen and NPP vegetation and land-use relationships and contribute to a more comprehensive landscape reconstruction (e.g., [118]). Future developments of this research will move in this direction and try to support both modern biodiversity awareness and cultural landscape conservation for sustainable regional resources management and exploitation. Future landscape management, resources exploitation, and restoration planning could benefit from the palaeoecological insights that this research provides, which take into account new factors, including both past and future environmental (climate) and social (history) changes.

Acknowledgments: This research was the PhD project of AF at the Earth System Science Doctorate School of the University of Modena and Reggio Emilia. The main investigation was carried out in collaboration with the Institute of Classical Archaeology of the University of Texas at Austin within the 'Metaponto project' (director: Prof. Joseph Coleman Carter), a vast integrated study of Greek colonies and rural populations of Basilicata. The indigenous sites were investigated in the framework of the multidisciplinary research lead by the Postgraduate School in Archaeological Heritage at Matera of the University of Basilicata (scientific officers: Profs. Massimo Osanna and Dimitris Roubis). The author is grateful to the anonymous reviewers for their constructive comments.

Conflicts of Interest: The author declares no conflict of interest.

References

1. Kareiva, P.; Watts, S.; McDonald, R.; Boucher, T. Domesticated Nature: Shaping Landscapes and Ecosystems for Human Welfare. *Science* **2007**, *316*, 1866–1869. [CrossRef] [PubMed]
2. Vitousek, P.M.; Mooney, H.A.; Lubchenco, J.; Melillo, J.M. Human domination of earth's ecosystems. *Science* **1997**, *277*, 494–499. [CrossRef]
3. Butzer, K.W. Environmental history in the Mediterranean world: Cross-disciplinary investigation of cause-and-effect for degradation and soil erosion. *J. Archaeol. Sci.* **2005**, *32*, 1773–1800. [CrossRef]
4. Mercuri, A.M. Genesis and evolution of the cultural landscape in central Mediterranean: The 'where, when and how' through the palynological approach. *Landsc. Ecol.* **2014**, *29*, 1799–1810. [CrossRef]
5. Mercuri, A.M.; Sadori, L.; Blasi, C. Editorial: Archaeobotany for cultural landscape and human impact reconstructions. *Plant Biosyst.* **2010**, *144*, 860–864. [CrossRef]
6. Mercuri, A.M.; Bandini Mazzanti, M.; Florenzano, A.; Montecchi, M.C.; Rattighieri, E. *Olea, Juglans* and *Castanea*: The OJC group as pollen evidence of the development of human-induced environments in the Italian peninsula. *Quat. Int.* **2013**, *303*, 24–42. [CrossRef]
7. Mercuri, A.M.; Sadori, L.; Uzquiano, P. Mediterranean and North-African cultural adaptations to mid-Holocene environmental and climatic changes. *Holocene* **2011**, *21*, 189–206. [CrossRef]
8. Kouli, K.; Masi, A.; Mercuri, A.M.; Florenzano, A.; Sadori, L. Regional vegetation histories: An overview of the pollen evidence from the Central Mediterranean. *Late Antiq. Archaeol.* **2015**, *11*, 69–82. [CrossRef]

9. Mercuri, A.M.; Florenzano, A.; Burjachs, F.; Giardini, M.; Kouli, K.; Masi, A.; Picornell-Gelabert, L.; Revelles, J.; Sadori, L.; Servera-Vives, G.; et al. From influence to impact: The multifunctional land-use in Mediterranean prehistory emerging from palynology of archaeological sites (8.0-2.8 ka BP). *Holocene*. in press.

10. Naveh, Z.; Whittaker, R.H. Structural and floristic diversity of shrublands and woodlands in northern Israel and other Mediterranean areas. *Vegetatio* **1979**, *41*, 171–190. [CrossRef]

11. Davis, F.W.; Goetz, S. Modelling vegetation pattern using digital terrain data. *Landsc. Ecol.* **1990**, *4*, 69–80. [CrossRef]

12. Carmel, Y.; Kadmon, R. Effects of grazing and topography on long-term vegetation changes in a Mediterranean ecosystem in Israel. *Plant Ecol.* **1999**, *145*, 243–254. [CrossRef]

13. Blondel, J.; Aronson, J. *Biology and Wildlife of the Mediterranean Region*; Oxford University Press: Oxford, UK, 1999; ISBN 9780198500360.

14. Joffre, R.J.; Vacher, C.; Long, G. The dehesa: An agrosilvopastoral system on the Mediterranean region with special reference to the Sirera Morena area of Spain. *Agrofor. Syst.* **1988**, *6*, 71–96. [CrossRef]

15. Callaway, R.M.; Davis, F.W. Vegetation dynamics, fire, and the physical environment in coastal central California. *Ecology* **1993**, *74*, 1567–1578. [CrossRef]

16. Seligman, N.G.; Perevolotsky, A. Has intensive grazing by domestic livestock degraded Mediterranean Basin rangelands? In *Plant-Animal interactions in Mediterranean-Type Ecosystems*; Arianoutsou-Faraggitaki, M., Groves, R.H., Eds.; Kluwer Academic Publishers: Dordrecht, The Netherlands, 1994; pp. 93–103. ISBN 9789401109086.

17. Amiri, F.; Ariapour, A.; Fadai, S. Effects of livestock grazing on vegetation composition and soil moisture properties in grazed and non-grazed range site. *J. Biol. Sci.* **2008**, *8*, 1289–1297. [CrossRef]

18. Díaz, S.; Lavorel, S.; McIntyre, S.; Falczuk, V.; Casanoves, F.; Milchunas, D.G.; Skarpe, C.; Rusch, G.; Sternberg, M.; Noy-Meyr, I.; et al. Plant trait responses to grazing—A global synthesis. *Glob. Chang. Biol.* **2007**, *13*, 313–341. [CrossRef]

19. Georgiadis, N.J.; Ruess, R.W.; McNaughton, S.J.; Western, D. Ecological conditions that determine when grazing stimulates grass production. *Oecologia* **1989**, *81*, 316–322. [CrossRef]

20. Feurdean, A.; Ruprecht, E.; Molnár, Z.; Hutchinson, S.M.; Hickler, T. Biodiversity-rich European grasslands: Ancient, forgotten ecosystems. *Biol. Conserv.* **2018**, *228*, 224–232. [CrossRef]

21. Pausas, J.G.; Bond, W.J. Humboldt and the reinvention of nature. *J. Ecol.* **2018**. [CrossRef]

22. Montalvo, J.; Casado, M.A.; Levassor, C.; Pineda, F.D. Species diversity patterns in Mediterranean grasslands. *J. Veg. Sci.* **1993**, *4*, 213–222. [CrossRef]

23. Biondini, M.E.; Manske, L. Grazing frequency and ecosystem processes in a northern mixed prairie, USA. *Ecol. Appl.* **1996**, *6*, 239–256. [CrossRef]

24. Pickup, G. Estimating the effects of land degradation and rainfall variation on productivity in rangelands: An approach using remote sensing and models of grazing and herbage dynamics. *J. Appl. Ecol.* **1996**, *33*, 819–832. [CrossRef]

25. Han, G.; Hao, X.; Zhao, M.; Wang, M.; Ellert, B.H.; Willms, W.; Wang, M. Effect of grazing intensity on carbon and nitrogen in soil and vegetation in a meadow steppe in Inner Mongolia. *Agric. Ecosyst. Environ.* **2008**, *125*, 21–32. [CrossRef]

26. Birks, H.J.B. Contributions of Quaternary palaeoecology to nature conservation. *J. Veg. Sci.* **1996**, *7*, 89–98. [CrossRef]

27. Willis, K.J.; Birks, H.J.B. What Is Natural? The Need for a Long-Term Perspective in Biodiversity Conservation. *Science* **2006**, *314*, 1261–1265. [CrossRef] [PubMed]

28. Willis, K.J.; Araújo, M.B.; Bennett, K.D.; Figueroa-Rangel, B.; Froyd, C.A.; Myers, N. How can a knowledge of the past help to conserve the future? Biodiversity conservation and the relevance of long-term ecological studies. *Philos. Trans. R. Soc. B* **2007**, *362*, 175–187. [CrossRef] [PubMed]

29. MacPherson, J. Applying Palaeoecology to Conservation: A long-term perspective for informed management of a fynbos nature reserve. *Plymouth Stud. Sci.* **2009**, *2*, 218–269.

30. Barnosky, A.D.; Hadly, E.A.; Gonzalez, P.; Head, J.; Polly, P.D.; Lawing, A.M.; Eronen, J.T.; Ackerly, D.D.; Alex, K.; Biber, E.; et al. Merging paleobiology with conservation biology to guide the future of terrestrial ecosystems. *Science* **2017**, *355*, eaah4787. [CrossRef]

31. Mercuri, A.M.; Marignani, M.; Sadori, L. Palaeoecology and long-term human impact in plant biology. *Plant Biosyst.* **2015**, *149*, 136–143. [CrossRef]

32. Beneš, J.; Mercuri, A.M. CEA 2018: The 14th Conference of Environmental Archaeology in Modena and the special issue of IANSA. *IANSA*. in press.
33. Galop, D.; Rius, D.; Cugny, C.; Mazier, F. A history of long-term human–environment interactions in the French Pyrenees inferred from the pollen data. In *Continuity and Change Cultural Adaptation to Mountain Environments Studies in Human Ecology and Adaptation*; Lozny, L.R., Ed.; Springer Science + Business Media: New York, NY, USA, 2013; Volume 7, pp. 19–30. ISBN 9781461457022.
34. Ejarque, A.; Julià, R.; Riera, S.; Palet, J.M.; Orengo, H.A.; Miras, Y.; Gascón, C. Tracing the history of highland human management in the Eastern Pre-Pyrenees: An interdisciplinary palaeoenvironmental study at the Pradell fen, Spain. *Holocene* **2009**, *19*, 1241–1255. [CrossRef]
35. Gauthier, E.; Bichet, V.; Massa, C.; Petit, C.; Vannière, B.; Richard, H. Pollen and non-pollen palynomorph evidence of medieval farming activities in southwestern Greenland. *Veg. Hist. Archaeobot.* **2010**, *19*, 427–438. [CrossRef]
36. Kouli, K. Plant landscape and land use at the Neolithic lake settlement of Dispilió (Macedonia, Northern Greece). *Plant Biosyst.* **2015**, *149*, 145–204. [CrossRef]
37. Bellotti, P.; Calderoni, G.; Dall'Aglio, P.L.; D'Amico, C.; Davoli, L.; Di Bella, L.; D'Orefice, M.; Esu, D.; Ferrari, K.; Bandini Mazzanti, M.; et al. Middle-to late-Holocene environmental changes in the Garigliano delta plain (Central Italy): Which landscape witnessed the development of the Minturnae Roman colony? *Holocene* **2016**, *26*, 1457–1471. [CrossRef]
38. Van Geel, B.; Buurman, J.; Brinkkemper, O.; Schelvis, J.; Aptroot, A.; van Reenen, G.; Hakbijl, T. Environmental reconstruction of a Roman Period settlement site in Uitgeest (The Netherlands), with special reference to coprophilous fungi. *J. Archaeol. Sci.* **2003**, *30*, 873–883. [CrossRef]
39. Florenzano, A.; Mercuri, A.M.; Carter, J.C. Economy and environment of the Greek colonial system in southern Italy: Pollen and NPPs evidence of grazing from the rural site of Fattoria Fabrizio (6th-4th cent. BC; Metaponto, Basilicata). *Ann. Bot.* **2013**, *3*, 173–181. [CrossRef]
40. Bowes, K.; Mercuri, A.M.; Rattigheri, E.; Rinaldi, R.; Arnoldus-Huyzendveld, A.; Ghisleni, M.; Grey, C.; MacKinnon, M.; Vaccaro, E. Peasant Agricultural Strategies in Southern Tuscany: Convertible Agriculture and the Importance of Pasture. In *The Economic Integration of Roman Italy: Rural Communities in a Globalising World*; de Haas, T.C.A., Tol, G., Eds.; Brill: Leiden, The Netherlands, 2017; Volume 404, pp. 170–199. ISBN 9789004345027.
41. Cremaschi, M.; Mercuri, A.M.; Torri, P.; Florenzano, A.; Pizzi, C.; Marchesini, M.; Zerboni, A. Climate change versus land management in the Po Plain (Northern Italy) during the Bronze Age: New insights from the VP/VG sequence of the Terramara Santa Rosa di Poviglio. *Quat. Sci. Rev.* **2016**, *136*, 153–172. [CrossRef]
42. Bosi, G.; Labate, D.; Rinaldi, R.; Montecchi, M.C.; Mazzanti, M.; Torri, P.; Riso, F.M.; Mercuri, A.M. A survey of the Late Roman period (3rd-6th century AD): Pollen, NPPs and seeds/fruits for reconstructing environmental and cultural changes after the floods in Northern Italy. *Quat. Int.* **2018**. [CrossRef]
43. Jouffroy-Bapicot, I.; Vannière, B.; Iglesias, V.; Debret, M.; Delarras, J.-F. 2000 Years of Grazing History and the Making of the Cretan Mountain Landscape, Greece. *PLoS ONE* **2016**, *11*, e0156875. [CrossRef]
44. Ejarque, A.; Miras, Y.; Riera, S. Pollen and non-pollen palynomorph indicators of vegetation and highland grazing activities obtained from modern surface and dung datasets in the eastern Pyrenees. *Rev. Palaeobot. Palynol.* **2011**, *167*, 123–139. [CrossRef]
45. Mazier, F.; Galop, D.; Brun, C.; Buttler, A. Modern pollen assemblages from grazed vegetation in the western Pyrenees, France: A numerical tool for more precise reconstruction of past cultural landscapes. *Holocene* **2006**, *16*, 91–103. [CrossRef]
46. Mazier, F.; Galop, D.; Gaillard, M.J.; Rendu, C.; Cugny, C.; Legaz, A.; Peyron, O.; Buttler, A. Multidisciplinary approach to reconstructing local pastoral activities—An example from the Pyrenean Mountains (Pays Basque). *Holocene* **2009**, *19*, 171–188. [CrossRef]
47. Cugny, C.; Mazier, F.; Galop, D. Modern and fossil non-pollen palynomorphs from the Basque mountains (western Pyrenees, France): The use of coprophilous fungi to reconstruct pastoral activity. *Veg. Hist. Archaeobot.* **2010**, *19*, 391–408. [CrossRef]
48. Ejarque, A.; Miras, Y.; Riera, S.; Palet, J.M.; Orengo, H.A. Testing microregional variability in the Holocene shaping of high mountain cultural landscapes: A palaeoenvironmental case-study in the eastern Pyrenees. *J. Archaeol. Sci.* **2010**, *37*, 1468–1479. [CrossRef]

49. Miras, Y.; Ejarque, A.; Orengo, H.A.; Riera, S.; Palet, J.M.; Poiraud, A. Prehistoric impact on landscape and vegetation at high altitudes: An integrated palaeoecological and archaeological approach in the eastern Pyrenees (Perafita valley, Andorra). *Plant Biosyst.* **2010**, *144*, 946–961. [CrossRef]

50. Mercuri, A.M.; Allevato, E.; Arobba, D.; Bandini Mazzanti, M.; Bosi, G.; Caramiello, R.; Castiglioni, E.; Carra, M.L.; Celant, A.; Costantini, L.; et al. Pollen and macroremains from Holocene archaeological sites: A dataset for the understanding of the bio-cultural diversity of the Italian landscape. *Rev. Palaeobot. Palynol.* **2015**, *218*, 250–266. [CrossRef]

51. Kleine, E.; Woldring, H.; Cappers, R.; Attema, P.; Delvigne, J. Il carotaggio del Lago Forano presso Alessandria del Carretto (Calabria, Italia). Nuovi dati sulla vegetazione olocenica e sulla storia dell'uso del suolo nella Sibaritide interna. In Proceedings of the Preistoria e Protostoria della Calabria: Scavi e Ricerche 2003, Atti delle giornate di studio, Pellaro, Reggio Calabria, Italy, 25–26 October 2003.

52. Di Rita, F.; Simone, O.; Caldara, M.; Gehrels, W.R.; Magri, D. Holocene environmental changes in coastal Tavoliere Plain (Apulia, southern Italy): A multiproxy approach. *Palaeogeogr. Palaeoclimatol. Palaeoecol.* **2011**, *310*, 139–151. [CrossRef]

53. Di Rita, F.; Lirer, F.; Bonomo, S.; Cascella, A.; Ferraro, L.; Florindo, F.; Insinga, D.D.; Lurcock, P.C.; Margaritelli, G.; Petrosino, P.; et al. Late Holocene forest dynamics in the Gulf of Gaeta (central Mediterranean) in relation to NAO variability and human impact. *Quat. Sci. Rev.* **2018**, *179*, 137–152. [CrossRef]

54. Joannin, S.; Brugiapaglia, E.; de Beaulieu, J.-L.; Bernardo, L.; Magny, M.; Peyron, O.; Goring, S.; Vannière, B. Pollen-based reconstruction of Holocene vegetation and climate in southern Italy: The case of Lago Trifoglietti. *Clim. Past* **2012**, *8*, 1973–1996. [CrossRef]

55. Mercuri, A.M.; Bandini Mazzanti, M.; Torri, P.; Vigliotti, L.; Bosi, G.; Florenzano, A.; Olmi, L.; Massamba N'siala, I. A marine/terrestrial integration for mid-late Holocene vegetation history and the development of the cultural landscape in the Po Valley as a result of human impact and climate change. *Veg. Hist. Archaeobot.* **2012**, *21*, 353–372. [CrossRef]

56. Maggi, R.; Nisbet, R.; Barker, G. *The Archeology of Pastoralism in Southern Europe*; Rivista di Studi Liguri; Istituto Internazionale di Studi Liguri-Museo Bicknell: Bordighera, Italy, 1990; Volume LVI.

57. Florenzano, A. Evolution of a Mediterranean Landscape as Shown by the Archaeo-Environmental Reconstruction of Lucanian sites. Ph.D. Thesis, Università di Modena e Reggio Emilia, Modena, Italy, 2013.

58. Florenzano, A.; Mercuri, A.M. Palynology of archaeological sites: The example of economy and human impact of the Metaponto area (6th-1st century BC). *Rendiconti Online* **2012**, *21*, 750–752.

59. Florenzano, A. Archaeobotany at Fattoria Fabrizio. In *The Chora of Metaponto 5: A Greek Farmhouse at Ponte Fabrizio*; Lanza Catti, E., Swift, K., Carter, J.C., Eds.; University of Texas Press: Austin, TX, USA, 2014; Volume 5, pp. 113–138. ISBN 9780292758643.

60. Florenzano, A. Archaeobotanical Analysis. In *The Chora of Metaponto 6: A Greek Settlement at Sant'Angelo Vecchio*; Silvestrelli, F., Edlund-Berry, I.E.M., Eds.; University of Texas Press: Austin, TX, USA, 2016; Volume 6, pp. 159–171. ISBN 9781477309476.

61. Florenzano, A. Palynological approach to reconstruct cultural landscape evolution: Case studies from South Italy. In Proceedings of the 2018 IEEE International Workshop on Metrology for Archaeology and Cultural Heritage, Cassino, Italy, 22–24 October 2018; ISBN 9781538652756.

62. Mercuri, A.M.; Florenzano, A.; Massamba N'siala, I.; Olmi, L.; Roubis, D.; Sogliani, F. Pollen from archaeological layers and cultural landscape reconstruction: Case studies from the Bradano Valley (Basilicata, southern Italy). *Plant Biosyst.* **2010**, *144*, 888–901. [CrossRef]

63. Florenzano, A.; Mercuri, A.M. Dal polline nei sedimenti alla ricostruzione del paesaggio e dell'economia di Torre di Satriano. In *Segni del Potere: Oggetti di lusso dal Mediterraneo nell'Appennino Lucano di età Arcaica*; Osanna, M., Vullo, M., Eds.; Osanna Edizioni: Venosa, Italy, 2013; pp. 163–168. ISBN 9788881674015.

64. Roubis, D.; Colacino, C.; Fascetti, S.; Pascale, S.; Pastore, V.; Sdao, F.; De Venuto, G.; Florenzano, A.; Mercuri, A.M.; Miola, A.; et al. The archaeology of ancient pastoral sites in the territory of Montescaglioso (4th–1st century BC). An interdisciplinary approach from the Bradano valley (Basilicata- southern Italy). *SIRIS* **2013**, *13*, 117–136.

65. Florenzano, A. La pastorizia nell'economia e nel modellamento del paesaggio mediterraneo. Esempi da siti archeologici del sud Italia. In *Storia e Archeologia Globale 2. I Pascoli, i Campi, il Mare. Paesaggi di altura e di Pianura in Italia dall'Età del Bronzo al Medioevo*; Cambi, F., De Venuto, G., Goffredo, R., Eds.; Edipuglia: Bari, Italy, 2015; pp. 245–252. ISBN 9788872287750.

66. Florenzano, A.; Marignani, M.; Rosati, L.; Fascetti, S.; Mercuri, A.M. Are Cichorieae an indicator of open habitats and pastoralism in current and past vegetation studies? *Plant Biosyst.* **2015**, *149*, 154–165. [CrossRef]

67. Florenzano, A.; Mercuri, A.M.; Rinaldi, R.; Rattighieri, E.; Fornaciari, R.; Messora, R.; Arru, L. The representativeness of *Olea* pollen from olive groves and the Late Holocene landscape reconstruction in central Mediterranean. *Front. Earth Sci.* **2017**, *5*, 85. [CrossRef]

68. Roubis, D. Ricerche archeologiche nell'abitato indigeno di Difesa San Biagio (Montescaglioso). In Proceedings of the Ricerche Sulla Casa in Magna Grecia e in Sicilia, Atti del Colloquio, Lecce, Italy, 22–24 June 1992; D'Andria, F., Mannino, K., Eds.; Congedo Editore: Galatina, Italy, 1996; pp. 235–253, ISBN 8880861506.

69. Osanna, M.; Roubis, D.; Sogliani, F. Ricerche archeologiche ad Altojanni (Grottole –MT) e nel suo territorio. Rapporto preliminare (2005–2007). *SIRIS* **2007**, *8*, 137–156.

70. Carter, J.C. *Living off the Chora: Diet and Nutrition at Metaponto*; Institute of Classical Archaeology, The University of Texas at Austin: Austin, TX, USA, 2003; ISBN 0970887957.

71. Carter, J.C. *Discovering the Greek Countryside at Metaponto*; The University of Michigan Press: Ann Arbor, MI, USA, 2006; ISBN 9780472114771.

72. Carter, J.V.; Swift, K. *The Chora of Metaponto 7: A Greek Sanctuary at Pantanello*; University of Texas Press: Austin, TX, USA, 2018; ISBN 9781477314234.

73. Osanna, M.; Colangelo, L.; Carollo, G. *Lo Spazio del Potere. La Residenza ad Abside, L'anaktoron, L'episcopio a Torre di Satriano*; Osanna Edizioni: Venosa, Italy, 2009; ISBN 9788881672790.

74. Blasi, C. *La Vegetazione d'Italia, Carta Delle Serie di Vegetazione, Scala 1:500.000*; Palombi & Partner S.r.l.: Roma, Italy, 2010; ISBN 98788606062909.

75. Di Pietro, R.; Fascetti, S.; Filibeck, G.; Blasi, C. Le serie di vegetazione della Regione Basilicata. In *La Vegetazione d'Italia, Carta Delle Serie di Vegetazione, Scala 1:500.000*; Blasi, C., Ed.; Palombi & Partner S.r.l.: Roma, Italy, 2010; pp. 375–389. ISBN 98788606062909.

76. De Capua, E.L.; Nigro, C.; Labriola, F. Boschi, biodiversità, territorio e variazioni ambientali. Interventi e attività della provincia di Matera. *Forest* **2005**, *2*, 110–129.

77. Florenzano, A.; Mercuri, A.M.; Pederzoli, A.; Torri, P.; Bosi, G.; Olmi, L.; Rinaldi, R.; Bandini Mazzanti, M. The significance of intestinal parasite remains in pollen samples from Medieval pits in the Piazza Garibaldi of Parma, Emilia Romagna, Northern Italy. *Geoarchaeology* **2012**, *27*, 34–47. [CrossRef]

78. Stockmarr, J. Tablets with spores used in absolute pollen analysis. *Pollen Spores* **1971**, *13*, 614–621.

79. Moore, P.D.; Webb, J.A.; Collinson, M.E. *Pollen Analysis*; Blackwell: London, UK, 1991; ISBN 9780632021765.

80. Beug, H.J. *Leitfaden der Pollen Bestimmung für Mittleleuropa und Angrenzende Gebeite*; Gustav Fisher Verlag: Stuttgart, Germany, 2004; ISBN 9783899370430.

81. Reille, M. *Pollen et Spores d'Europe et d'Afrique du Nord*, 2nd ed.; Laboratoire de Botanique Historique et Palynologie: Marseille, France, 1999; ISBN 2950717535.

82. Van Geel, B. Non-pollen palynomorphs. In *Tracking Environmental Change Using Lake Sediments, Vol. 3, Terrestrial, Algal, and Siliceous Indicators*; Smol, J.P., Birks, H.J.B., Last, W.M., Eds.; Kluwer: Dordrecht, The Netherlands, 2001; pp. 99–119. ISBN 9781402006814.

83. Mudie, P.J.; Marret, F.; Rochon, A.; Aksu, A.E. Non-pollen palynomorphs in the Black Sea corridor. *Veg. Hist. Archaeobot.* **2010**, *19*, 531–544. [CrossRef]

84. Miola, A. Tools for Non-Pollen Palynomorphs (NPPs) analysis: A list of Quaternary NPP types and reference literature in English language (1972–2011). *Rev. Palaeobot. Palynol.* **2012**, *186*, 142–161. [CrossRef]

85. Mazier, F. Modélisation de la Relation Entre Pluie Pollinique Actuelle, Végétation et Pratiques Pastorales en Moyenne Montagne (Pyrenées et Jura). Application Pour L'interprétation des Données Polliniques Fossiles. Ph.D. Thesis, U.F.R. des Sciences et Techniques, Université de Franche Comté, Besançon, France, 2006.

86. Grimm, E.D. *TGView*; Illinois State Museum, Research and Collection Center: Springfield, MA, USA, 2004.

87. Mercuri, A.M. Human influence, plant landscape evolution and climate inferences from the archaeobotanical records of the Wadi Teshuinat area (Libyan Sahara). *J. Arid Environ.* **2008**, *72*, 1950–1967. [CrossRef]

88. Mercuri, A.M. Plant exploitation and ethnopalynological evidence from the Wadi Teshuinat (Tadrart Acacus, Libyan Sahara). *J. Archaeol. Sci.* **2008**, *35*, 1619–1642. [CrossRef]

89. Dimbleby, G.W. *The Palynology of Archaeological Sites*; Academic Press: London, UK, 1985; ISBN 9780122164804.

90. Hall, S.A. Deteriorated pollen grains and the interpretation of quaternary pollen diagrams. *Rev. Palaeobot. Palynol.* **1981**, *32*, 193–206. [CrossRef]

91. Horowitz, A. *Palynology of Arid Lands*; Elsevier: Amsterdam, The Netherlands, 1992; ISBN 9780444882776.

92. Traverse, A. *Sedimentation of Organic Particles*; Cambridge University Press: Cambridge, UK, 2005; ISBN 9780521675505.

93. Davis, O.K.; Anderson, R.S. Pollen in packrat (Neotoma) middens: Pollen transport and the relationship of pollen to vegetation. *Palynology* **1987**, *11*, 185–198. [CrossRef]

94. Davis, O.K.; Buchmann, S.L. Insect sources of pollen clumps in archeological sites in Southwestern U.S.A.: Ground-nesting bees and mites. *AASP Contrib. Ser.* **1994**, *29*, 63–74.

95. Robbins, E.I.; Cuomo, M.C.; Haberyan, K.A.; Mudie, P.J.; Chen, Y.Y.; Head, E. Chapter 27. Fecal pellets. In *Palynology: Principles and Applications*; Jansonius, J., McGregor, D.C., Eds.; American Association of Stratigraphic Palynologists Foundation: Salt Lake City, UT, USA, 1996; Volume 3, pp. 1085–1097. ISBN 9780931871030.

96. Grenfell, H.R. Probable fossil zygnemataceian algal spore genera. *Rev. Palaeobot. Palynol.* **1995**, *84*, 201–220. [CrossRef]

97. Medeanic, S. Freshwater algal palynomorph records from Holocene deposits in the coastal plain of Rio Grande do Sul, Brazil. *Rev. Palaeobot. Palynol.* **2006**, *141*, 83–101. [CrossRef]

98. Miola, A.; Bondesan, A.; Corain, L.; Favaretto, S.; Mozzi, P.; Piovan, S.; Sostizzo, I. Wetlands in the Venetian Po Plain (northeastern Italy) during the Last Glacial Maximum: Interplay between vegetation, hydrology and sedimentary environment. *Rev. Palaeobot. Palynol.* **2006**, *141*, 53–81. [CrossRef]

99. Di Rita, F.; Magri, D. Holocene drought, deforestation, and evergreen vegetation development in the central Mediterranean: A 5,500 year record from Lago Alimini Piccolo, Apulia, southeast Italy. *Holocene* **2009**, *19*, 295–306. [CrossRef]

100. Sadori, L.; Narcisi, B. The postglacial record of environmental history from Lago di Pergusa, Sicily. *Holocene* **2001**, *11*, 655–670. [CrossRef]

101. Mercuri, A.M.; Bandini Mazzanti, M.; Florenzano, A.; Montecchi, M.C.; Rattighieri, E.; Torri, P. Anthropogenic Pollen Indicators (API) from archaeological sites as local evidence of human-induced environments in the Italian peninsula. *Ann. Bot.* **2013**, *3*, 143–153. [CrossRef]

102. Jalut, G.; Dedoubat, J.J.; Fontugne, M.; Otto, T. Holocene circum-Mediterranean vegetation changes: Climate forcing and human impact. *Quat. Int.* **2009**, *200*, 4–18. [CrossRef]

103. Noë, R.; Blom, C.W.P.M. Occurrence of three *Plantago* species in Coastal dune grasslands in relation to pore-volume and organic matter content of the soil. *J. Appl. Ecol.* **1981**, *19*, 177–182. [CrossRef]

104. Brun, C. Anthropogenic indicators in pollen diagrams in eastern France: A critical review. *Veg. Hist. Archaeobot.* **2011**, *20*, 135–142. [CrossRef]

105. Blackford, J.J.; Innes, J.B. Linking current environments and processes to fungal spore assemblages: Surface NPM data from woodland environments. *Rev. Palaeobot. Palynol.* **2006**, *141*, 179–187. [CrossRef]

106. Graf, M.T.; Chmura, G.L. Development of modern analogues for natural, mowed and grazed grasslands using pollen assemblages and coprophilous fungi. *Rev. Palaeobot. Palynol.* **2006**, *141*, 139–149. [CrossRef]

107. Baker, A.G.; Bhagwat, S.A.; Willis, K.J. Do dung fungal spores make a good proxy for past distribution of large herbivores? *Quat. Sci. Rev.* **2013**, *62*, 21–31. [CrossRef]

108. Davis, O.K.; Shafer, D. *Sporormiella* fungal spores, a palynological means of detecting herbivore density. *Palaeogeogr. Palaeoclimatol. Palaeoecol.* **2006**, *237*, 40–50. [CrossRef]

109. Hillbrand, M.; Hadorn, P.; Cugny, C.; Hasenfratz, A.; Galop, D.; Haas, J.N. The palaeoecological value of *Diporotheca rhizophila* ascospores (Diporothecaceae, Ascomycota) found in Holocene sediments from Lake Nussbaumersee, Switzerland. *Rev. Palaeobot. Palynol.* **2012**, *186*, 62–68. [CrossRef]

110. Le Bailly, M.; Bouchet, F. Ancient dicrocoeliosis: Occurrence, distribution and migration. *Acta Trop.* **2010**, *115*, 175–180. [CrossRef] [PubMed]

111. Bartolome, J.; Franch, J.; Plaixats, J.; Seligman, N.G. Diet selection by sheep and goats on Mediterranean heath-woodland range. *J. Range Manag.* **1998**, *51*, 383–391. [CrossRef]

112. Allen, J.R.M.; Watts, W.A.; McGee, E.; Huntley, B. Holocene environmental variability–the record from Lago Grande di Monticchio, Italy. *Quat. Int.* **2002**, *88*, 69–80. [CrossRef]

113. Russo Ermolli, E.; Di Pasquale, G. Vegetation dynamics of south-western Italy in the last 28 kyr inferred from pollen analysis of a Tyrrhenian Sea core. *Veg. Hist. Archaeobot.* **2002**, *11*, 211–219. [CrossRef]

114. Di Rita, F.; Fletcher, W.J.; Aranbarri, J.; Margaritelli, G.; Lirer, F.; Magri, D. Holocene forest dynamics in central and western Mediterranean: Periodicity, spatio-temporal patterns and climate influence. *Sci. Rep.* **2018**, *8*, 8929. [CrossRef] [PubMed]

115. Caroli, I.; Caldara, M. Vegetation history of Lago Battaglia (Eastern Gargano coast, Apulia Italy) during the Middle-Late Holocene. *Veg. Hist. Archaeobot.* **2007**, *16*, 317–327. [CrossRef]

116. Tinner, W.; van Leeuven, J.F.N.; Colombaroli, D.; Vescovi, E.; van der Knaap, W.O.; Henne, P.D.; Pasta, S.; D'Angelo, S.; La Mantia, T. Holocene environmental and climatic changes at Gorgo Basso, a coastal lake in southern Sicily, Italy. *Quat. Sci. Rev.* **2009**, *28*, 1498–1510. [CrossRef]

117. Foster, D.R.; Schoonmaker, P.K.; Pickett, S.T.A. Insights from palaeoecology to community ecology. *Trends Ecol. Evol.* **1990**, *5*, 119–122. [CrossRef]

118. Davis, B.A.S.; Zanon, M.; Collins, P.; Mauri, A.; Bakker, J.; Barboni, D.; Barthelmes, A.; Beaudouin, C.; Bjune, A.E.; Bozilova, E.; et al. The European Modern Pollen Database (EMPD) project. *Veg. Hist. Archaeobot.* **2013**, *22*, 521–530. [CrossRef]

Article

The Impact of Late Holocene Flood Management on the Central Po Plain (Northern Italy)

Filippo Brandolini * and Mauro Cremaschi

Earth Sciences Department "Ardito Desio"—Università degli Studi di Milano, Milan 20133, Italy; mauro.cremaschi@unimi.it
* Correspondence: fibrandolini@gmail.com; Tel.: +39-025-031-5525

Received: 22 September 2018; Accepted: 28 October 2018; Published: 31 October 2018

Abstract: Fluvial environments have always played a crucial role in human history. The necessity of fertile land and fresh water for agriculture has led populations to settle in floodplains more frequently than in other environments. Floodplains are complex human–water systems in which the mutual interaction between anthropogenic activities and environment affected the landscape development. In this paper, we analyzed the evolution of the Central Po Plain (Italy) during the Medieval period through a multi-proxy record of geomorphological, archaeological and historical data. The collapse of the Western Roman Empire (5th century AD) coincided with a progressive waterlogging of large floodplain areas. The results obtained by this research shed new light on the consequences that Post-Roman land and water management activities had on landscape evolution. In particular, the exploitation of fluvial sediments through flood management practices had the effect of reclaiming the swamps, but also altered the natural geomorphological development of the area. Even so, the Medieval human activities were more in equilibrium with the natural system than with the later Renaissance large-scale land reclamation works that profoundly modified the landscape turning the wetland environment into the arable land visible today. The analysis of fluvial palaeoenvironments and their relation with past human activities can provide valuable indications for planning more sustainable urbanized alluvial landscapes in future.

Keywords: flood management; wetland; land use change; landscape transformation; resilience; late Holocene; medieval age

---

1. Introduction

Floodplains are preferred areas for human settlements due to their suitability for agriculture activities, and many studies substantiate the interpretation of floodplains as complex human–water systems [1–3]. Indeed, water and land management activities have altered fluvial landscape development to create cultivable land, while simultaneously protecting communities from the risk of flooding events, which are still the most common natural disasters worldwide [4,5].

In Europe, the reciprocal interaction between fluvial environments and human activities has been documented since the Neolithic [6,7], when fluvial landscapes were first altered for agriculture purposes. Today, floodplains are densely cultivated, and the modern European countryside is principally derived from the human landscape modification that occurred in medieval times.

In the Early Medieval Age (6th–9th centuries AD) few cases of large-scale anthropogenic land and water management activities are known (es. *Fossa Carolina* [8,9]) compared to later centuries. Even though there are some differences between the northern and central Europe and the Mediterranean region, the anthropogenic reshaping of the natural environment was principally a result of overall population growth. Between the 10th and the 13th centuries AD, concurrent with the Medieval Warm Period climate phase [10–12], the European population grew substantially, almost tripling (in Northern and Central Italy, the urban population doubled), and increasing the demand for cultivated lands [13–16]. Cereals became a more significant constituent in the average diet and in the agrarian regime compared to the centuries before, leading populations to reconfigure the medieval natural landscape for agricultural purposes. In creating new land for cultivation and settlement, the European communities triggered a massive landscape transformation through woodland clearance, arable intensification, the development of irrigation systems and the drainage of wetlands. Land reclamations works profoundly modified many European regions: the peatlands in the Netherlands [17–20]; the coastal marshlands in the UK [13,21] and in North Frisia (Germany) [22] and the alluvial wetlands in the Po Valley (Northern Italy) (Figure 1) [17]. In the latter, specifically the Emilia–Romagna region (Central Po Plain), the earliest evidence of attempts to clear the forests and drain the wetlands is mentioned in historical documents from the late 8th century [23,24], but only from the 10th to the 13th centuries were land and water management activities actually carried out widely. The proponents of those activities were mostly monasteries [25–27] such as Nonantola, Santa Giulia, Mirandola, San Benedetto in Polirone [28], and also local lords who contributed to colonizing new farmland by forcing rural people to live into new fortified settlements (*incastellamento* process [29–31]).

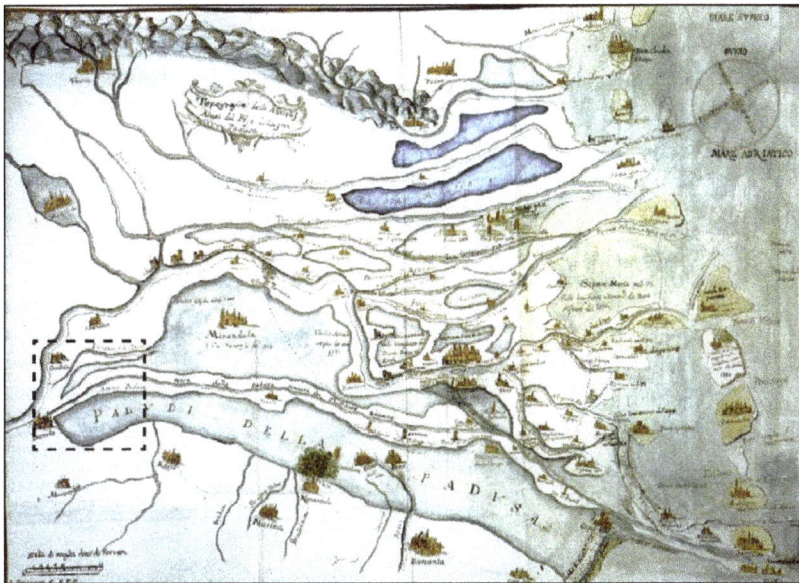

Figure 1. The wetlands in the Po Plain before the Renaissance land reclamation works, 1570 AD. Library of the Univerisità degli Studi di Bologna [32]. The dashed line indicates the research area.

According to historical maps and documents the land reclamation of the Central Po Plain reached its peak during the Renaissance (15th–16th centuries AD) [33] and continued in the Modern Age (17th–18th centuries AD), with the last marsh areas only being reclaimed in the 20th century AD: channels and drainage system are still active and allow the Po Valley to be drained and be cultivatable [34].

During the last two centuries, the height of the Po river embankments have been raised significantly to protect urban areas, and the river has become increasingly controlled, but the debate around future sustainable flood management activities is ongoing. In particular, with steadily increasing embankment heights, the potential flood depth increases, which in turn increases the flood damage if a failure occurs. The heightening of embankments, indeed, represents a component of the so-called 'levee effect' [1,35]; flood defenses actually increase overall vulnerability, as protection from regular flooding reduces perceptions of risk and encourages inappropriate development in alluvial floodplains [3]. A possible correct strategy to reduce the flood hazard when considering very large or extreme events consists of exploiting areas outside of the main embankments with the 'Room-for-River' approach [36,37]. In the case of the Po River, for example, the main embankments are already very tall along large sections of the river, and their further raising is not environmentally sustainable [2].

The environmental evolution of the Central Po Plain and its relationship to human activities has been thoroughly studied for prehistoric periods and the Roman Era [38–44]; however, these human–environment interactions in subsequent centuries are less apparent. The post-Roman Era represents a crucial moment in the Late Holocene development of the Po Plain when a large portion of the well-organized Roman countryside turned into vast alluvial wetlands [45]; the Roman drainage system collapsed, and the inhabitants had to deal with living in a waterlogged environment.

This study aims to detect and disentangle the mutual interaction between anthropogenic activities and environmental changes that occurred in the post-Roman Era integrating geomorphological analysis with historical and archaeological data. The application of a multidisciplinary approach enables the quantification of the impact of human activities on the development of the wetland environment in Central Po Plain during the Middle Ages.

2. Study Area: Geomorphological and Historical Contexts

The research area is located in the Central Po Valley, south of the Po River, a few kilometers north of the Tuscan–Emilian Apennine margin, between the cities of Parma and Reggio Emilia (Figure 2). Since the Pleistocene, the Apennine watercourses shaped the landscape developing alluvial megafans with their sediments (Figure 3). In the study area, the distal parts of Holocene alluvial fans are characterized by a telescopic shape, a result of subsequent aggradation (Aggradation indicates the increase in land elevation, typically in a river system, due to the deposition of sediment.)—entrenchment phases due to the alternation of glacial-interglacial periods; each aggradational cycle causes an incision on the top of the previous fan, while a new fan prograded in a more distal position [41]. In the study area, the alluvial ridges of the Enza, Crostolo and Tresinaro, emerging from the flat floodplain, flank depression areas known in fluvial geomorphology as backswamps, which are characterized by deposits of fine silts and clays deposited after flood events [46].

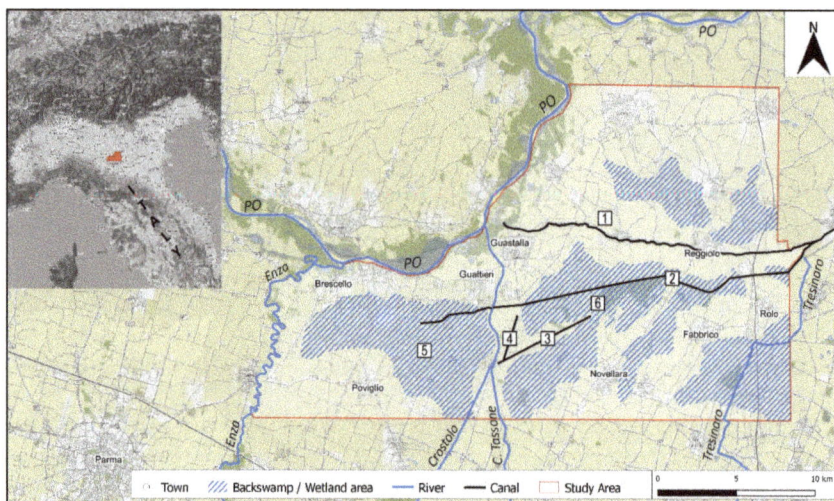

Figure 2. Location of the study area. 1—Tagliata Canal; 2—Parmigiana Canal; 3—*Crustulus Vetus*; 4—Camporainero area; 5—Valle di Gualtieri backswamp; 6—Valle di Novellara backswamp.

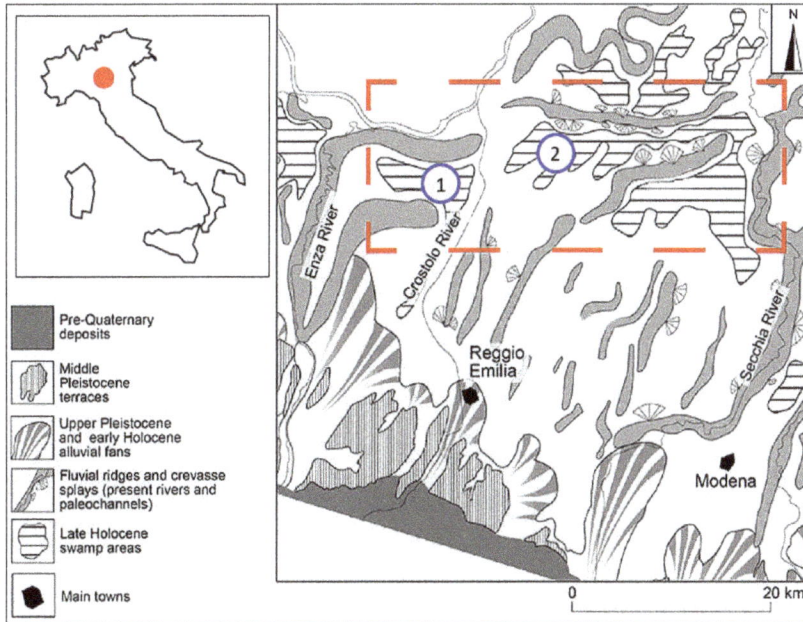

Figure 3. Schematic geomorphic map of the Po plain in the Emilia region (adapted from [47–49]). The red dashed line highlights the study area: 1—Valle di Gualtieri; 2—Valle di Novellara.

The landscape evolution of this portion of the Central Po Plain has a long-standing connection with human activities. Protohistoric human–environmental interactions have been widely investigated [39–41,49,50] but the first historical large-scale land and water management project in the study area is dated to the 2nd century BC. At that time, the Romans colonized the Po Valley and profoundly modified the natural landscape dividing the cultivated land into square fields by a regular grid of roads and ditches [38,51,52]. After the collapse of the Roman Empire (5th century AD), lack of maintenance of the irrigation systems [24] associated with a cooling climate phase (i.e., the so-called Migration Period or the Dark Age Cold Period; [10,11,53] led to the progressive waterlogging of the Po Valley and the natural depressions on the right side of the Po River turned in two vast swamp basins known as Valle di Gualtieri and Valle di Novellara (Figure 3). In particular, the Roman water management augmented the Holocene sedimentary process of Apennine watercourse with consequent aggradation of the river beds. Therefore, in the post-Roman period, channel diversions affected the southern tributaries of the Po River, especially the Crostolo Creek that flowed in the floodplain depressions: vast areas became marshy and Roman sites, and centuriation tracks were often buried under fluvial and palustrine deposits [42,45]. The waterlogging process of the area continued until the 10th century AD influencing human sustenance and settling practices [25,54]. According to historical–archaeological data the wetlands were exploited for fishing as well as for water transport by boat [55,56] while the early medieval sites were settled on the fluvial ridges, in topographically higher and strategic positions relative to the surrounding swampy meadows [24,30,31,57]. In the study area, channelization and reclamation works started in the 10th century AD [58–62] and increased between the 12th and the 13th centuries [25,26,63], promoted both by monasteries [28,64] and by private landowners [65].

Large-scale ground reclamation began during the Renaissance and is known in the literature as Bonifica Bentivoglio, after the name of its promoter, Cornelio Bentivoglio, Lord of Guastalla. The Bonifica Bentivoglio drainage system reclaimed a significant portion of the swampy meadows and was renovated and updated many times in the Modern Age; between 17th and 19th centuries, many homesteads were settled in the reclaimed farmland. The land reclamation of this portion of the

Central Po Plain continued for centuries after the Renaissance and was completed only in the 20th century AD [33,34,66], resulting in an entirely artificial landscape.

3. Materials and Methods

This study has been performed using a multi-proxy record to analyze the Late Holocene transformations of the Central Po Plain: geomorphological tools were combined with archaeological data and historical documents.

The geomorphological literature on the Central Po Valley [45,47,48,67,68] constituted a useful starting point in association with a 1-m DEM (produced from high-resolution LiDAR) provided by the national geodatabase [69]. The elevation checkpoints (m.a.s.l.) provided by the regional geodatabase [70] have been elaborated with the software QGIS (2.18) to draw contour lines with an equidistance of 0.5 m and to create the final Digital Terrain Model (DTM) of the research area (Figure 4). Moreover, to enable the interpretation of the geomorphological features detected by the DTM, a 3D model has been created by the software QGIS plugin Qgis2threejs exaggerating the DTM Z-value ×50.

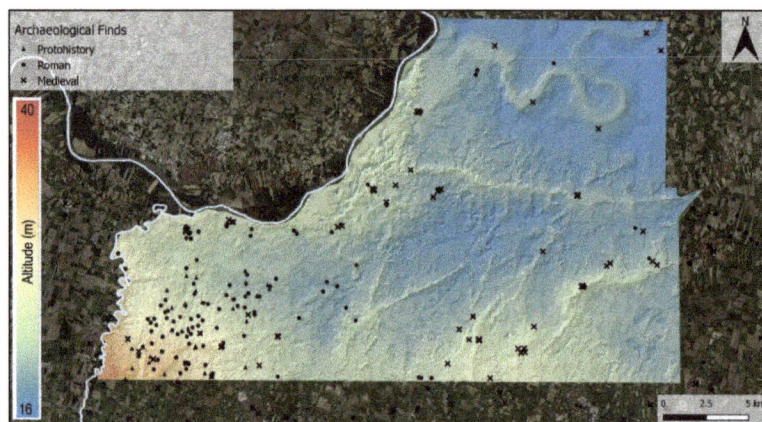

Figure 4. Digital Terrain Model elaborated with software QGIS implemented with archaeological records. In the north of the mapped area, the meandering geomorphological feature corresponds to the so-called *Po Morto* or *Dead Po*, active in Roman Times.

The regional soil map has been applied to the 3D model to detect the maximum extension of the medieval swamps before the Renaissance land reclamation works, and our interpretation is based on the different types of sediments and their concentrations.

The landforms detected have been dated and contextualized according to archaeological and historical records. The archaeological evidence (sites and materials) has been compiled from various records [48,51,62,71,72], a regional web database [73] and terrain surveys. The 18th and 19th century transcriptions of medieval accounts and parchments [58–60,74] served as an advantageous starting point in the analysis of the historical documents along with the 20th-century editions of medieval documents kept at the Parma and Reggio Emilia National Historical Archives [75–77]. Additional information has been collected at the Novellara Historical Archives, including from the archive fund "Cavamenti Acque (1495–1931)" [78]. All of the archaeological (sites and materials) and historical (place names) data have been organized according to epochs in different layers using QGIS software.

4. Results and Discussion

The multidisciplinary approach allowed us to shed new light on anthropogenic activities related to land and water management in the post-Roman landscape. In medieval times, humans exploited the palustrine environment (Palustrine environment refers to any (no-tidal) inland wetland. Wetlands within this category include inland marshes, swamps and floodplains.) as a resource, until the demand for more arable land led to land reclamation. The impact of medieval human activities on the landscape resulted in altering the natural shape of the area through the development of anthropogenic geomorphological features.

The results will be discussed in three categories related to the backswamps limits, the Tagliata Canal and the Crostolo River, respectively.

4.1. Backswamps Limits

As mentioned above, the collapse of the Roman Empire was associated with a period of climatic instability [79] that led to the progressive waterlogging of the Po Valley; frequent flood events have been reported in historical chronicles between 6th and 7th centuries AD [80]. In the study area, the floodplain on the right side of the Po River turned in two vast backswamps; the Valle di Gualtieri and the Valle di Novellara (hereafter, the two *Valli*). The backswamps are the lower area of floodplains, poorly drained, and where only finer material accumulates during a flooding event [81]. This palustrine environment served as a resource for silvopastoral practice, especially in the Early Medieval period. Historical documents [75,76,82] report that local communities exploited the Valli swamps for fishing and transport by boat; in sale-purchase agreements "*piscationes, paludes et lacus*" (i.e., fisheries, marshes, ponds) were traded, as well as any other goods such as fields or vineyards. Despite this central role in the economy of the period, in historical documents, there is no precise information about the swamps' geographical limits, perhaps because their physical boundaries constantly fluctuated during the Middle Ages climatic variations.

To determine accurate limits for the two Valli, a 3D model of the area (Figure 5) has been extrapolated exaggerating the DTM elevation attribute ×50. Superimposing the regional soil map (Geoportale Emilia Romagna) on the 3D model shows a relationship between the concentration of clayey and silt-clayey soils and the lowest areas detected in the DTM (Figure 6). These data are entirely compatible with the geomorphological definition of back swamps as low flat standing areas where fine sediments settle after a flood event. According to the model rendered, the maximum extension of the Valle di Gualtieri and Valle di Novellara covered an area of ≈250 Km^2 (Figure 7). The medieval archaeological records [30,31,51,52] seems to further confirm the swamps limits as determined by the 3D model; all of the archaeological finds and the historical toponyms [59,75] are distributed surrounding the two Valli. The wetlands had an economic potential, but one that was far below the full agricultural potential of the same land after the reclamation that occurred starting the 10th century AD. Further palaeoecological and geoarchaeological analyses will be useful to provide more information about the character of the wetland environment in the study area.

Figure 5. The 3D model: altitude checkpoints attribute ×50.

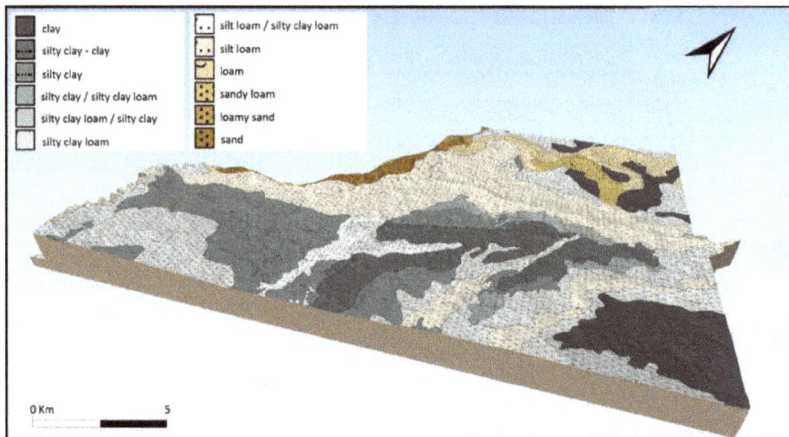

Figure 6. Application of the soil map [70] to the 3D model. Grey and Dark-Grey areas correspond to fine sediments settled after flood events.

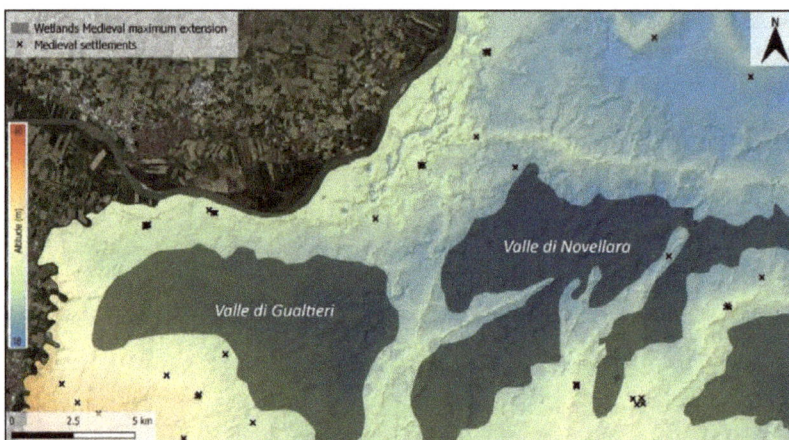

Figure 7. Maximum extension of wetlands before the Renaissance large-scale land reclamation project.

4.2. Tagliata Canal

The most valuable data about anthropogenic activities in the medieval environment concern the northern limit of the Valle di Novellara. According to the DTM, this backswamp is delimited in the north by the ridge of the so-called Tagliata Canal. In scientific literature [41,43,47,67,68] the Tagliata Canal is considered a Proto-historic Po ridge characterized by crevasse splays on both sides (Figure 8); however, geomorphological, archaeological, and historical data suggest a new interpretation. First, the distribution of archaeological evidence in the study area shows an absence of Bronze Age and Roman Era findings (both sites and materials) (Figure 9). It is likely that the subsequent reworking of the fluvial landscape has obscured pre-medieval archaeological evidence. The 13th-century chronicler Fra Salimbene de Adam reported that, between the two towns of Guastalla and Reggiolo, there used to be a waterlogged area that was drained only after the excavation of the Tagliata Canal (Table 1, I). Thus, the accretion of Tagliata Canal ridge occurred after the collapse of the Roman Empire, not before.

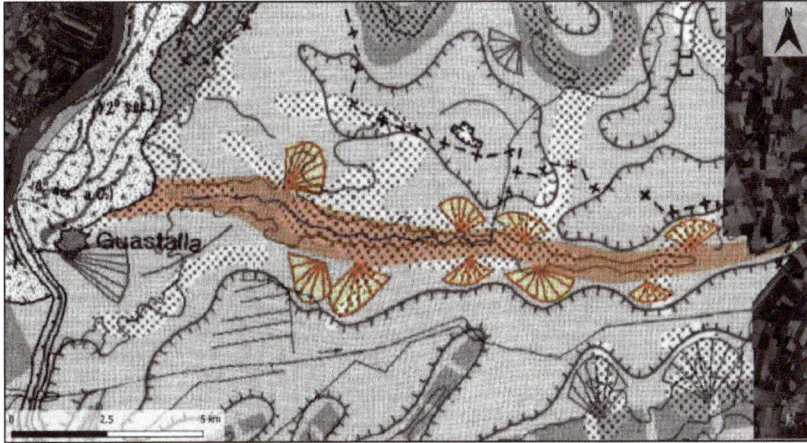

Figure 8. The Tagliata Canal in the geomorphological map of the Po Plain edited in 1997 [47]. The Tagliata Canal is considered a Proto-historic Po ridge characterized by crevasse splays on both sides.

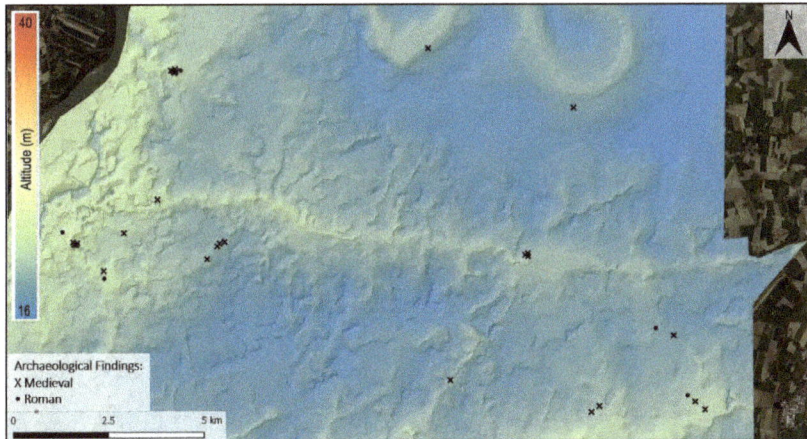

Figure 9. The distribution of archaeological sites and materials in the study area shows an absence of Bronze Age and Roman Era findings suggesting that the Tagliata Canal ridge developed in medieval times.

Moreover, historical documents [59,60,74] report that the Tagliata Canal was artificially excavated in 1218 for commercial purposes. The new canal enabled a bypass of the Po River from Guastalla to Reggiolo for the city of Cremona in order to avoid paying commercial fees to the city of Mantova, which controlled the local waterways.

The medieval chroniclers, also, report that the construction of Tagliata Canal had negative implications for the environment with frequent floods in the surrounding farmland. In 1269, for example, the chronicler Fra Salimbene reported a severe flood event (Table 1, II) and many other floods occurred between the 13th and 14th centuries AD [60].

The geomorphological evidence of those medieval floods should be attested via the crevasse splay detected near *Fangaia* and *Villarotta*, two villages that were settled on the Tagliata ridge after the canal excavation. Nevertheless, these two villages' place names are reminiscent of flooding vocabulary (i.e., *Fangaia* from *"Fango"* → mud, refers to alluvial sediments; or *Villarotta* from *Rotta* → breach levee, refers to a crevasse splay event) (Figure 10).

Table 1. Historical quotes about the environmental transformation of the research area in medieval times.

No.	Object	Chronicler	Medieval Italian	English Translation
I	Tagliata Canal	Fra Salimbene de Adam (13th century AD) [60] p. 89	*"Tra Guastalla e Reggiolo era una stesa di terreno paludoso le cui acque incanalate nel detto cavo e asciugato il territorio, si conquistarono alla coltivazione ubertosissime campagne [. . .]"*	"Between Guastalla and Reggiolo there used to be a waterlogged area whose waters were channelled in the canal (Tagliata): the land was drained and turned into very fertile farmland".
II	Tagliata Canal	Fra Salimbene de Adam (13th century AD) [60] p. 93	*"Questa Tagliata impaludò larga zona di terreni, distrusse e sommerse molte ville, e dove prima di aveva abbondanza di frumento e di vino, ora si ha copia di pesci di diverse specie"*	"The Tagliata canal flooded a wide area, razed many houses, and in place of wheat fields and vineyards now there is an abundance of fishes of different species".
III	Tagliata Canal	Ireneo Affò (18th century AD) [58] Vol. 8, p. 233	*"[. . .] La comunità da tempo memorabile aveva diritto di rompere gli argini quando menavano acque torbide perché spandere si potessero nelle valli. E molti uomini testificarono il mirabile effetto che ne era seguito, accennando de campi allora coltivabili nel luogo dei quali a loro memoria solevano i pescatori andare con le barche [. . .]"*	"Since time immemorial, the community (of Guastalla) was allowed to breach the levees when the watercourses carried turbid water to spread them in the wetlands. Many people testified to the admirable effect that followed, and they said that now cultivable fields replaced places in which fishermen were used to going by boat"
IV	Crostolo River	Ireneo Affò (18th century AD) [58] Vol. 8, p. 240	*"[. . .] fu loro conseguenza permesso di fare un Cavo in Camporainero dal Crostolo [. . .] lungo la Via di Roncaglio per cui tirar nelle valli più agevolmente tali acque"*	"(the city council of Guastalla) allowed to excavated a canal from the Crostolo river to the Camporaneiro area, along the Roncaglio Road, in order to divert the (turbid) waters in the wetlands more easily".

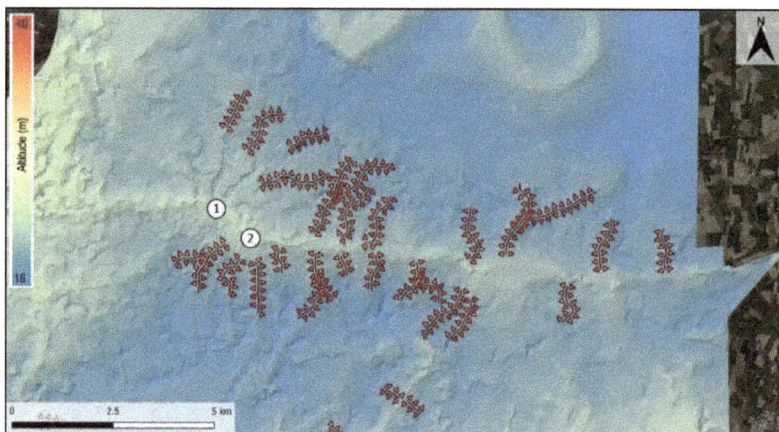

Figure 10. Geomorphological analysis of the Tagliata Canal ridge: crevasse splays and land-fill ridges. 1—Villarotta; 2—Fangaia.

The geomorphological analyses provide more nuanced information about the shape of the Tagliata Canal. In the DTM, the morphology of Tagliata Canal ridge seems to be more complicated than what is represented in the geomorphological map of The Po Plain [47]. The crevasse splays show unusual elongated small ridges not compatible with natural fluvial crevasse splays. Information in the historical documents provide a possible explanation for their genesis: the chronicler Affo' [58] reports that since the 13th century AD, people of Guastalla were allowed to breach the artificial levees of rivers and canals in a situation of high, muddy discharge (Table 1, III). This practice had the effect of infilling the swamps with sediments, thus creating new farmland, and the elongated shape of the unusual landforms could be the results of those practices. Although such flood management practices have never been reported in geomorphological and geoarchaeological studies on Central Po Plain so far, these few cases noted in the literature could support our interpretation. Recently, medieval archaeological excavations in the Ligurian Apennine reported that watercourse sediments and colluvial deposits had been exploited to reclaim palustrine environments [83,84]. In the medieval sites of Mogge di Ertola (Genova—Italy) and Torrio (Piacenza—Italy), marshes were turned into terraces by applying a technique called *"colmata di monte*": the sediments carried by mountain watercourse were managed to fill palustrine areas and create new arable land and pasture [85]. The process of breaching levees described in Affo' [58], is very similar to the "*colmata di monte*" reported in the Apennine archaeological sites. The elongated shape of the Canal Tagliata crevasse splays are the results of medieval flood management practices intended to reclaim the backswamp of Valle di Novellara; we decided to define these unique anthropogenic geomorphological features as Land-Fill Ridge (or *dosso per colmata*) (Figure 10).

Additionally, the land-fill ridge detected in this study serves as an interesting comparison with the warping practices that occurred in England in the 18th century AD. Research in Humberland Levels [86,87] and the lower Trent Valley [88] wetlands has highlighted the considerable degree to which a combination of natural alluviation and anthropogenic warping deposits have concealed palaeoland surfaces and the archaeological record of the two regions. The warping practices have been carried out since the 18th century AD and consist of the artificial diversion of fluvial sediments in a wetlands area. Warping was conducted mainly to fertilize large areas: where the peats were too acidic, warp deposits served to mask unproductive wetlands with a light, well-drained silty or silty-clay soil. Anthropogenic warping practices also aimed to reduce the impact of the spring tides which had left large areas of the region waterlogged for most of the year. The land to be warped was first enclosed by embankments, then a regular network of small canals (often still visible in aerial images as crop marks) ensured the rapid and even distribution of flood water throughout the compartment, creating a uniform deposit [86,88]. Even though the process that generates land-fill ridges and warps is quite similar (i.e., the artificial exploitation of substantial silt and clay load carried in suspension by canals and rivers), in the research area, there is no historical or geoarchaeological evidence of both embankments and fertilizing practices in alluvial wetlands.

4.3. Crostolo River

In the post-Roman era, the Crostolo River diversion channel caused the waterlogging of a large area of the Roman countryside; the river flowed into the Valli backswamps turning the farmland into a palustrine environment (Figure 11). In medieval times, human enterprise on the Crostolo River was concentrated in the portion of the watercourse that crossed the city of Reggio Emilia. Here, to prevent flooding hazards, the river was artificially diverted outside the city walls [49].

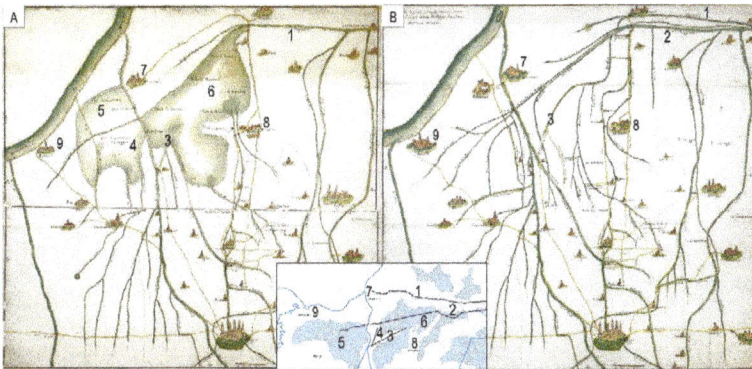

Figure 11. Medieval historical maps (18th-century copies) that show the environmental situation of the study area before (**A**), (for more details, see also Figure 12) and after (**B**), (for more details, see also Figure 13) the Renaissance wetland reclamation. The numbers in the schematic box in the center of the image helps to orientate: 1—Tagliata Canal; 2—Parmigiana Canal; 3—Crustulus Vetus; 4—Camporainero area; 5—Valle di Gualtieri backswamp; 6—Valle di Novellara backswamp; 7—Town of Guastalla; 8—Town of Novellara; 9—Town of Gualtieri. (AsMo 52 and 53, XVIIIsec. Modena National Historical Archive—"Congragazione delle Acque e delle Strade, Reggio e Reggiano") [89].

In the study area, the DTM highlights two linear Crostolo ridges, probably the result of anthropogenic activities (Figure 14).

According to Affo' [58], when the practice of breaching river and canal levees was forbidden, the city council of Guastalla allowed the excavation of a channel to divert fluvial sediments from the Crostolo River to the backswamps. The geomorphological linear ridges detected in the DTM are likely to be the result of this large-scale flood management activities. The project of the Guastalla city council likely intended to not spread the fluvial sediments in different areas (land-fill ridges) but to concentrate the muddy river discharge in the wetland to create more arable land. The medieval parchments reported by Affo' [58] gives more details about this channel, stating that it was excavated in an area called "*Camporainero*" along the "*Via di Roncaglio*"; both the place names are still used and correspond to the area of the Crostolo straight ridge-oriented NNE (Figure 14). Furthermore, a couple of historical maps (17th century AD copies of Late Medieval originals) help in understanding the geomorphological evolution of the area.

As shown in the historical map (Figures 11A and 12), *Camporainero* is a portion of the Valle di Gualtieri wetland, and the Crostolo River was artificially diverted into the Po River using the Early Medieval canal called *Fossa di Roncaglio*. The orientation of the NNE ridge match with the historical description provided by the chronicles. In this map the NNE ridge is not indicated, although it is detected in the DTM: probably the original project consisted of the diversion of the Crostolo River using the *Camporainero* Canal as well as the *Fossa di Roncaglio* Canal. The Renaissance project was probably adapted and modified because *Camporainero* Canal ridge is still recognizable, while the *Fossa di Roncaglio* has been replaced by the Crostolo River. On the other hand, the linear ridge-oriented NE (Figure 14) has always been identified in the available literature [48,62] as the Crostolo paleochannel active in Roman times and called *Crustulus Vetus* (or *Crostolo Vecchio* → Old Crostolo). In this study, both geomorphological and historical–archaeological proxies support a new interpretation. First, the ridge leads directly to the Valle Novellara backswamp and it ends in the Roman farmland: here some road and ditches of the centuriation grid are still visible in aerial images [65]. It is not possible that the river watercourse passed here in Roman times. Moreover, along this linear ridge, Roman archaeological finds have never been reported.

Moreover, historical maps dated to the 17th and 18th centuries AD [78] indicate that, along this ridge, a canal called *Fossa Alessandrina* and was excavated and a church was constructed (the still standing Chiesa di San Bernardino). In Roman times the Crostolo River is likely to have been flanked by artificial levees, and to have flowed into the Po River, north of the medieval Tagliata Canal (*Po Morto*, see Figure 4). The results of this study support the idea that the Crostolo River in the post-Roman era flowed directly in the Valle di Novellara backswamp developing the fluvial ridge oriented to NE in the DTM. Later, in the 16th century, as recorded in medieval documents, the river was artificially diverted into an area called *Camporainero* to reclaim wetlands and create new cultivable land for the community of Guastalla. The result is the NNE oriented fluvial ridge detected in the DTM (Figure 14).

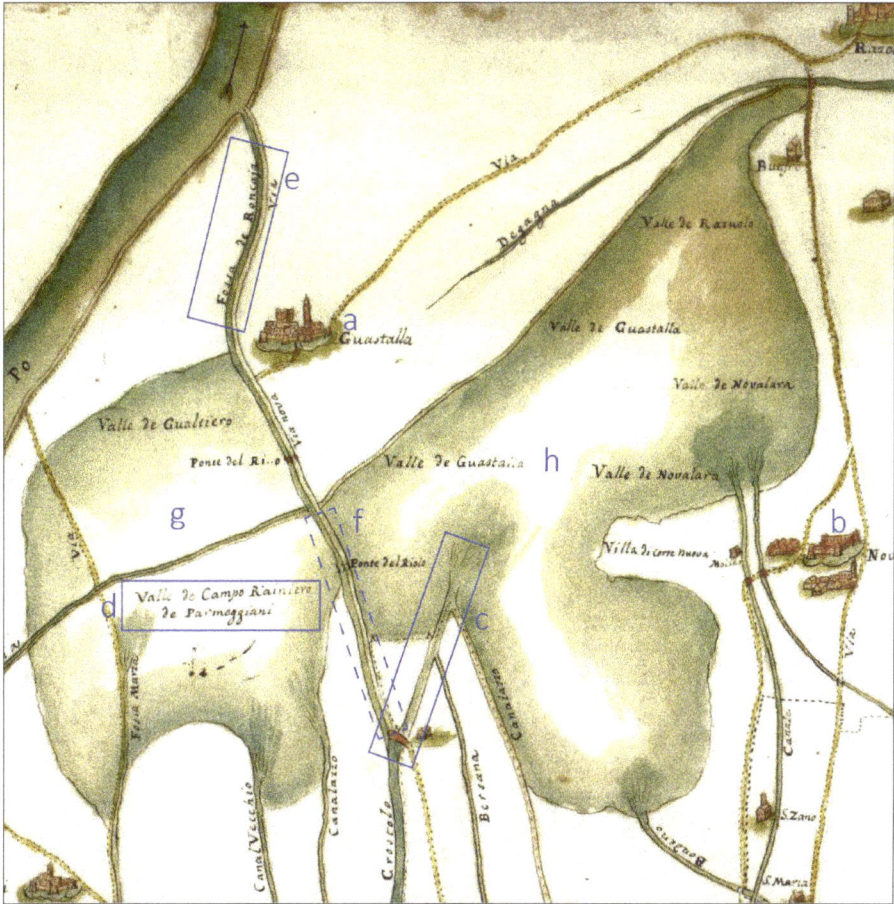

Figure 12. Detail of the medieval historical map that shows the landscape before the Renaissance land reclamation project (Figure 11A). (AsMo 52, XVIIIsec. Modena National Historical Archive—"Congragazione delle Acque e delle Strade, Reggio e Reggiano) [89]. This map shows the project of artificially diversion on the Crostolo River from the Valle di Novellara to the Po Plain through the Early Medieval *Fossa di Roncaglio* Canal. a—Town of Guastalla; b—Town of Novellara; c—the so-called *Crustulus Vetus*, still active at that time; d—the wetland area called *Camporainero*; e—*Fossa di Roncaglio* Canal; f—the area of the NNE Crostolo ridge, developed to fill the *Camporainero* area with Crostolo sediments; g—Valle di Gualtieri wetland; h—Valle di Novellara wetland.

Figure 13. Detail of the medieval historical map which shows the landscape before the Renaissance land reclamation project (Figure 11B). (AsMo 53, XVIIIsec. Modena National Historical Archive—"Congragazione delle Acque e delle Strade, Reggio e Reggiano) [89]. This map shows the study area after the Renaissance Bonifica Bentivoglio land reclamation project. a—Town of Guastalla; b—Town of Novellara; c—*Crostolo Vecchio* or *Crustulus Vetus*, the medieval Crostolo watercourse that flowed into the Valle di Novellara wetland; d—The artificial diversion of the Crostolo River in the Po River; there are no more indications of both *Camporainero* and *Roncaglio*; e—Drained area in place of the Valle di Gualtieri wetland; f—Drained area in place of Valle di Novellara wetland; g—*Botte Bentivoglio* hydraulic device.

The developing of the Camporainero Canal ridge shows similarities to the human management of the crevasses spays reported in the lowlands of the Adige River (Northern Italy) [90]. The first attempts to manage crevasses splays in this area are dated to 12th century AD, and they were finalized at lowering flood hazard and allowing discharge for water mills and navigation through the control of the water intake from the Adige River into the Castagnaro and Malopera Rivers [90].

The Crostolo river was artificially diverted in the Po River only after 1576 during the large-scale land reclamation works called *Bonifica Bentivoglio*: this project included the excavation of a canal-oriented EW (Parmigiana Canal) to drain the two Valli backswamps and the construction of a hydraulic device that allowed this canal to pass under the Crostolo River. The *Bonifica Bentivoglio* drastically changed the natural medieval landscape by turning the swamps into farmland and constraining the rivers by artificial embankments (Figures 11 and 15).

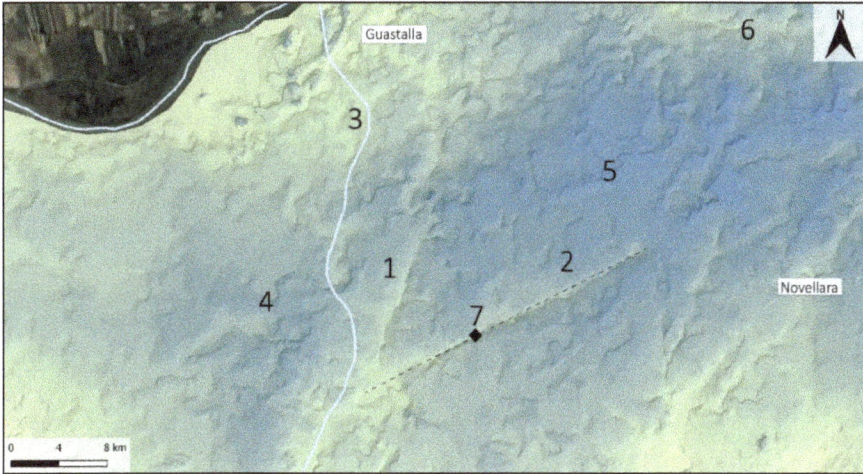

Figure 14. Digital Terrain Model (DTM) of the Crostolo River medieval ridges. 1—NEE Crostolo ridge; 2—NE Crostolo ridge, called *Crustulus Vetus*; 3—Modern Crostolo River course; 4—Valle di Gualtieri wetland area; 5—Valle di Novellara wetland area; 6—Tagliata Canal ridge; 7—The dashed line indicates the modern Age *Fossa Alessandrina* Canal, the black square represents the S. Bernardino Church.

Figure 15. The crossing between the Crostolo River (1) with the *Parmigiana Canal* (2) in the modern completely drained countryside. The white square highlights the 17th century AD structure of the *Botte Bentivoglio* hydraulic device (3).

5. Conclusions

This study sheds new light on the evolution of the Central Po Plain landscape in the Late Holocene.

The dynamic climatic conditions that occurred after the collapse of the Western Roman Empire led to a significant change in the well-organized Roman countryside, turning large floodplain areas in swamps. In the Early Medieval Age (6th–9th centuries AD) the natural landscape development seems to not have been impacted any significant human-induced changes. On the contrary, the local communities adapted themselves to the post-Roman palustrine environment settling in positions of higher elevation around the swamps limits and using them for navigation as well as for silvopastoral sustenance practices. In contrast, contemporaneous with the Medieval Climate Anomaly (10th–13th century AD), anthropogenic activities altered the landscape to fulfil the new socio-economical needs.

In the research area, the most evident anthropogenic geomorphological features developed in medieval times is the Tagliata Canal ridge. This canal represents a relevant landscape-modifying agent both for natural (flood events) and anthropogenic (land-fill ridges) causes. The exploitation of fluvial sediment to reclaim wetland areas affected the Crostolo River watercourse, as well. The result of human flood management practices was the development of new cultivable land in place of alluvial wetlands. This human-induced landscape transformation is likely not to have happened abruptly, leaving time for the natural environment to adapt to the new anthropogenic landforms. Richard Hoffmann [15] states that "the ecological concept of sustainability is a dynamic equilibrium between human activities in nature and the ability of the natural system to respond to those activities." According to this definition, medieval flood management represented a more sustainable land reclamation technique than the Renaissance large-scale projects that deeply modified the landscape with channelization and artificial levees, turning the palustrine environment into the modern countryside.

Nevertheless, the medieval flood management activities reported in this study constitute an example of the modern "Room-for-River" strategy [36,37]. Indeed, the primary purpose of the medieval practices was to create new arable land, but at the same time, the artificial diversion of fluvial sediments enables the control of floodwaters avoiding the risk of inundation, for example, in urbanized areas or productive farmland. Similar practices have been reported in Northeast Italy [90] were the medieval human management of crevasses splays aimed to control the Adige River discharge to reduce the flood hazard. Understanding the past anthropogenic effect on the landscape is essential for its future sustainable management [91], especially in the Central Po Plain, where the debate around future sustainable solutions to reduce flood hazards is ongoing [3] and the continued raising of embankments is not environmentally sustainable [2].

Author Contributions: Conceptualization, F.B. and M.C.; Data curation, F.B.; Formal analysis, F.B.; Funding acquisition, M.C.; Investigation, F.B.; Methodology, F.B. and M.C.; Project administration, M.C.; Resources, M.C.; Supervision, M.C.; Validation, F.B. and M.C.; Writing—original draft, F.B.; Writing—review and editing, F.B. and M.C.

Funding: This research was carried out in the framework of the SUCCESSO-TERRA Project funded by PRIN-MIUR (project PRIN20158KBLNB, Principal Investigator: M. Cremaschi).

Acknowledgments: The authors thank Nicki Whitehouse (University of Plymouth) and Benjamin Gearey (University College Cork) for their suggestions about a possible comparison between the land-fill ridges and warping practices. Also, the authors thank also to Rachel Kulick (University of Toronto) that reviewed the English of the manuscript. Finally, the authors are grateful to the anonymous reviewers for their constructive criticism and comments.

Conflicts of Interest: The authors declare no conflict of interest.

References

1. Merz, B.; Hall, J.; Disse, M.; Schumann, A. Fluvial flood risk management in a changing world. *Nat. Hazards Earth Syst. Sci.* **2010**, *10*, 509–527. [CrossRef]
2. Castellarin, A.; Di Baldassarre, G.; Brath, A. Floodplain Management Strategies for Flood Attenuation in the River Po. *River Res. Appl.* **2011**, *27*, 1037–1047. [CrossRef]

3.   Domeneghetti, A.; Carisi, F.; Castellarin, A.; Brath, A. Evolution of flood risk over large areas: Quantitative assessment for the Po river. *J. Hydrol.* **2015**, *527*, 809–823. [CrossRef]

4.   Dankers, R.; Feyen, L. Flood hazard in Europe in an ensemble of regional climate scenarios. *J. Geophys. Res.* **2009**, *114*, 1–16. [CrossRef]

5.   Akter, T.; Quevauviller, P.; Eisenreich, S.J.; Vaes, G. Impacts of climate and land use changes on flood risk management for the Schijn River, Belgium. *Environ. Sci. Policy* **2018**, *89*, 163–175. [CrossRef]

6.   Feulner, F. The Late Mesolithic Bark Floor of the Wetland Site of Rüde 2, Schleswig-Holstein, Germany. *J. Wetl. Archaeol.* **2011**, *11*, 109–119. [CrossRef]

7.   Ollivier, V.; Fontugne, M.; Hamon, C.; Decaix, A.; Hattè, C.; Jalabadz, M. Neolithic water management and flooding in the Lesser Caucasus (Georgia). *Quat. Sci. Rev.* **2018**, *197*, 267–287. [CrossRef]

8.   Hausmann, J.; Zielhofer, C.; Werther, L.; Berg-Hobohm, S.; Dietricha, P.; Heymann, R.; Werban, U. Direct push sensing in wetland (geo)archaeology: High-resolution reconstruction of buried canal structures (Fossa Carolina, Germany). *Quat. Int.* **2018**, *473 Pt A*, 21–36. [CrossRef]

9.   Kirchner, A.; Zielhofer, C.; Werther, L.; Schneider, M.; Linzen, S.; Wilken, D.; Wunderlich, T.; Rabbel, W.; Meyer, C.; Schmidt, J.; et al. A multidisciplinary approach in wetland geoarchaeology: Survey of the missing southern canal connection of the Fossa Carolina (SW Germany). *Quat. Int.* **2018**, *473*, 3–20. [CrossRef]

10.   Büntgen, U.; Tegel, W.; Nicolussi, K.; Mc Cormick, M.; Frank, D.; Trouet, V.; O Kaplan, J.; Herzig, F.; Heussner, K.U.; Wanner, H.; et al. 2500 Years of European Climate Variability and Human Susceptibility. *Science* **2011**, *331*, 578–582. [CrossRef] [PubMed]

11.   Christiansen, B.; Ljungqvist, F.C. The extra-tropical Northern Hemisphere temperature in the last two millennia: Reconstructions of low-frequency variability. *Clim. Past* **2012**, *8*, 765–786. [CrossRef]

12.   Mensing, S.; Tunno, I.; Sagnotti, L.; Florindo, F.; Noble, P.; Archer, C.; Zimmerman, S.; Pavón-Carrasco, F.J.; Cifani, G.; Passigli, S.; et al. 2700 years of Mediterranean environmental change in central Italy: A synthesis of sedimentary and cultural records to interpret past impacts of climate on society. *Quat. Sci. Rev.* **2015**, *116*, 72–94. [CrossRef]

13.   Rippon, S. Adaptation to a changing environment: The response of marshland communities to the late medieval 'crisis'. *J. Wetl. Archaeol.* **2001**, *1*, 15–39. [CrossRef]

14.   Malanima, P. Urbanisation and the Italian economy during the last millennium. *Eur. Rev. Econ. Hist.* **2005**, *9*, 99–102. [CrossRef]

15.   Hoffmann, R.C. *An Environmental History of Medieval Europe*; Cambridge University Press: Cambridge, UK, 2014; ISBN 9781139050937.

16.   Mensing, S.; Tunno, I.; Cifani, G.; Passigli, S.; Noble, P.; Archer, C.; Piovesan, G. Human and climatically induced environmental change in the Mediterranean during the Medieval Climate Anomaly and Little Ice Age: A case from central Italy. *Anthropocene* **2016**, *15*, 49–59. [CrossRef]

17.   Curtis, D.R.; Campopiano, M. Medieval land reclamation and the creation of new societies: Comparing Holland and the Po Valley, c.800–c.1500. *J. Hist. Geogr.* **2014**, *44*, 93–108. [CrossRef]

18.   Groenewoudt, B.; Van Doesburg, D. Peat People. On the Function and Context of Medieval Artificial Platforms in a Coastal Wetland, Eelder- and Peizermaden, The Netherlands. *J. Wetl. Archaeol.* **2018**, *18*, 77–96. [CrossRef]

19.   Willemsen, J.; van't Veer, R.; van Geel, B. Environmental change during the medieval reclamation of the raised-bog area Waterland (The Netherlands): A palaeophytosociological approach. *Rev. Palaeobot. Palynol.* **1996**, *94*, 75–86. [CrossRef]

20.   Pals, J.P.; van Dierendonck, M.C. Between flax and fabric: Cultivation and processing of flax in a mediaeval peat reclamation settlement near midwoud (Prov. Noord Holland). *J. Archaeol. Sci.* **1988**, *15*, 237–251. [CrossRef]

21.   Gardiner, M.; Hartwell, B. Landscapes of Failure: The Archaeology of Flooded Wetlands at Titchwell and Thornham (Norfolk), and Broomhill (East Sussex). *J. Wetl. Archaeol.* **2006**, *6*, 137–160. [CrossRef]

22.   Hadler, H.; Vött, A.; Newig, J.; Emde, K.; Finkler, C.; Fischer, P.; Willershäuser, T. Geoarchaeological evidence of marshland destruction in the area of Rungholt, present-day Wadden Sea around Hallig Südfall (North Frisia, Germany), by the Grote Mandrenke in 1362 AD. *Quat. Int.* **2018**, *473*, 37–54. [CrossRef]

23.   Brühl, C. (Ed.) *Codice Diplomatico Longobardo*; III, no. 41; Istituto Storico Italiano per il Medio Evo: Rome, Italy, 1973.

24. Squatriti, P. *Water and Society in Early Medieval Italy AD 400–1000*; Cambridge University Press: Cambridge, UK, 1998; ISBN 0521522064.
25. Fumagalli, V. *L'uomo e L'ambiente nel Medioevo*; Editori Laterza: Roma, Italy, 1999; ISBN 884206954X.
26. Fumagalli, V. *Città e Campagna nell'Italia Medievale*; Pàtron Editore: Bologna, Italy, 1985; ISBN 978-8855519328.
27. Magnusson, R. *Water Technology in the Middle Ages: Cities, Monasteries, and Waterworks after the Roman Empire*; JHU Press: Baltimore, MD, USA, 2001; ISBN 080186626X.
28. Ambrosini, C.; De Marchi, P.M. (Eds.) *Uomini e Acque a San Benedetto Po—Il Governo del Territorio tra Passato e Futuro, in Atti del Convegno (Mantova-San Benedetto Po, 10–12 maggio 2007)*; All'Insegna del Giglio: Firenze, Italy, 2010; ISBN 9788878144224.
29. Hodges, R. *Dark Age Economics: A New Audit*; Bloomsbury Academic: Bristol, UK, 2012; ISBN 9780715636794.
30. Settia, A.A. *Proteggere e Dominare, Fortificazioni e Popolamento nell'Italia Medievale*; Viella Libreria Editrice: Roma, Italy, 1999; ISBN 9788883346071.
31. Settia, A.A. *Castelli e Villaggi nell'Italia Padana*; Liguori Editore: Napoli, Italy, 1984; ISBN 8820712113.
32. Wikimedia Commons. Available online: https://commons.wikimedia.org/wiki/File:Paludes1570.jpg (accessed on 13 March 2018).
33. Gabbi, B. *La Bonifica Bentivoglio-Enza. Antologia Documentaria Sulle Acque*; Reggio Emilia: Parma, Italy, 2001; ISBN 881031299.
34. Saltini, A. *Dove L'uomo Separò la Terra Dalle Acque: Storia Delle Bonifiche dell'Emilia-Romagna*; Reggio Emilia: Parma, Italy, 2005; ISBN 8881034336.
35. Tobin, G.A. The Levee Love Affair: A Stormy Relationship. *J. Am. Water Resour. Assoc.* **1995**, *31*, 359–367. [CrossRef]
36. Vis, M.; Klijn, F.; De Bruijn, K.M.; Van Buuren, M. Resilience strategies for flood risk management in The Netherlands. *Int. J. River Basin Manag.* **2003**, *1*, 33–40. [CrossRef]
37. Silva, W.; Klijn, F.; Dijkman, J. Room for the Rhine Branches in The Nederlands. In *What the Research Has Taught Us*; Delft Hydraulics Report R3294; RIZA Report: Delft, The Netherlands, 2001; p. 31.
38. Bottazzi, G. Gli agri centuriati di Brixellum e di Tannetum. In *L'Emilia in età Romana*; Ricerche di Topografia Antica: Modena, Italy, 1987; pp. 149–191.
39. Cremaschi, M.; Pizzi, C.; Valsecchi, V. Water management and land use in the terramare and a possible climatic co-factor in their abandonment: The case study of the terramara of Poviglio Santa Rosa (northern Italy). *Quat. Int.* **2006**, *151*, 87–98. [CrossRef]
40. Mele, M.; Cremaschi, M.; Giudici, M.; Lozej, A.; Pizzi, C.; Bassi, A. The Terramare and the surrounding hydraulic structures: A geophysical survey of the Santa Rosa site at Poviglio (Bronze Age, northern Italy). *J. Archaeol. Sci.* **2013**, *40*, 4648–4662. [CrossRef]
41. Cremaschi, M.; Nicosia, C. Sub-Boreal Aggradation along the Apennines margin of the Central Po plain: Geomorphological and geoarchaeological aspects. *Gèomorphologie* **2012**, *2*, 156–174. [CrossRef]
42. Cremaschi, M.; Mercuri, A.M.; Torri, P.; Florenzano, A.; Pizzi, C.; Marchesini, M.; Zerboni, A. Climate change versus land management in the Po Plain (Northern Italy) during the Bronze Age: New insights from the VP/VG sequence of the Terramara Santa Rosa di Poviglio. *Quat. Sci. Rev.* **2016**, *136*, 153–172. [CrossRef]
43. Ravazzi, C.; Marchetti, M.; Zanon, M.; Perego, R.; Quirino, T.; Deaddis, M.; De Amicis, M.; Margaritora, D. Lake evolution and landscape history in the lower Mincio River valley, unravelling drainage changes in the central Po Plain (N-Italy) since the Bronze Age. *Quat. Int.* **2013**, *288*, 195–205. [CrossRef]
44. Mercuri, A.M.; Montecchi, M.C.; Pellacani, G.; Florenzano, A.; Rattighieri, E.; Cardarelli, A. Environment, human impact and the role of trees on the Po plain during the Middle and Recent Bronze Age: Pollen evidence from the local influence of the terramare of Baggiovara and Casinalbo. *Rev. Palaeobot. Palynol.* **2015**, *218*, 231–249. [CrossRef]
45. Marchetti, M. Environmental changes in the central Po Plain (northern Italy) due to fluvial modifications and anthropogenic activities. *Geomorphology* **2002**, *44*, 361–373. [CrossRef]
46. Charlton, R. *Fundamentals of Fluvial Geomorphology*; Routledge: New York, NY, USA, 2007; ISBN 978-0415334549.
47. Castiglioni, G.B.; Ajassa, R.; Baroni, C.; Biancotti, A.; Bondesan, A.; Bondesan, M. *Carta Geomorfologica Della Pianura Padana. 3 Fogli Alla Scala 1:250.000*; IRIS: Firenze, Italy, 1997.
48. Cremaschi, M.; Bernabò Brea, M.; Tirabassi, J.; Dall'Aglio, P.L.; Baricchi, W.; Marchesini, A.; Nepoti, S. L'evoluzione del settore centromeridionale della valle padana, durante l'età del bronzo, l'età romana e l'età altomedievale, geomorfologia ed insediamenti. *Padusa* **1980**, *16*, 5–25.

49. Cremaschi, M.; Storchi, P.; Perego, A. Geoarchaeology in urban context: The town of Reggio Emilia and river dynamics during the last two millennia in Northern Italy. *Geoarchaeology* **2018**, *33*, 52–66. [CrossRef]

50. Bernabò Brea, M.; Cremaschi, M. *Il Villaggio Piccolo Della Terramara di Santa Rosa di Poviglio: Scavi 1987–1992*; IRIS: Firenze, Italy, 2004; ISBN 8860450144.

51. Bottazzi, G.; Bronzoni, L.; Mutti, A. *Carta Archeologica del Comune di Poviglio 1986–1989*; IRIS: Poviglio, Italy, 1995.

52. Settis, S.; Pasquinucci, M. *Misurare La Terra: Centuriazione e Coloni nel Mondo Romano*; Franco Cosimo Panini: Modena, Italy, 1984; ISBN 88-7686-014-2.

53. Lamb, H.H. *Climate, History and the Modern World*; Routledge: London, UK, 1995; ISBN 9780415127356.

54. Montanari, M. *L'alimentazione Contadina Nell'alto Medioevo*; Liguori Publications: Napoli, Italy, 1983; ISBN 8820707748.

55. Calzolari, M. La navigazione interna in Emilia Romagna tra VIII e XIII secolo. In *Vie del Commercio in Emilia Romagna Marche*; Alfieri, N., Ed.; Silvana: Cinisello Balsamo, Italy, 1991; pp. 115–124. ISBN 9788836603053.

56. Racine, P. Poteri medievali e percorsi fluviali nell'Italia padana. *Quaderni Storici* **1986**, *61*, 9–32.

57. Brandolini, F.; Trombino, L.; Sibilia, E.; Cremaschi, M. Micromorphology and site formation processes in the Castrum Popilii Medieval Motte (N Italy). *J. Archaeol. Sci. Rep.* **2018**, *20*, 18–32. [CrossRef]

58. Affò, I. *Istoria di Guastalla*; Costa: Parma, Italy, 1786.

59. Tiraboschi, G. *Dizionario Topografico Degli Stati Estensi*; tipogr. Camerale: Modena, Italy, 1824.

60. Cantarelli, C. *Cronaca di fra Salimbene Parmigiano Dell'ordine dei Minori*; Luigi Battei: Parma, Italy, 1882.

61. Pasquali, G. L'azienda curtense e l'economia dei secoli VI–XI. In *Uomini e Campagne nell'Italia Medievale*; Cortonesi, A., Ed.; Laterza: Roma, Italy, 2003; pp. 19–20. ISBN 9788842066682.

62. Mancassola, N. Uomini e acque nella pianura reggiana durante il Medioevo (Secoli IX-XIV). In *Acque e Territorio nel Veneto Medievale*; Canzian, D., Simonetti, R., Eds.; Viella: Roma, Italy, 2012; pp. 115–132. ISBN 9788867281473.

63. Fumagalli, V. *L'alba del Medioevo*; L'Edizione Del Mulino: Bologna, Italy, 2004; ISBN 978-88-15-25387-3.

64. Rao, R. *I Paesaggi dell'Italia Medievale*; Routledge: Roma, Italy, 2016; ISBN 8843077759.

65. Brandolini, F.; Cremaschi, M. Valli-Paludi nel Medioevo: Il rapporto tra uomo e acque nella Bassa Pianura Reggiana. Le bonifiche "laiche" per colmata. In *VIII Congresso Nazionale di Archeologia Medievale. Pré-Tirages (Matera, 12–15 Settembre 2018)*; Sogliani, F., Gargiulo, B., Annunziata, E., Vitale, V., Eds.; All'Insegna del Giglio: Firenze, Italy, 2018; Volume 2, pp. 72–78. ISBN 9788878148673.

66. Mori, A. *Le Antiche Bonifiche Della Bassa Reggiana*; La Bodoniana: Parma, Italy, 1923.

67. Castaldini, D. Evoluzione della rete idrografica centropadana in epoca protostorica e storica. In *Atti del Convegno Nazionale di Studi—Insediamenti e Viabilità Nell'alto Ferrarese Dall'età Romana al Medioevo, Cento 1987*; ITA: Ferrara, Italy, 1989; pp. 115–134.

68. Castiglioni, G.B.; Pellegrini, G.B. *Note Illustrative della Carta Geomorfologica Della Pianura Padana*; Supplementi di Geografia Fisica e Dinamica Quaternaria; Comitato Glaciologico Italiano: Torino, Italy, 2001; Volume 4, pp. 328–421.

69. Geoportale Nazionale. Available online: http://www.pcn.minambiente.it/mattm/ (accessed on 24 February 2018).

70. Geoportale Emilia Romagna Region. Available online: http://geoportale.regione.emilia-romagna.it/it (accessed on 13 March 2018).

71. Degani, M. *Carta Archeologica Della Carta d'Italia al 1:100.000*; Foglio 74 (Città e Provincia di Reggio Emilia); LS Olschki: Firenze, Italy, 1974.

72. Baricchi, W. (Ed.) *Insediamento Storico e Beni Culturali Bassa Pianura Reggiana: Comuni di Boretto, Brescello, Fabbrico, Gualtieri, Guastalla, Luzzara, Novellara, Poviglio, Reggiolo, Rolo*; Reggio Emilia: Parma, Italy, 1989.

73. Castelli dell'Emilia-Romagna: Censimento e Schedatura. Available online: http://geo.regione.emilia-romagna.it/schede/castelli/index.jsp (accessed on 18 April 2018).

74. Affò, I. *Storia di Parma*; Carmignani: Parma, Italy, 1792.

75. Drei, G. *Le Carte Degli Archivi Parmensi dei Secoli X–XI*; Reggio Emilia: Parma, Italy, 1924.

76. Torelli, P. *Le Carte Degli Archivi Reggiani: Fino al 1050*; Reggio Emilia: Parma, Italy, 1924.

77. Torelli, P.; Gatta, F.S. *Le Carte degli Archivi Reggiani 1051–1060*; Reggio Emilia: Parma, Italy, 1938.

78. Angiolini, E.; Torresan, S. *Archivio Storico del Comune di Novellara*; Inventario; Comune di Novellara: Novellara, Italy, 2015.

79. Larsen, L.; Vinther, B.; Briffa, K.; Melvin, T.; Clausen, H.; Jones, P.; Siggaard-andersen, M.; Hammer, C.; Eronen, M.; Grudd, H.; et al. New ice core evidence for a volcanic cause of the A.D. 536 dust veil. *Geophys. Res. Lett.* **2008**, *35*, L04708. [CrossRef]
80. Squatriti, P. The Floods of 589 and Climate Change at the Beginning of the Middle Ages: An Italian Microhistory. *Speculum* **2010**, *85*, 799–826. [CrossRef] [PubMed]
81. Macphail, R.I.; Goldberg, P. *Practical and Theoretical Geoarchaeology*; Blackwell Publishing: Oxford, UK, 2006; ISBN 0632060441.
82. Iannacci, L.; Mezzetti, M.; Modesti, M.; Zuffrano, A. *Chartae Latinae Antiquiores. Facsimile-Edition of the Latin Charters. Ninth Century. Part XCI-Italy LXIII, 2nd Series; Reggio Emilia: Firenze, Italy*; Urs Graf Verlag: Zurich, Switzerland, 2012; ISBN 9783859512337.
83. Moreno, D. Miglioramenti agrari sullo spartiacque Trebbia—Aveto. Tracce di "colmate di monte" di età post-medievale. In *La Natura Della Montagna*; Cevasco, R., Ed.; Gallucci: Setri Levante, Italy, 2013; pp. 424–439. ISBN 88-97264-22-0.
84. Moreno, D. Improving land with water: Evidence of historical "colmate di monte" in the Trebbia-Aveto watershed. In *Proceedings of 2nd Workshop on Environmental History and Archaeology*; All'Insegna del Giglio: Montebruno, Italy, 2002.
85. Cevasco, A.; Cevasco, R. "Montagne che libbiano" e zone umide colmate? Il "lago" di Torrio (Val d'Aveto, Ferriere, Pc). In *La Natura Della Montagna*; Cevasco, R., Ed.; Gallucci: Sestri Levante, Italy, 2013; pp. 440–448. ISBN 88-97264-22-0.
86. Lillie, M.; Weir, D. Alluvium and warping in the Humberhead Levels: The identification of factors obscuring palaeo land surfaces and the archaeological record. In *Wetland Heritage of the Humberhead Levels*; Van de Noort, R., Ellis, S., Eds.; University of Hull: East Riding of Yorkshire, UK, 1997; pp. 191–218. ISBN 0859581926.
87. Mansell, L.; Whitehouse, N.; Gearey, B.; Barratt, P.; Roe, H. Holocene floodplain palaeoecology of the Humberhead Levels; implications for regional wetland development. *Quat. Int.* **2014**, *341*, 91–109. [CrossRef]
88. Lillie, M.; Weir, D. Alluvium and warping in the lower Trent valley. In *Wetland Heritage of the Ancholme and Lower Trent Valleys—An Archaeological Survey*; Van de Noort, R., Ellis, S., Eds.; University of Hull: East Riding of Yorkshire, UK, 1998; pp. 103–122. ISBN 0859581934.
89. Adani, G.; Badini, G.; Baricchi, W.; Pellegrini, M.; Pozzi, F.M.; Spaggiari, A. *Vie D'acqua nei Ducati Estensi*; Cassa di Risparmio di Reggio Emilia, Amilcare Pizzi Editore: Cinisello Balsamo/Milano, Italy, 1990; ISBN 9788837055455.
90. Mozzi, P.; Piovan, S.; Corrò, E. Long-term drivers and impacts of abrupt river changes in managed lowlands of the Adige river and northern PO delta (Northern Italy). *Quat. Int.* **2018**, in press. [CrossRef]
91. Merz, B.; Aerts, J.; Arnbjerg-Nielsen, K.; Baldi, M.; Becker, A.; Bichet, A.; Blöschl, G.; Bouwer, L.M.; Brauer, A.; Cioffi, F.; et al. Floods and climate: Emerging perspectives for flood risk assessment and management. *Nat. Hazards Earth Syst. Sci.* **2014**, *14*, 1921–1942. [CrossRef]

sustainability

MDPI

Article

Forest Landscape Change and Preliminary Study on Its Driving Forces in Ślęża Landscape Park (Southwestern Poland) in 1883–2013

Piotr Krajewski [1,*], **Iga Solecka** [1] **and Karol Mrozik** [2]

1   The Faculty of Environmental Engineering and Geodesy, Department of Spatial Economy, Wroclaw University of Environmental and Life Sciences, 50-375 Wroclaw, Poland; iga.solecka@upwr.edu.pl
2   The Faculty of Environmental Engineering and Spatial Management, Institute of Land Improvement, Environmental Development and Geodesy, Poznan University of Life Sciences, 60-637 Poznan, Poland; kmrozik@up.poznan.pl
*   Correspondence: piotr.krajewski@upwr.edu.pl; Tel.: +48-660-709-433

Received: 11 October 2018; Accepted: 29 November 2018; Published: 30 November 2018

Abstract: Changes in forest landscapes have been connected with human activity for centuries and can be considered one of the main driving forces of change from a global perspective. The spatial distribution of forests changes along with the geopolitical situation, demographic changes, intensification of agriculture, urbanization, or changes in land use policy. However, due to the limited availability of historical data, the driving forces of changes in forest landscapes are most often considered in relation to recent decades, without taking long-term analyses into account. The aim of this paper is to determine the level and types of landscape changes and make preliminary study on natural and socio-economic factors on changes in forest landscapes within the protected area, Ślęża Landscape Park, and its buffer zone using long-term analyses covering a period of 140 years (1883–2013). A comparison of historical and current maps and demographic data related to three consecutive periods of time as well as natural and location factors by using the ArcGIS software allows the selected driving forces of forest landscape transformations to be analyzed. We took into account natural factors such as the elevation, slope, and exposure of the hillside and socio-economic drivers like population changes, distances to centers of municipalities, main roads, and built-up areas.

Keywords: driving forces; landscape change; landscape dynamics; forest landscape; land use; land cover; landscape change index

---

1. Introduction

On a global scale, human beings have been the main driving force for the transformation of the Earth's surface for several hundred years [1]. People have significantly influenced landscape changes. Transformations have occurred to meet the needs of society and its individual units. Today, we can see the effects of many historical changes in the landscape. Over the last few decades, these changes have intensified due to strong socio-economic changes, including changes in agriculture, industry, or transport [2]. These transformations are particularly evident in the countries in Central and Eastern Europe, where profound political and socio-economic changes have taken place [3]. The significance of the changes that followed the fall of communism have been particularly emphasized [4,5]. Changes resulting from the enlargement of the European Union in 2004 have also been analyzed [6]. Both traditional agricultural landscapes [7–9], urban and industrial [10–13], have undergone strong transformations, as well as landscapes with high value for tourists [14,15]. Changes in forest landscapes and those in the immediate vicinities of forests [16–19] are particularly noticeable. The results of palaeoenvironmental studies [20], used also as primary source for the reconstruction of forest cover [21], have more and more often become the starting point for research on long-term landscape changes. So far, in many cases, landscape change studies have been limited only to identifying the sizes and types of transformations, ignoring the identification of factors that could have a significant impact on the landscape. Meanwhile, understanding the phenomena that lie behind a specific transformation is crucial in the context of conservations and sustainable landscape management [22]. Understanding the cause and effect relationship is one of the ideas on which the analysis of driving forces of landscape changes is based [23]. The knowledge of the causes allow the main driving forces and categories of phenomena that have helped to shape the landscape to be classified. The reason for many changes, especially in suburban areas [24,25], is urbanization, and this is causing more and more intensive transformations. As the result of this, the percentage of urbanized landscapes is increasing at the expense of natural and semi-natural landscapes. Other causes of changes in landscapes include the intensification of agriculture, the succession of forests in abandoned areas, increased demand for service areas, the development of renewable energy sources, and the creation of protected areas [26]. The land use policy of local authorities is also a frequent driving force of landscape change [27]. This often leads to the degradation of historically shaped landscapes that are part of the local cultural heritage which should be protected elements in the land use policy [28].

The acceleration of landscape changes in recent decades has also been noticed by the Council of Europe. To limit negative changes, the European Landscape Convention was adopted in 2000 [29]. The signatories of this convention, including Poland, have recognized the landscape as being an important part of quality of life and a key element of the well-being of society of which every person has the responsibility to protect, shape, and plan for. However, the Polish government has not yet implemented the provisions of the convention to Polish law despite the obligations resulting from the ratification in 2004. That is why no landscape assessment on a whole country scale has been done, and thus, it has not been possible to identify the forces and pressures that have caused the transformations as well as their levels of intensity. Methods for determination of the landscape change index have been developed [30]. However, these studies have been conducted locally and not in the context of the whole country. Knowledge in this field is particularly important where the purposes and functions of particular areas change frequently as the result of adapting space to current social and economic needs. Areas of tourist investment are particularly located within or close to areas that are attractive in terms of aesthetic appeal and that are often protected by one of the landscape protection forms. An example of this type of area in Poland is landscape parks—areas with high landscape value that cover both forest areas as well as areas of arable land, rural settlements, and sometimes, also small towns. In the context of this form of protection, the necessity of monitoring and planning changes in the landscape has been emphasized [31,32].

When conducting research on landscape changes, we should always be aware that the condition of the landscape which we are currently observing is a physiognomic reflection and synthesis of changes caused by a number of different factors. This has become the basis for the concept of the driving forces for landscape change [23,33], also known as drivers [34] or key processes [35]. These are the factors that have caused noticeable changes in the landscape and have significantly influenced the direction of its further transformation. Driving forces are divided into five basic groups: socio-economic, political, technological, cultural, and natural/spatial [23]. Analyses of the causes of changes in the landscape may concern different spatial scales and different time periods. The results of analyses often include the identification of the main actors of change (people/institutions) which have had a decisive influence on the change of landscape [36,37]. This research trend has been considered in many European countries. Analyses of driving forces have concerned, in particular, landscapes in Switzerland [38], Germany [39], Slovakia [40,41], the Mediterranean landscapes [42,43], and across Europe [44]. In the context of forest areas, analyses of driving forces have usually also included entire countries or regions [45–47]. Smaller areas have rarely been analyzed [48]. Although, in previous years, interest in research on the driving forces of landscape changes has increased significantly, there is still a lack of examples of this type of research in some European Union countries, including Poland [49]. The case study of the Ślęża

Landscape Park and its buffer zone, where the protected landscape and important elements of the Lower Silesia's cultural heritage in Poland have been analyzed, will supplement the knowledge of the character of forest landscape change in this part of Europe which has not been recognized well in this aspect. Preliminary results of driving forces studies will give a good basis for further analysis of cause and result relationship between identified factors and observed phenomena for a full understanding of the dynamics of changes.

To meet the needs of the research, we assumed that forest landscape is a landscape perceived by people to be a mosaic of various land cover types dominated by forests and which provide goods and services related to forests. The main goal of our research was to determine the level and types of landscape changes and make preliminary study on the dependencies between basic driving forces—natural and socio-economic—and changes in forest landscapes over a long-term period of about 140 years (1883–2013). To simplify the way that the boundaries of forest landscapes were defined, we chose the Ślęża Landscape Park and its buffer zone as a case study area. We assumed that goods and services related to the forest landscape of the Ślęża Massif were limited to the surrounding protected area and the buffer zone. However, we would like to stress that the manuscript is mainly focused on spatiotemporal analyses of forest landscape change which have a great importance for analysis of its driving forces. The results of preliminary study on driving forces of forest landscape change presented in this manuscript is the first study related to area of research and gives a good basis for further studies on the correlation of each factor with the observed types of changes in forest landscapes. Analyses of other possible driving forces like land use policies and local legal regulations are planned in the second stage of research.

In order to achieve our intended goal, we performed comparative analyses of land cover maps, demographic and location data available in the analyzed periods, together with a summary of historical events, which we supplemented with an analysis of natural factors in ArcGIS software. To identify and understand the level and character of forest landscape changes and its driving forces in the Ślęża Landscape Park and its buffer zone, we tried to answer four questions:

1. What have the quantitative changes within the forest and non-forest landscapes over a period of 140 years, divided into three periods of time, covering approximately 40–50 years, been?
2. In which time period was the landscape change index (LCI), which determines the level of change in landscapes, more intensive? Is it possible to calculate the index based on historical data?
3. What have the main types of forest and non-forest landscape transformations been in the studied area?
4. What are the characteristics of identified landscape changes in the context of natural factors such as elevation, slope, and hillside exposure as well as socioeconomic factors such as population changes, distances to built-up areas, main roads, and capitals of municipalities?

2. Materials and Methods

2.1. Case Study Area

The area of our research was the Ślęża Landscape Park and its buffer zone (Figure 1). The park is one of 12 protected areas of this type in Lower Silesia with a total area of 158.07 km², including a buffer zone. The research area covers the part of six communes located around the highest hill of Przedgórze Sudeckie—Ślęża (717 m a.s.l.). The research area is located about 40 km southwest of the capital of the region, Wrocław. The landscape park consists of three main parts. Together, the three parts of park occupy an area of 76.78 km². They are connected to each other with an area of 81.29 km²—a buffer zone around the landscape park which protects it against external threats resulting from human activity.

The Ślęża Landscape Park was created on 8 June 1988 to preserve the landscape of the Ślęża Massif, including preservation of the local character and scale of development in historically shaped settlement units and undeveloped spaces in the open forest and meadow landscape and to protect the area's diverse natural, geological, and geomorphological value [50]. Currently, the area within the limits of the Ślęża Landscape Park is covered mainly by forests (about 62%), which occupy the slopes of Ślęża, Radunia, and the surrounding hills. The lower slopes and flatlands are predominantly used for agricultural purposes. There is one city located in the Northern part of the landscape park. The remaining units of the settlement system are villages that are evenly distributed around the hills among the agricultural and forest areas. In total, areas covered with forest within the boundaries of the landscape park and its buffer zone currently occupy over 33%, and areas not covered with forest occupy 67%, of which 59% is agricultural land, 7% is built-up areas, and 1% is other areas.

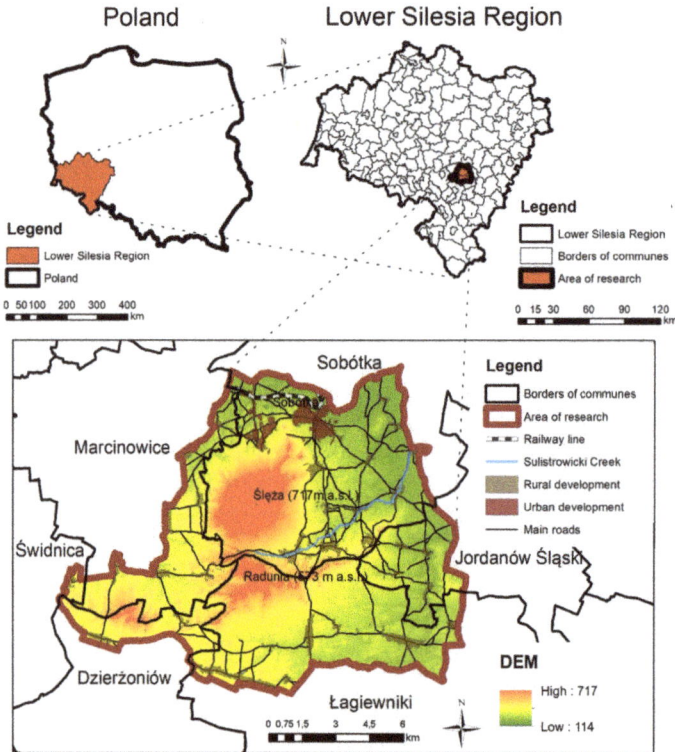

Figure 1. Location and digital elevation model of the research area.

Before analyzing the driving forces of landscape changes within the research area, it should be mentioned that the Ślęża mountain has been a special place for the inhabitants of Lower Silesia since ancient times. Raised above the surrounding flat areas, it was a place of pagan sun worship of the local tribes which lived at the foot of the mountain in the later Bronze Age and early Iron Age (1300 BC–500 BC). Today, the Ślęża Massif is a protected area as a landscape park, and partly as a Natura 2000 area. It enjoys great popularity among tourists as well as among the residents of the largest city in the region, Wrocław. The surrounding landscape is a factor that attracts the rich inhabitants of the city who build their second homes there. Due to the rich deposits of granite, it is also an area that has been associated with mining for years.

In terms of types of forest in the park, there is a division into two parts [51]. The top of Ślęża mountain is overgrown with *Galio odorati-Fagetum* and *Luzulo pilosae-Fagetum* and is protected as a nature reserve. However, these forest associations are preserved in a small part of their potential area of occurrence. Paleobotanical studies revealed that the primary vegetation for entire area of Ślęża Massif was beech and oak forest. The lower parts, today occupied mainly by arable land and built-up areas, were dominated by oak-hornbeam forest [52]. Today, in most of the study area, these types of forest have been replaced by the monoculture of *Picea abies* with a dense forest stand that is often devoid of undergrowth and was formed in the 19th century. In the Northern part, there is also a fragment of an old and well-preserved *Tilio platyphyllis-Acerion pseudoplatani* with rich forest undergrowth. The peak of the Radunia mountain is overgrown with *Potentillo albae-Quercetum* with a dominance of *Quercus petraea*. On the top, there is sparse forest with an unusual dwarf form of this species. However, at the foot of hills, there are numerous orchards, mostly of *Cerasus* sp. Some areas, formerly used for agriculture, have undergone spontaneous forest succession.

2.2. Identification of Changes in Forest Landscapes

To make it possible to identify landscape changes in last 140 years, it was necessary to select the proper archival source materials that would have a similar content and give detailed information about land cover. Source maps should have also been published at the same scale, which ensured a similar rate of data generalization. Four series of maps published in 1883–1889, 1936–1938, 1977, and 2013 at 1:25,000 scale, containing information of land cover, were selected from available source materials. The basis for choosing the source maps was availability of good quality map sheets covering the whole research area and the similarity of land cover types. The full information of collected maps is included in Table 1. Five sheets of maps from the oldest considered series published in 1883–1889, and next published in 1936–1938 (Topographische Karte—Messtischblatt) refer to the period under German rule. The maps were made in the field and are characterized by very high detail and extraordinary land cover mapping accuracy. They are one of the best sources of information on land cover transformation over the last 140 years and are commonly used in landscape change analyses [53]. Polish land cover maps from 1977 and 2013 were developed on the basis of aerial photos supplemented with field verification.

The availability of source maps was a direct cause of identification of landscape changes in three periods covering approximately 40–50 years: 1883(89)–1936(38), 1936(38)–1977, and 1977–2013. Publication dates in parentheses represent the newest map from the series. There were no more published series of land cover maps at the same scale that would cover the whole area of the study. The lack of data sources made it impossible to increase the number of analyzed periods of time. Selected intervals were long enough to identify forest transformations at archival maps and covered different political and economic conditions which can be considered as underlying driving forces of landscape change in the study area. Chronological time intervals allowed us to identify and compare data and find some trends of landscape transformations in the last 140 years. All the maps contain basic land cover classes—such as forest (young and mature), meadows, agricultural land, urban and rural development, water areas, mining areas, etc.—appear consistently over all time periods. The changes between these land cover classes represent the most landscape transformations within research area. Therefore, it gives enough information and comparable data from each time interval.

Georeferences were given for all historical maps used in the analysis. As a ground control points, we chose the least volatile elements in the landscape like the tops of hills, the intersections of main roads, and preserved historical buildings. The newest database of topographic objects from 2013 was the reference layer for the remaining maps. A rubber sheeting method of transformation called 'spline' was used to calibrate the maps. It gave the best calibration results when ground control points are important and exact registration is required [54]; however, it causes significant distortions of the maps [55] that can be overcome by using a dense grid of ground control points.

The selected four series of source maps were the basis for preparing four land cover maps showing most common elements of the landscape. Prepared maps made it possible to compare data in ArcMap 10.2.2. Land cover was divided into two groups—areas covered with forest and areas not covered with forest. In addition, the maps show the major roads and railways. Full classification of land cover types is included in Table 2.

Table 1. Collected cartographic materials on a scale of 1:25,000 used to analyze changes in forest landscapes.

Map Name	Type of Data	Land Cover Classes	Map Sheets	Date of Release	Source of Data
Topographische Karte (Messtischblatt)	raster data	forest, plantations and forest nurseries, meadows, pastures, arable land, wetland, marshland, fallow land, water areas, mining areas, urban and rural areas	Zobten Jordansmühl Weizenrodau Lauterbach Mörschelwitz	1883 1889 1885 1883 1884	Library of Wroclaw University
Topographische Karte (Messtischblatt)	raster data	forest, plantations and forest nurseries (not observed), meadows, pastures, arable land, wetland, marshland, fallow land, water areas, mining areas, urban and rural areas	Zobten am Berge Jordansmühl Weizenrodau Lauterbach Rosenborn	1938 1938 1936 1936 1937	Library of Wroclaw University
Topographic map of Poland	raster data	mature forest, young forest, bushes, orchards, meadows, arable land, wetland, water areas, mining areas, urban and rural areas	Sobótka, Mysłaków, Dzierżoniów, Jordanów Śląski, Kobierzyce	1977	Head Office of Geodesy and Cartography
Database of Topographic Objects	vector data	mature forest, young forest, bushes, orchards, meadows, pastures, arable land, wetland, fallow land, water areas, mining areas, landfill, urban and rural areas	Sobótka, Mysłaków, Dzierżoniów, Jordanów Śląski, Kobierzyce	2013	Head Office of Geodesy and Cartography

Table 2. Classification of land cover types and subtypes.

Land Cover type	Land Cover Subtype	Description of Class
Forest area (FA)	Mature forest area (MFA)	Areas covered with forest except young forest areas
	Young forest area (YFA)	Reforested woodland areas, plantations and forest nurseries
Non-forest area (NFA)	Water area (WA)	Water reservoirs and main watercourses
	Agricultural land (AL)	Arable land, meadows, pastures, bushes and orchards
	Rural development area (RA)	Built-up and recreational areas within all villages
	Urban development area (UA)	Built-up and recreational areas in the city of Sobótka
	Mining area (MA)	Existing quarries and material storage areas
	Other area (OA)	Other unclassified land cover classes, such as wetland, marshland, fallow land, landfills, and forest parking lots

The next step was to create a database containing the area of each type of land cover type and its percentage share in relation to the entire analyzed area in each of the time periods. The basis for this was land cover maps prepared for each time period. In order to calculate the landscape change index (LCI) for each period, it was necessary to specify parameters that indicate changes in the percentage share of areas covered by each of land cover type. It was calculated with

$$CA_i = 100 \times (A_{t+1} - A_t)/TA \tag{1}$$

where CA_i represents changes in the percentage share of areas covered by each land cover type in relation to the total area of research (%); A_{t+1} represents the area covered with each type of land cover during the time interval $t + 1$ (ha); A_t represents the area covered with each type of land cover during the time interval t (ha); and TA represents the total research area (ha).

This allowed us to determine the landscape change index (LCI) for each of the time intervals. The LCI is defined as the absolute values of change in the land cover types that have the greatest impact on the shape of the landscape, assuming that both increases and decreases in these values cause changes in the landscape [30]. The index was calculated for each period of time by multiplying a factor of one-half by the sum of the absolute values of change in percentage share of areas covered by each land type cover in relation to the whole analyzed area. Summing the absolute values of change of each land cover type essentially doubled the index, so the LCI included a factor of one-half to reflect the actual level of change. LCI was calculated with

$$LCI_t = \frac{1}{2} \times \sum_{i=1}^{n} |CA_i| \qquad (2)$$

where LCI_t represents the landscape change index in each time interval; and $|CA_i|$ represents the absolute value of change in percentage share of the areas covered by each land cover type in relation to the total research area.

To identify the nature and scale of changes related to forest area all polygons which were detected as including changes by the clip tool in ArcGIS software (tool used to cut out a piece of features representing land cover classes from one of analyzed period of time by using the features from another period of time) were analyzed in terms of type of transformation and total and average area. Identification of changes related to forests allowed us to create a classification of types and subtypes of changes in forest landscapes that took place in the Ślęża Landscape Park in the analyzed time intervals. This classification includes three types of change and a total of nine subtypes of change related to forest landscapes. Other possible types and subtypes of change not listed in the classification below were not observed in the research area. Full classification of types and subtypes of forest landscape changes is included in Table 3.

Table 3. Classification of types and subtypes of forest landscape changes.

Type of Forest Landscape Change	Subtype of Forest Landscape Change	Code of Change
Transformations within forest landscapes (temporary deforestation, maturation of forest)	Change from a mature forest area into a young forest area (MFA-YFA)	A1
	Change from a young forest area into a mature forest (YFA-MFA)	A2
Transformation from forest landscapes into non-forest landscapes (permanent deforestation)	Change from a mature forest area into agricultural land (MFA-AL)	B1
	Change from a mature forest area into a rural development area (MFA-RA)	B2
	Change from a mature forest area into an urban development area (MFA-UA)	B3
	Change from a mature forest area into a mining area (MFA-MA)	B4
Transformation of non-forest landscapes into forest landscapes (afforestation)	Change from agricultural land into a young forest area (AL-YFA)	C1
	Change from agricultural land into a mature forest area (AL-MFA)	C2
	Change from a mining area to a mature forest area (MA-MFA)	C3

2.3. Driving Forces of Changes in Forest Landscapes

Natural driving forces associated with topographic conditions may largely determine type of landscape change [36]. Therefore, three of such kind factors (elevation, slope, and hillside exposure) were taken into account. Based on reference data [51,56], variables like natural disasters and type of soils were considered insignificant and not considered in this study. Socioeconomic factors also play important roles in landscape changes [38]. Therefore, we selected four factors for consideration—human demographic changes, distances to main roads, and capitals of municipalities and built-up areas. The analyzed a group of topographic factors including data obtained on the basis of analyses of the digital elevation model (DEM) with a mesh resolution of 1×1 m obtained on the basis of aerial laser scanning (LIDAR) with an average height error of 0.2 m. The digital elevation model was the basis for constructing maps of possible natural drivers of change, such as elevation, slope, and hillside exposure, by using ArcMAP 10.2.2 software with the 3D Analyst extension. However, it should be emphasized that these factors can be considered as an indirect driving forces of change because they determine only the accessibility of the area. Construction of new communication routes as well as better and better construction equipment enables changes in higher located areas.

The second group, socio-economic factors, included human demographic changes which were determined from the population censuses of 1885, 1941, 1978, and 2011 for individual villages within the boundaries of the study area. The distances to the main roads (national, province, and county), the five nearest municipalities (Sobótka, Marcinowice, Jordanów Śląski, Łagiewniki, and Dzierżoniów) and borders of built-up areas were calculated from the axes of roads, and the central parts of villages and towns which were obtained on from the land cover maps developed in the first stage of research for different periods of time. The geopolitical situation in particular periods of time—as well as the main historical, cultural, and technological events—were also described as a background to the research results. The main data sources were literature and chronicles of events. The complete list of analyzed driving forces and data sources is shown in Table 4.

Table 4. Analyzed driving forces of changes in forest landscapes.

Group of Driving Forces	Type of Driving Force	Data Source	Variables Used to Quantify Factors
Natural	Slope grade Hillside exposure Elevation	Map of slope grade from the digital elevation model (DEM) Map of the hillside exposure from the DEM Map of elevation from the DEM	Slope (%) Hillside exposure (Flat, N, NE, E, SE, S, SW, W, NW) Elevation (m a.s.l.)
Socioeconomic	Human demographic changes Distance to the main roads Distance to the capital of municipality The distance to the built-up area	Population censuses from 1885, 1941, 1978, and 2011 Land cover maps from four time periods Land cover maps from four time periods Land cover maps from four time periods	Number of inhabitants in each village (pcs.) Distances to national, provincial, and county roads (km) Distance to the center of the capital of the 5 nearest municipalities (km) Distance to the nearest border of built-up area (km)
Political	Political events—national and local	Chronicles of events, literature	-
Cultural	Cultural events, social changes	Chronicles of events, literature	-
Technological	Changes in crop technology and forest management	Chronicles of events, literature	-

Next step was to determine the characteristics of each observed type of change within the forest landscapes in each of the time intervals in the context of selected possible natural and socioeconomic driving forces. Each polygon that underwent transformation was described in terms of its dominant slope, the exposure of the hillside, elevation, human demographic changes of the nearest village and city, and the distances to the nearest major towns (with headquarter of commune council), main roads (national, provincial and county roads), and built-up areas. Then polygons were classified for the appropriate classes in terms of elevation (50 m intervals), slopes (10 % intervals), hillside exposure (according to the sides of the world), distances from municipalities (2 km intervals), distances from main roads (500 m intervals), and distances from built-up areas (250 m intervals). Last step was to count the polygons separately for each type of landscape transformation specified in the classification (A1–A2, B1–B4, C1–C3).

It should be noted that used methods may be the source of some errors resulting from the process of identification of past landscape transformations connected especially with mapping of changes. One should also be aware that the archival source materials, despite being published at the same scale, differ in terms of their purpose (administrative, military maps, etc.), used signs, or degree of data generalization. Since the basis for mapping changes were archival maps from different time periods the analysis had the potential to contain some errors resulting from the inaccuracy of the georeferencing of individual maps or the method of drawing polygons. To minimize the amount of erroneous data, we eliminated all polygons that symbolized a change in an area smaller than 0.1 ha Analyses of driving forces behind landscape changes have also some limitations because they do not indicate cause and effect relationships between selected driving forces and identified landscape transformations. Therefore, the correlation analyses between each factor and landscape changes are necessary in the next stage of research. Data collected at the initial stage of driving forces analyses can only be the basis of further studies for full understanding processes shaping the landscape.

3. Results

3.1. Changes in Forest and Non-Forest Land Cover in the Years 1883–2013

Changes in the area of land cover classes is one of the main indicators of landscape dynamics. Analyses of four prepared land cover maps (Figure 2) show that the area covered with forest has systematically increased in particular periods, mainly at the expense of the area used for agriculture. There were also significant transformations within the forest area—some parts were temporarily deforested and some of the young forests reached a mature age. No young forests were found in 1936(38) within the research area, but the plantation and forest nurseries are mentioned in the source map legend. The area of rural, urban development, and mining also systematically increased. Changes of the area of land cover classes in each period are presented in Figure 3.

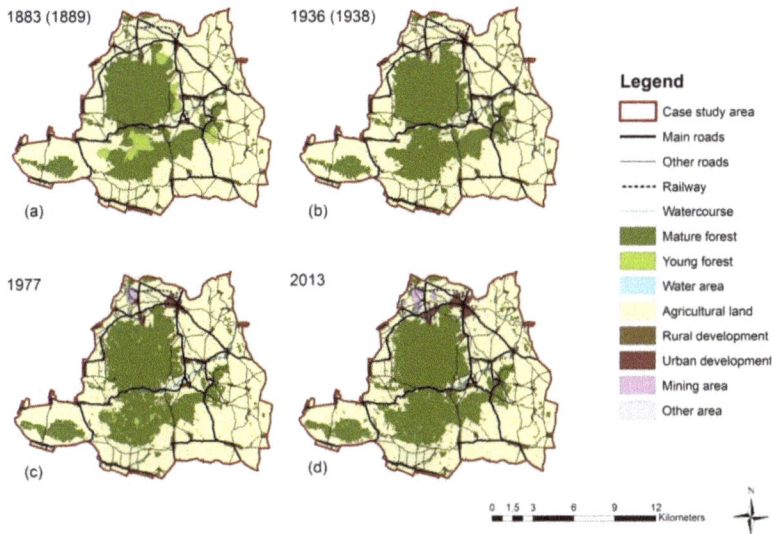

Figure 2. Changes in the land cover of the study area between 1883(89) and 2013: (**a**) map of land cover in 1883(89); (**b**) map of land cover in 1936(38); (**c**) map of land cover in 1977; (**d**) map of land cover in 2013.

In the first of the analyzed time intervals, 1883(89)–1936(38), there was a slight increase in the area covered with forests, from 29.32% to 29.75%. However, at that time, there were transformations within the forest structure—significant areas of young forests occupying 3.34% of the study area had already developed a mature structure. The largest increase (2.22%) of forest area was recorded between 1936(38) and 1977. In 2013, the forest area reached a value of 33.11%. Among the areas not covered with the forest, there was a systematic increase in the area of rural development, with the exception of the period 1936(38)–1977 when part of the rural area was included in the city limits of Sobótka. The mining area also systematically increased from 1936(38) to 2013. This was related to the expansion of one of the largest granite mines in Lower Silesia, located in the Northern part of the study area. Agricultural areas decreased systematically from 65.86% in 1883(89) to 58.38% in 2013. The largest decreases occurred in the periods 1936(38)–1977 and 1977–2013. Changes in the area of land cover classes are shown in Figure 4

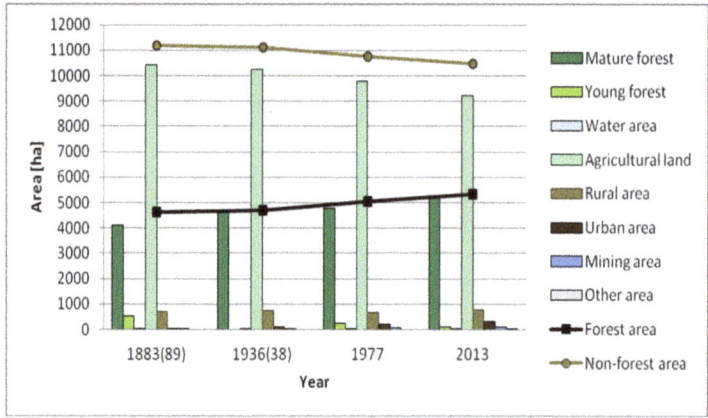

Figure 3. The areas of land cover classes between 1883(89) and 2013.

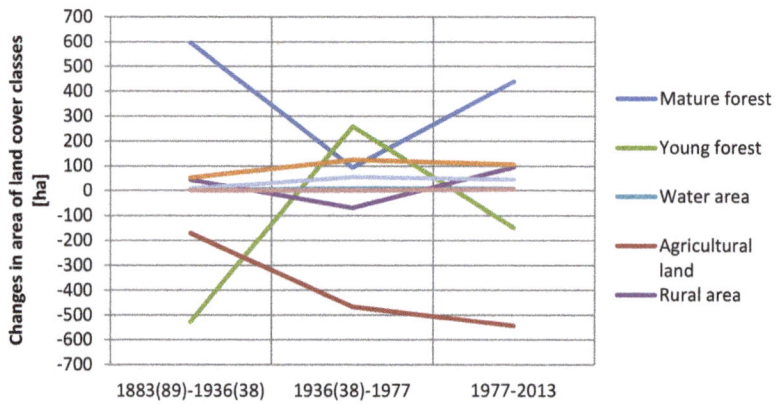

Figure 4. Land cover changes in analyzed time intervals.

The largest changes within the areas covered by the forest occurred in the period 1883(89)–1936(38). The landscape change index level (LCI) for this period of time was more than twice as high as the index for the period 1977–2013 and more than three times higher than in 1936(38)–1977 (Table 5). Significantly lower dynamics of changes concerned areas not covered by forest in the period 1883(89)–1936(38). The LCI in this period was almost three times lower than in the subsequent analyzed periods. It can be observed that the LCI for the whole area in the period 1883(89)–1936(38) is comparable to the indicator for the period 1977–2013. In the first of these periods, the changes concerned mainly area covered with forest and in the second, they mainly concerned non-forest area.

Table 5. Dynamics of changes in land cover in the research area between 1883(89) and 2013.

Time Interval	Indicator	Forest Area				Non-Forest Area			
		MFA [1]	YFA [2]	WA [3]	AL [4]	RA [5]	UA [6]	MA [7]	OA [8]
1883(89)–1936(38)	CA (%)	3.77	−3.34	0.01	−1.08	0.28	0.33	0.04	0
	LCI		3.55				0.87		
1936(38)–1977	CA (%)	0.59	1.63	0.04	−2.95	−0.45	0.79	0.35	0
	LCI		1.11				2.29		
1977–2013	CA (%)	2.78	−0.94	0.04	−3.44	0.60	0.67	0.28	0.02
	LCI		1.86				2.53		

[1] Mature forest area; [2] young forest area; [3] water area; [4] agricultural land; [5] rural area; [6] urban area; [7] mining area; [8] other area.

An analysis of landscape transformations showed a total of 1302 polygons with a total area of 2236.1 ha that have changed since 1883(89). In the period 1883(89)–1936(38), the vast majority of recorded transformations (more than 60%) were changes from areas originally used for agriculture to areas of mature forest. The largest area of change concerned transformations of young forest to mature forest. However, new areas of young forests were not detected in 1936(38). About half of the transformations in 1936(38)–1977 were connected with the abandonment of agricultural land process as well as forest succession. Temporary and permanent deforestation was detected in 22% of changes. The average area of permanent deforestation was two times lower than that of temporary deforestation. The third of the analyzed time periods, 1977–2013, was characterized by the largest total number of changes, but the area of change decreased in comparison with previous analyzed periods. There were many temporary deforestations within the areas of mature forest and permanent deforestation in order to acquire new areas for agricultural activity. A greater area of deforestation was also recorded in order to acquire new areas for development and mining. The full list of landscape transformations types is presented in Table 6 and Figure 5.

Table 6. Types and subtypes of forest landscape changes between 1883(89) and 2013; NP: number of polygons; AS: average size of polygon; TA: total area of change

Type of Change [1]	Subtype of Change [2]	1883(89)–1936(38)			1936(38)–1977			1977–2013		
		NP (pcs)	AS (ha)	TA (ha)	NP (pcs)	AS (ha)	TA (ha)	NP (pcs)	AS (ha)	TA (ha)
A	A1	0	0.0	0.0	112	2.1	235.2	129	0.6	82.1
	A2	9	57.5	517.7	0	0	0	116	2.1	248.8
B	B1	10	3.5	34.9	106	1.0	108.4	93	1.2	112.0
	B2	3	0.6	1.8	0	0	0	24	0.4	9.2
	B3	0	0.	0.	0	0	0	5	0.6	3.0
	B4	7	0.3	1.8	2	3.4	6.7	7	1.8	12.3
C	C1	0	0	0	15	0.9	14.1	26	0.8	19.6
	C2	44	2.6	116.0	243	1.6	396.2	328	0.9	310.1
	C3	0	0	0	14	0.2	2.5	9	0.4	3.7
Total number/area of changes		73	-	672.2	492	-	763.1	737	-	800.8

[1] A: temporary deforestation, maturation of forest; B: permanent deforestation; C: afforestation; [2] A1: change of mature forest area into young forest area; A2: change of young forest area into mature forest; B1: change of mature forest area into agricultural land; B2: change of mature forest area into rural development area; B3: change of mature forest area into urban development area; B4: change of mature forest area into mining area; C1: change of agricultural land into young forest area; C2: change of agricultural land into mature forest area; C3: change of mining area to mature forest area.

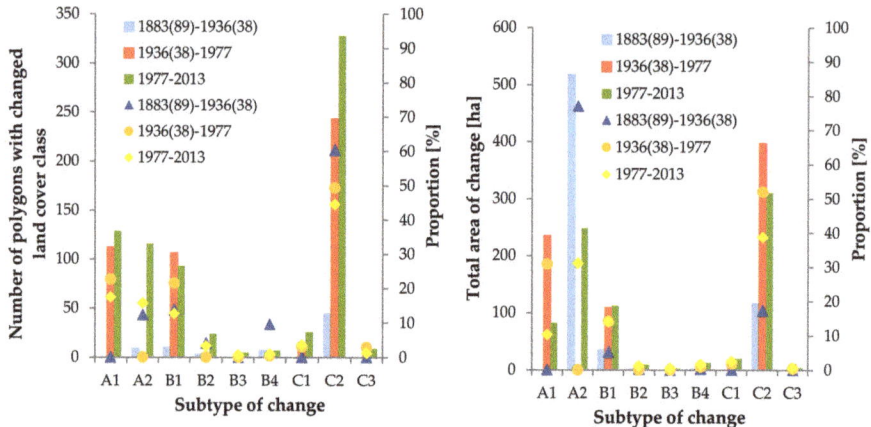

Figure 5. Area and number of different landscape changes in analyzed time intervals.

3.2. Driving Forces of Changes in Forest Landscapes between 1883(89) and 2013

3.2.1. Natural Driving Forces of Landscape Change

In the first analyzed period, 1883(89)–1936(38), most of the changes concerned areas located at an average elevation of 250 to 300 m a.s.l. There were almost no changes at elevations higher than 400 m a.s.l. Transformations within forest areas (type A) were mainly observed in the 200–250 m a.s.l range, changes of forest landscapes into non-forest landscapes (type B) at elevations lower than 200 m a.s.l and changes of non-forest landscapes into forest landscapes (type C) in the ranges from 200 to 300 m a.s.l. In the next period, 1936(38)–1977, a reduction in the percentage of changes which concerned transformations at elevations less than 200 m a.s.l was noticeable. The percentage of transformations at elevations above 350 m increased as a result of changes within forest landscapes (type A). The largest number of such changes occurred mainly at elevations from 250 to 350 m a.s.l. Transformations from forest landscapes into non-forest landscapes (type B) and from non-forest landscapes into forest landscapes (type C) dominated at elevations from 200 to 300 m a.s.l. Most changes in the third analyzed period, 1977–2013, were in the ranges from 200 to 250 m a.s.l. and from 250 to 300 m a.s.l., while in areas below 200 m a.s.l., the changes occurred slightly more frequently than in the previous period. On the other hand, the percentage of changes observed at elevations higher than 350 m a.s.l. increased. When we look at the types of changes, it can be observed that transformations within forest areas (type A) dominated at 300–350 m a.s.l., while a larger percentage of forest landscape transformations into non-forest ones (type B) was recorded at lower elevations, from 200 to 250 m a.s.l.

The second analyzed factor was the average slope of the changed area. In each of the analyzed periods, about 65% of observed changes concerned areas with an average slope in the range of 1–10%, while 31% of landscape transformations took place on slopes with an average grade of 20–30%. In the context of dominant hillside exposure the percentage share of changes was at a similar level in all analyzed periods of time. Among the areas where flat surface is not dominant, most changes were observed on the Northern and Northeastern exposures of the hillside, and the least number of changes were observed on the western, northwestern, and southwestern exposures of the hillside (see Figure 6).

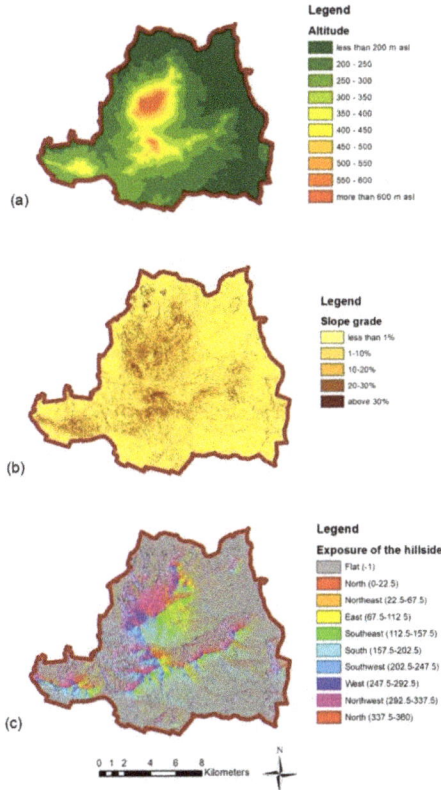

Figure 6. Maps of the analyzed natural driving forces of landscape change: (**a**) map of elevation; (**b**) map of slope grade; (**c**) map of hillside exposure.

3.2.2. Socioeconomic Driving Forces of Landscape Change

The group of socioeconomic driving forces of landscape changes includes, among others, population changes [57] and distances to roads, well-known facilities, places, or local service centers [58]. In the context of human population changes, we can observe that total number of people in the study area decrease slightly from 17,727 in 1885 to 16,339 in 2013. The number of people living in urban and rural areas has changed significantly. Only 2344 people lived in Sobótka in 1885, while 15,383 people lived in rural areas. The number of residents grew slightly to 17,819 in 1941. Human population of the city of Sobótka increased to 3412 inhabitants. The number of inhabitants in the rural part of the study area significantly decreased to 1978 as a result of population loss and mass displacement of native inhabitants in the post-war period. A total of 6043 people lived in Sobótka city, and 9729 lived in the rural areas. The population of the city of Sobótka increased to 7030 people, and the number of inhabitants of rural areas decreased to 9369 people in 2013. The changes in the human population in the study area are shown in Figure 7.

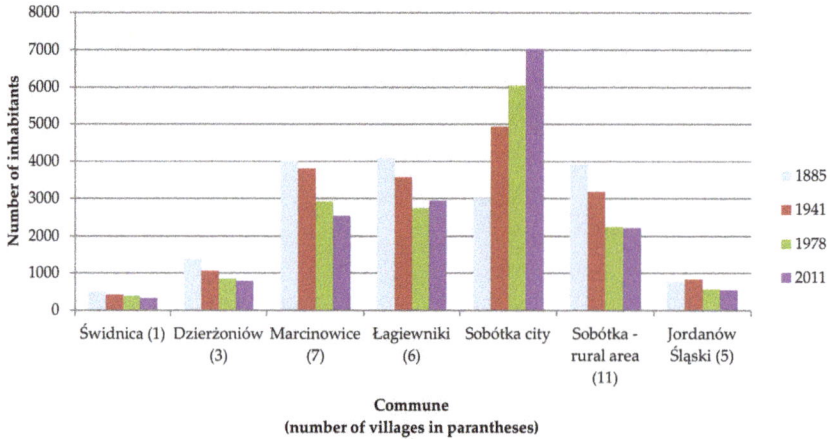

Figure 7. Population changes in communes of Ślęża Landscape Park (source: population censuses from 1885, 1941, 1978, 2011).

Most of the transformations occurred within the Sobótka precinct where Ślęża mountain is located, and within Jaźwina, Tąpadła, and Słupice precinct where Radunia mountain is located. A comparison of this indicator with population changes showed that less changes were observed in areas where the number of inhabitants in particular periods did not change significantly. The largest numbers of changes were identified in precincts where population changes were more noticeable.

The next analyzed factor from socioeconomic group of driving forces was the distance of areas that have changed from the centers of the surrounding municipalities (Figure 8). In all analyzed periods, the highest percentage of change concerned areas located at a distance of 6–8 km from the center of a municipality. The fewest changes were observed at distances of less than 2 km from the center of a municipality because only one municipality is located within the research area. The largest differences between the analyzed periods were noticed in the case of changes in locations at a distance of more than 8 km from the center of a municipality. Changes in forest areas (type A) were most frequently found at distances of 6–8 km and more than 8 km from the center of a municipality. The percentage of changes at a distance of more than 8 km increased in each subsequent period. Changes of forest landscapes into non-forest landscapes (type B) were observed to have occurred at a similar level at distances of 4–6 and 6–8 km from the center of a municipality. In the case of transformations of non-forest landscapes into forest landscapes (type C), the vast majority concerned locations at distances of 6–8 km from a municipality.

The second factor was the distance from a main road. In this context, changes were most frequently observed in all analyzed periods at distances of up to 500 m and from 500 to 1000 m from main roads. However, when we compared the periods, a systematic increase in the percentage of change at greater distances from the main roads was noticeable. For example, the percentage of change at a distance of 1500–2000 m increased from 4.11% in 1883(89) to 10.85% in 2013. The percentage of changes observed at a distance of 500–1000 m from the main road decreased from 1883(89) to 2013 more than 10%. The dominance of changes within forest-covered areas (type A) characteristically occurred at distances of 500 to 1500 m from the main road, although at larger distances, an increase in the change effect was noticeable. Transformations from forest landscapes into non-forest ones (type B) were recorded mainly at a distance of less than 500 m and from non-forest to forest landscapes (type C) also at a distance of 500 to 1000 m.

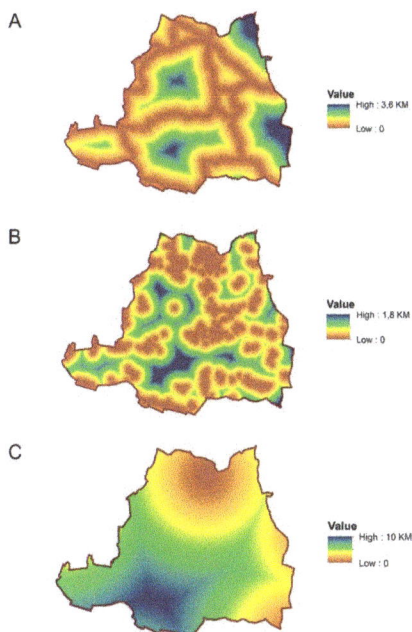

Figure 8. Maps of the analyzed socioeconomic driving forces of landscape change: (**A**) map of the distance to a main road; (**B**) map of the distance to a built-up area; (**C**) map of the distance to the center of a municipality.

The last element of the analysis was the distance of observed changes from built-up areas. In this aspect, the ratio of locations with changes located less than 250 m from built-up areas (about 30%) was maintained at a similar level in all periods. On the other hand, the percentage of changes at locations at a distance of 250–500 m decreased systematically from 1883(89) to 2013. At the same time, the number of transformations located at distances greater than 750 m from built-up areas increased. Most changes in forest areas (type A) were observed at a distance of 750–1000 m from built-up areas. The vast majority of forest landscape transformations into non-forest landscapes (type B) covered areas less than 250 m from built-up areas. Similarly, in the case of changes from non-forest into forest landscapes (type C), the highest percentage of change was observed at distances of less than 250 m, as well as at distances of 250 to 500 m from built-up areas.

4. Discussion

Land cover analyses showed that the area covered by forests within the present limits of Ślęża Landscape Park increased from 29.32% at the end of the 19th century to 33.80% in 2013. Szymura et al. indicate that forest area increased of the entire Sudeten area from 30.4% in the 18th century to 36.4% in the 20th century [56]. The same situation has also been observed in other mountainous areas in Poland. An example of such process is the Beskid Mountains [59]. However, despite the increase in the forest area, the percentage of forest-covered area still remains smaller than in other mountain and foothill areas [60]. At the same time, it is higher than that in lowland areas [45]. This is the effect of the rather low elevation of the area, which favors agricultural land use; in particular, it includes areas below 200 m a.s.l.

The level of changes in forest and non-forest landscapes in particular periods, as indexed by the landscape change index, is difficult to compare with other research results, because studies using this index have not been used to analyze historical data—only contemporary data [41]. Considering the nature of the indicator used, based on the relative data obtained from cartographic analyses, a comparison with other areas would be possible only if the same source materials were used. It should be assumed that the accuracy of the data increases when more advanced methods can be used for the interpretation and processing of field or aerial data. Therefore, there was a difference in the number of changes identified from the prepared land cover maps. It grew in successive periods. In the first analyzed time interval, only 73 polygons representing changes in land cover were identified; in the second period, 492; and in the third, 737. Because of differences in the ability to obtain data of the same scale and degree of data generalization, the results of research should be interpreted with the use of relative rather than absolute data [61]. The approach of mapping the interactions between factors influencing urban development and environmental resources, which was used in the presented paper, corresponds with the latest studies in the field of socio-environmental vulnerability [62].

The basic factor which determines land cover within research area are topographical conditions. The slopes of Ślęża and surrounding hills have been covered with forests for years, and settlement areas are developing in the foothills, while arable land dominate in flat areas. Weak soil is mentioned in literature as the basic driving force behind changes in forest landscapes [63]; however, as shown by Szymura et al. [56], in the Sudety area, the soil type has not had a significant impact on landscape changes. Analyses of the driving forces of changes in forest landscapes have mainly concerned the Carpathian area [19,60,64], with its Polish part investigated less frequently [60,65]. There has been little research on the area of the Sudetes and its foothills. Generally, there has been a lack of research in Poland to allow comparisons of obtained results with other research. This was confirmed by Plieninger et al. [49] who showed that among 144 analyzed articles which identified driving forces of landscape change across Europe, only four referred to case studies located in Poland. In neighboring Germany, Czech Republic, and Slovakia, the number of such studies is much higher.

The analysis of the selected natural driving forces of landscape changes revealed that the topographical conditions of identified landscape transformations have changed over time. Changes related to forest area (type A) in the first analyzed period concerned elevations of 200–250 m a.s.l.; in the next, they occurred mainly at elevations of 250–350 m a.s.l.; and in the third period, they began to dominate at elevations of 300–350 m a.s.l. Landscape changes were identified at higher elevations in successive periods. Transformations of forest landscapes into non-forest landscapes (type B) and from non-forest to forest landscapes (type C) dominated at elevations of 200–300 m a.s.l. in all periods. In the lower-lying areas, these changes have been less frequent due to the increased area of arable land. The landscape remains basically unchanged at elevations greater than 350 m a.s.l. Areas with a slope from 1 to 10% and those in the north and Northeastern exposures of hillsides which are characterized by less sunlight are more susceptible to change. A smaller terrain slope makes landscape transformation easier, especially in terms of newly built-up areas as well as in areas of deforestation. On the northern slopes, these changes remain less visible due to the shorter lighting time.

The analysis of selected socioeconomic driving forces of landscape change showed that the stabilization of the population in the analyzed period helped to reduce the number of landscape changes. In areas where changes in the population were greater, more landscape changes were also observed. Transformations of forest-covered areas were the most common in the areas farthest away from the center of a municipality—at distances of 6–8 km and over 8 km. The other types of changes most commonly occurred at distances of 4–6 km and 6–8 km from the center of a municipality. The distances to the main road and built-up area also played important roles in the distribution of change types. Transformations within forest landscapes usually occurred at distances of 500–1500 m from main roads and 500–1250 m from built-up areas. Relatively close locations were preferred for the transportation of timber. Changes from forest landscapes into non-forest landscapes dominated at distances of less than 500 m from main roads and less than 250 m from built-up areas which is connected mainly with the locations of new buildings being close to existing ones with good access to roads. Changes of non-forest landscapes into forests were commonly identified at distances of

0–1000 m from major roads and at distance of 0–500 m from building areas. This was connected with the progressive process of succession close to the forest border.

In order to fully understand the changes that took place in forest landscapes during the discussed period of time, it was important to consider changes in the broader context of the political environment, cultural, and technological events. That is why we analyzed the chronicles of events and literature of those that could have influenced landscape changes. The political and technological driving forces of landscape changes from 1883(89)–1936(38) included changes in forest management policy. After a period of very strong forest exploitation in the 18th century and the beginning of the 19th century, the demand for wood in Prussia decreased significantly. Additionally, as a result of the inflow of capital from war contributions, the sale of state forests stopped. Gradually, due to the development of modern forestry sciences, which occurred at the end of the 19th century, the level of forest cover increased. The reason for this was a favorable tax policy, as a result of which large areas of land of poor quality, unsuitable for agriculture, were afforested. The use of appropriate tax rates indirectly protected forest landscapes, especially in mountainous areas [66]. An important technological event was a new railway line from Wrocław to Sobótka that was built in 1885. It influenced the development of the city in the Northern part, but also enabled the intensification of the extraction of minerals [67]. In the second analyzed period, after the end of the Second World War, the geopolitical situation changed completely. Lower Silesia became a part of Poland. The population was displaced—people from the Eastern part of pre-war Poland came to the research area. The top of Ślęża mountain has been protected since 1954 as a nature reserve. In 1973, two neighboring rural areas were included in the city limits of Sobótka. The political system in Poland changed in 1989—the period of communism ended after 50 years. An important event was the establishment of a landscape park of the entire area of the Ślęża Massif in 1988. In 2004, Poland joined the European Union, which was another indirect cause of change [30]. Due to the subsidies for afforested, poor quality, arable land within the study area, new forest areas have begun to appear quickly.

5. Conclusions

This study of the area covering the present borders of Ślęża Landscape Park related to the landscape changes in the period 1883(89)–2013 successfully measured the levels of landscape change in the three analyzed periods of time covering approximately 40–50 years. The analyses showed that the level of change in forest-covered areas was the highest in 1883(89)–1936(38) and in 1977–2013 for non-forest areas. Moreover, results of the studies on the landscape change index (LCI) show that it can be successfully used to determine the level of historical landscape change. Our results revealed that the percentage of area covered by forest has systematically grown since the end of the 19th century, mainly at the expense of agricultural land. The most frequently identified changes occurred in forest-covered areas and were connected with the temporary deforestation, maturation of forests and afforestation, especially transformations from agricultural land to mature forest area.

The study revealed that landscape transformations inside forest-covered areas have been located at higher elevations in successive periods while other drivers, like the slope grade and exposure of the hillside, have been constant. Analyses of socioeconomic driving forces of changes showed that they may be considered as one of the causes of landscape changes, but more studies are required. It should be noted that research may contain some errors, which can be the result of inaccuracies in the mapping of changes, and limitations related to the lack of analysis of cause and effect relationships between driving forces and landscape changes. However, this research provides valuable insight into the landscape changes of the last 140 years, introduces a new tool for assessing the level of historical landscape transformation (LCI), and indicates the direction of further study on driving forces behind landscape change of Ślęża Landscape Park.

Author Contributions: Methodology, P.K.; Data curation, all authors; Formal analysis, P.K. and I.S.; Investigation: P.K.; Writing—Original Draft Preparation, P.K., Writing—Review & Editing, P.K. and K.M.

Funding: This research was co-funded by the Polish National Science Centre, grant number 2013/09/D/HS4/01858. We are also grateful to the Ministry of Science and Higher Education (MNiSW, Poland) for supporting open access publishing in the framework of the Statutory funds of Department of Land Management, Wroclaw University of Environmental and Life Sciences.

Acknowledgments: The authors would like to give special thanks to the administration of the Lower Silesian Association of Landscape Parks for providing data for the analysis. Pre-print version of the manuscript titled "Driving Forces behind Forest Landscape Change in Ślęża Landscape Park (Southwester Poland) in 1883–2013" was initially submitted to the MDPI Platform www.preprints.org.

Conflicts of Interest: The authors declare no conflict of interest.

References

1. Vitousek, P.M.; Mooney, H.A.; Lubchenco, J.; Melillo, J.M. Human Domination of Earth's Ecosystems. *Science* **1997**, *277*, 494–499. [CrossRef]

2. Antrop, M. Landscape change and the urbanization process in Europe. *Landsc. Urban Plan.* **2004**, *67*, 9–26. [CrossRef]

3. Bičík, I.; Kupková, L.; Jeleček, L.; Kabrda, J.; Štych, P.; Janoušek, Z.; Winklerová, J. Land Use Changes in the Czech Republic 1845–2010: Socio-Economic Driving Forces. Springer: Basel, Switzerland, 2015; ISBN 978-3-319-17671-0.

4. Łowicki, D. Land use changes in Poland during transformation: Case study of Wielkopolska region. *Landsc. Urban Plan.* **2008**, *87*, 279–288. [CrossRef]

5. Skokanová, H.; Falťan, V.; Havlíček, M. Driving forces of main landscape change processes from past 200 years in Central Europe—differences between old democratic and post-socialist countries. *Ekológia (Bratislava)* **2016**, *35*, 50–65. [CrossRef]

6. Heffner, K. Zmiany przestrzenne na obszarach wiejskich w Polsce w okresie transformacji i po wejściu do Unii Europejskiej. In *Obszary wiejskie—wiejska przestrzeń i ludność, aktywność społeczna i przedsiębiorczość*; Heffner, K., Klemens, B., Eds.; Studia KPZK PAN: Warsaw, Poland, 2016; pp. 12–27.

7. Bičik, I.; Jeleček, L.; Štěpánek, V. Land-use changes and their social driving forces in Czechia in the 19th and 20th centuries. *Land Use Policy* **2001**, *18*, 65–73. [CrossRef]

8. Peña, J.; Bonet, A.; Bellot, J.; Sánchez, J.R.; Eisenhuth, D.; Hallett, S.; Aledo, A. Driving forces of land-use change in a cultural landscape of Spain. In *Modelling Land-use Change*; Springer: Dordrecht, The Netherlands, 2007; pp. 97–116. ISBN 978-1-4020-6484-5.

9. Liesovsky, J.; Bezak, P.; Spulerova, J.; Liesovsky, T.; Koleda, P.; Dobrovodska, M.; Bürgi, M.; Gimmi, U. The abandonment of traditional agricultural landscape in Slovakia—Analysis of extent and driving forces. *J. Rural Stud.* **2015**, *37*, 75–84. [CrossRef]

10. Hersperger, A.M.; Bürgi, M. Driving forces of landscape change in the urbanizing Limmat Valley, Switzerland. In *Modelling Land-Use Change*; Springer: Dordrecht, The Netherlands, 2007; pp. 45–60, ISBN 978-1-4020-6484-5.

11. Tokarczyk-Dorociak, K.; Kazak, J.; Szewrański, S. The Impact of a Large City on Land Use in Suburban Area—The Case of Wrocław (Poland). *J. Ecol. Eng.* **2018**, *19*, 89–98. [CrossRef]

12. Zewdie, M.; Worku, H.; Bantider, A. Temporal Dynamics of the Driving Factors of Urban Landscape Change of Addis Ababa During the Past Three Decades. *Environ. Manag.* **2018**, *61*, 132–146. [CrossRef]

13. Idczak, P.; Mrozik, K. Periurbanisation—Evidence from Polish metropolitan areas. *Econ. Environ. Stud.* **2018**, *18*, 183–202. [CrossRef]

14. Przybyła, K.; Kulczyk-Dynowska, A. Transformations of Tourist Functions in Urban Areas of the Karkonosze Mountains. *IOP Conf. Ser. Mater. Sci. Eng.* **2017**, *245*, 072001. [CrossRef]

15. Krajewski, P. Assessing changes in high-value landscape—Case study of the municipality of Sobotka in Poland. *Polish J. Environ. Stud.* **2017**, *26*, 2603–2610. [CrossRef]

16. Saura, S.; Martín-Queller, E.; Hunter, M.L. Forest landscape change and biodiversity conservation. In *Forest Landscapes and Global Change*; Azevedo, J., Perera, A., Pinto, M., Eds.; Springer: New York, NY, USA, 2014; pp. 167–198, ISBN 978-1-4939-0953-7.

17. Choi, J.; Lee, S.; Ji, S.Y.; Jeong, J.-C.; Lee, P.S.-H. Landscape Analysis to Assess the Impact of Development Projects on Forests. *Sustainability* **2016**, *8*, 1012. [CrossRef]

18. Wu, Z.; Ge, Q.; Dai, E. Modeling the Relative Contributions of Land Use Change and Harvest to Forest Landscape Change in the Taihe County, China. *Sustainability* **2017**, *9*, 708. [CrossRef]

19. Kruhlov, I.; Thom, D.; Chaskovskyy, O.; Keeton, W.S.; Scheller, R.M. Future forest landscapes of the Carpathians: Vegetation and carbon dynamics under climate change. *Reg. Environ. Chang.* **2018**, *18*, 1555–1567. [CrossRef]

20. Mercuri, A.M.; Marignani, M.; Sadori, L. Palaeoecology and long-term human impact in plant biology, Plant Biosystems. *Plant Biosyst.* **2015**, *149*, 136–143. [CrossRef]

21. Zanon, M.; Davis, B.A.S.; Marquer, L.; Brewer, S.; Kaplan, J.O. European Forest Cover During the Past 12,000 Years: A Palynological Reconstruction Based on Modern Analogs and Remote Sensing. *Front. Plant Sci.* **2018**, *9*, 253. [CrossRef] [PubMed]

22. Antrop, M. Sustainable landscapes: contradiction, fiction or utopia? *Landsc. Urban Plan.* **2006**, *75*, 187–197. [CrossRef]

23. Bürgi, M.; Hersperger, A.M.; Scheenberger, N. Driving forces of landscape change—Current and new directions. *Landsc. Ecol.* **2004**, *19*, 857–868. [CrossRef]

24. Hersperger, A.M.; Bürgi, M. Going beyond landscape change description: Quantifying the importance of driving forces of landscape change in a Central Europe case study. *Land Use Policy* **2009**, *26*, 640–648. [CrossRef]

25. Lin, X.; Wang, Y.; Wang, S.; Wang, D. Spatial differences and driving forces of land urbanization in China. *J. Geogr. Sci.* **2015**, *25*, 545–558. [CrossRef]

26. Plieninger, T.; Bieling, C. (Eds.) Connecting cultural landscapes to resilience. In *Resilience and the Cultural Landscape: Understanding and Managing Change in Human-shaped Environments*; Cambridge University Press: New York, NY, USA, 2012; pp. 3–26, ISBN 978-1-107-02078-8.

27. Mrozik, K.; Bossy, M.; Zaręba, K. Polityka przestrzenna gmin wiejskich na tle zmian zagospodarowania przestrzennego wynikających z suburbanizacji. *Rocznik Ochrona Środowiska* **2012**, *14*, 761–771.

28. Solecka, I.; Raszka, B.; Krajewski, P. Landscape analysis for sustainable land use policy: A case study in the municipality of Popielów, Poland. *Land Use Policy* **2018**, *75*, 116–126. [CrossRef]

29. Council of Europe. European Landscape Convention, Florence, 2000. ETS No. 176. Available online: http://conventions.coe.int/Treaty/en/Treaties/Html/176.htm. (accessed on 20 June 2018).

30. Krajewski, P. Landscape change index as a tool for spatial analysis. *J. IOP Mater. Sci. Engin.* **2017**, *245*, 072014. [CrossRef]

31. Kistowski, M. *Wybrane Aspekty Zarządzania Ochroną Przyrody w Parkach Krajobrazowych*; Uniwersytet Gdański: Gdańsk, Poland, 2004; ISBN 83-7326-188-5.

32. Krajewski, P. Problemy planistyczne na terenach parków krajobrazowych w sąsiedztwie Wrocławia na przykładzie Ślężańskiego Parku Krajobrazowego. *Res. Papers Wroclaw Univ. Econ.* **2014**, *367*, 147–154. [CrossRef]

33. Brandt, J.; Primdahl, J.; Reenberg, A. Rural land-use and dynamic forces—Analysis of 'driving forces' in space and time. In *Land-use Changes and Their Environmental Impact in Rural Areas in Europe*; Krönert, R., Baudry, J., Bowler, I.R., Reenberg, A., Eds.; UNESCO: Paris, France, 1999; pp. 81–102, ISBN 1-85070-047-8.

34. Seabrook, L.; McAlpine, C.; Fensham, R. Cattle, crops and clearing: Regional drivers of landscape change in the Bridglow Belt, Queensland, Australia, 1840–2004. *Landsc. Urban Plan.* **2006**, *78*, 373–385. [CrossRef]

35. Marucci, D. Landscape history as a planning tool. *Landsc. Urban Plan.* **2000**, *49*, 67–81. [CrossRef]

36. Scheenberger, N.; Bürgi, M.; Hersperger, A.M.; Ewald, K.C. Driving forces and rates of landscape change as a promising combination for landscape change research—An application on the northern fringe of the Swiss Alps. *Land Use Policy* **2007**, *24*, 349–361. [CrossRef]

37. Hersperger, A.M.; Gennaio, M.; Verburg, P.H.; Bürgi, M. Linking land change with driving forces and actors: Four conceptual models. *Ecol. Soc.* **2010**, *15*, 1–17. [CrossRef]

38. Bürgi, M.; Straub, A.; Gimmi, U.; Salzmann, D. The recent landscape history of Limpach Valley, Switzerland: Considering three empirical hypotheses on driving forces of landscape change. *Landsc. Ecol.* **2010**, *25*, 287–297. [CrossRef]

39. Bieling, C.; Plieninger, T.; Schaich, H. Patterns and causes of land change: Empirical results and conceptual considerations derived from a case study in the Swabian Alb, Germany. *Land Use Policy* **2013**, *35*, 192–203. [CrossRef]

40. Lieskovský, J.; Kanka, R.; Bezák, P.; Štefunková, D.; Petrovič, F.; Dobrovodská, M. Driving forces behind vineyard abandonment in Slovakia following the move to a market-oriented economy. *Land Use Policy* **2013**, *32*, 356–365. [CrossRef]

41. Krajewski, P. Landscape changes in selected suburban area of Bratislava (Slovakia). In *Landscape and Landscape Ecology: Proceedings of the 17th International Symposium on Landscape Ecology*; Halada, L., Baca, A., Boltizar, M., Eds.; Institute of Landscape Ecology, Slovak Academy of Sciences: Bratislava, Slovakia, 2016; pp. 110–117, ISBN 978-80-89325-28-3.

42. Serra, P.; Pons, X.; Sauri, D. Land-cover and land-use change in a Mediterranean landscape: A spatial analysis of driving forces integrating biophysical and human factors. *Appl. Geogr.* **2008**, *28*, 189–209. [CrossRef]

43. Geri, F.; Amici, V.; Rocchini, D. Human activity impact on the heterogeneity of Mediterranean landscape. *Appl. Geogr.* **2010**, *30*, 370–379. [CrossRef]

44. Bürgi, M.; Bieling, C.; Von Hackwitz, K.; Kizos, T.; Liesovsky, J.; Martin, M.G.; McCarthy, S.; Müller, M.; Plieninger, T.; Printsmann, A. Processes and driving forces in changing cultural landscapes across Europe. *Landsc. Ecol.* **2017**, *32*, 2097–2112. [CrossRef]

45. Szabo, P. Driving forces of stability and change in woodland structure: A case-study from the Czech lowlands. *For. Ecol. Manag.* **2010**, *259*, 650–656. [CrossRef]

46. Loran, C.; Munteanu, C.; Verburg, P.H.; Schmatz, D.R.; Bürgi, M.; Zimmermann, N.E. Long-term change in drivers of forest cover expansion: An analysis for Switzerland (1850–2000). *Reg. Environ. Chang.* **2017**, *17*, 2223–2235. [CrossRef]

47. Mwangi, H.M.; Lariu, P.; Julich, S.; Patil, S.D.; McDonald, M.A.; Feger, K.-H. Characterizing the Intensity and Dynamics of Land-Use Change in the Mara River Basin, East Africa. *Forests* **2018**, *9*, 8. [CrossRef]

48. Frayer, J.; Müller, D.; Sun, Z.; Munroe, D.K.; Xu, J. Processes Underlying 50 Years of Local Forest-Cover Change in Yunnan, China. *Forests* **2014**, *5*, 3257–3273. [CrossRef]

49. Plieninger, T.; Draux, H.; Fagerholm, N.; Bieling, C.; Bürgi, M.; Kizos, T.; Kuemmerle, T.; Primdahl, J.; Verburg, P.H. The driving forces of landscape change in Europe: A systematic review of the evidence. *Land Use Policy* **2016**, *57*, 204–214. [CrossRef]

50. Regulation of the Lower Silesian Voivode dated 4 April 2007 regarding Ślęża Landscape Park. Available online: http://oi.uwoj.wroc.pl/dzienniki/Dzienniki_2007/Dz_U_Nr_94.pdf (accessed on 20 June 2018).

51. Krajewski, P. *Ślęża Landscape Park*; Lower Silesian Association of Landscape Park: Wroclaw, Poland, 2012; ISBN 978-83-63166-03-8.

52. Matuszkiewicz, W.; Faliński, J.B.; Kostrowicki, A.S.; Matuszkiewicz, J.M.; Olaczek, R.; Wojterski, T. *Potential Natural Vegetation of Poland. General Map 1:300,000*; IGiPZ PAN: Warszawa, Poland, 1995.

53. Baude, M.; Meyer, B.C. Changes of landscape structure and soil production function since the 18th century in north-west Saxony. *J. Environ. Geogr.* **2012**, *3*, 11–23.

54. Affek, A. Georeferencing of historical maps using GIS, as exemplified by the Austrian Military Surveys of Galicia. *Geographia Pol.* **2013**, *86*, 375–390. [CrossRef]

55. Jaskulski, M.; Łukasiewicz, D.; Nalej, M. Comparison of methods for historical map transformation. *Ann. Geomatics* **2013**, *11*, 41–57.

56. Szymura, T.H.; Murak, S.; Szymura, M.; Raduła, M.W. Changes in forest cover in Sudety Mountains during the last 250 years: Patterns, drivers, and landscape-scale implications for nature conservation. *Acta Societatis Botanicorum Poloniae* **2018**, *87*, 1–14. [CrossRef]

57. Song, K.; Wang, Z.; Du, J.; Liu, L.; Zeng, L.; Ren, C. Wetland Degradation: Its Driving Forces and Environmental Impacts in the Sanjiang Plain, China. *Environ. Manag.* **2014**, *54*, 255–271. [CrossRef] [PubMed]

58. Corona, R.; Galicia, L.; Palacio-Prieto, J.L.; Bürgi, M.; Hersperger, A.M. Local deforestation patterns and driving forces in a tropical dry forest in two municipalities of southern Oaxaca, Mexico (1985–2006). *Investigaciones Geográficas* **2016**, *91*, 86–104. [CrossRef]

59. Sobala, M.; Rahmonov, O.; Myga-Piątek, U. Historical and contemporary forest ecosystem changes in the Beskid Mountains (southern Poland) between 1848 and 2014. *iForest Biogeosci. For.* **2017**, *10*, 939–947. [CrossRef]

60. Szabó, P. Changes in woodland cover in the Carpathian Basin. In *Human Nature: Studies in Historical Ecology and Environmental History*; Szabó, P., Hédl, R., Eds.; Institute of Botany of the ASCR: Brno, Czech Republic, 2008; pp. 106–115.

61. Plit, J. Analiza historyczna jako źródlo informacji o środowisku przyrodniczym. *Problemy Ekologii Krajobrazu* **2006**, *16*, 217–226.

62. Szewrański, S.; Świąder, M.; Kazak, J.K.; Tokarczyk-Dorociak, K.; Van HooF, J. Socio-environmental vulnerability mapping for environmental and flood resilience assessment: The case of ageing and poverty in the city of Wroclaw, Poland. *Integr. Environ. Assess. Manag.* **2018**, *14*, 592–597. [CrossRef] [PubMed]

63. Wulff, M.; Rujner, H. A GIS-based method for the reconstruction of the late eighteenth century forest vegetation in the Prignitz region (NE Germany). *Landsc. Ecol.* **2011**, *26*, 153–168. [CrossRef]

64. Kozak, J. Forest cover change in the Western Carpathians in the past 180 years: A case study in the Orawa region in Poland. *Mt. Res. Dev.* **2003**, *23*, 369–375. [CrossRef]

65. Munteanu, C.; Kuemmerle, T.; Boltiziar, M.; Butsic, V.; Gimmi, U.; Lúboš, H.; Kaim, D.; Király, G.; Konkoly-Gyuró, É.; Kozak, J.; et al. Forest and agricultural land change in the Carpathian region—A meta-analysis of long-term patterns and drivers of change. *Land Use Policy* **2014**, *38*, 685–697. [CrossRef]

66. Nyrek, A. *Kultura użytkowania gruntów uprawnych, lasów i wód na Śląsku od XV do XX wieku*; Wydawnictwo Uniwersytetu Wrocławskiego: Wrocław, Poland, 1992.

67. Staffa, M.; Mazurski, K.R.; Czerwiński, J.; Pisarski, G. *Słownik geografii turystycznej Sudetów. Tom 20. Masyw Ślęży, Równina Świdnicka, Kotlina Dzierżoniowska*; Wydawnictwo I-Bis: Wrocław, Poland, 2005.

sustainability

MDPI

Article

Responses of Vegetation Cover to Environmental Change in Large Cities of China

Kai Jin [1], Fei Wang [1,2,3,]* and Pengfei Li [2,4]

1 Institute of Soil and Water Conservation, Northwest A&F University, Yangling 712100, China; jinkai-2014@outlook.com
2 Institute of Soil and Water Conservation, Chinese Academy of Sciences and Ministry of Water Resources, Yangling 712100, China; forest@nwafu.edu.cn
3 University of Chinese Academy of Sciences, Beijing 100049, China
4 College of Geomatics, Xi'an University of Science and Technology, Xi'an 710054, China
* Correspondence: wafe@ms.iswc.ac.cn; Tel.: +86-29-8701-2411

Received: 3 December 2017; Accepted: 15 January 2018; Published: 20 January 2018

Abstract: Vegetation cover is crucial for the sustainability of urban ecosystems; however, this cover has been undergoing substantial changes in cities. Based on climate data, city statistical data, nighttime light data and the Normalized Difference Vegetation Index (NDVI) dataset, we investigate the spatiotemporal variations of climate factors, urban lands and vegetation cover in 71 large cities of China during 1998–2012, and explore their correlations. A regression model between growing-season NDVI (G-NDVI) and urban land proportion (PU) is built to quantify the impact of urbanization on vegetation cover change. The results indicate that the spatiotemporal variations of temperature, precipitation, PU and G-NDVI are greatly different among the 71 cities which experienced rapid urbanization. The spatial difference of G-NDVI is closely related to diverse climate conditions, while the inter-annual variations of G-NDVI are less sensitive to climate changes. In addition, there is a negative correlation between G-NDVI trend and PU change, indicating vegetation cover in cities have been negatively impacted by urbanization. For most of the inland cities, the urbanization impacts on vegetation cover in urban areas are more severe than in suburban areas. But the opposite occurs in 17 cities mainly located in the coastal areas which have been undergoing the most rapid urbanization. Overall, the impacts of urbanization on G-NDVI change are estimated to be −0.026 per decade in urban areas and −0.015 per decade in suburban areas during 1998–2012. The long-term developments of cities would persist and continue to impact on the environmental change and sustainability. We use a 15-year window here as a case study, which implies the millennia of human effects on the natural biotas and warns us to manage landscapes and preserve ecological environments properly.

Keywords: vegetation cover; urbanization; climate change; NDVI; cities; China

1. Introduction

Over a half of the world's population lives in urban areas which suffer from environmental problems (e.g., air pollution and ecosystem degradation) [1,2]. As one of the most important parts in the earth land ecosystem, vegetation provides a wide range of social and environmental services to urban life, and benefits the sustainability of urban ecosystems [2–4]. For example, urban vegetation can sequestrate carbon [5], regulate microclimates [6], improve air quality [7], preserve biodiversity [8], conserve soil and water and mitigate nature disasters [9]. However, palaeoecological and palaeoenvironmental records show that vegetation and ecosystem have been obviously influenced by human disturbances and climate change in long-term landscape evolution [10–12]. Remarkably, vegetation cover within and around urban areas has experienced obvious transformation during the past decades, which has significantly influenced the sustainability of urban ecosystems [1].

Generally, studies on vegetation variations at macro scales (e.g., city scale and regional scale) are conducted based on the remote sensing technique. Normalized difference vegetation index (NDVI) and vegetation net primary productivity (NPP) derived from remote sensing datasets are main indicators of vegetation activities [13]. NPP is the rate of atmospheric carbon uptake by vegetation, which is a useful proxy of vegetation ecosystem services ability [14]. Based on MODIS-1 km NPP products, Fu et al. [15] showed a total NPP loss of approximately 167 × 106 g·C from 2001 to 2006 in Guangzhou city, China, while the spatiotemporal pattern of NPP showed obvious variations in local areas. A different approach is to use NDVI to estimate fractional vegetation cover [16], or directly applied to investigate the changes of vegetation cover [17,18]. Based on GIMMS NDVI datasets, Eastman et al. [19] indicated that the changing trend of NDVI during 1982–2011 exhibited significant spatial differences over global land surfaces. Differences in spatiotemporal variations of vegetation could be largely attributed to the effects of climate factors [10], anthropogenic activities [20], and their interactions [15,21].

Precipitation and temperature has been demonstrated to be the key climate factors for plant growth and vegetation development in the literatures [22,23]. Vegetation cover in humid regions was generally higher than that in arid and semi-arid regions [10,24]. But the response of vegetation to precipitation and temperature change were varied in temporal patterns [23,25,26]. For example, Zhou et al. [27] demonstrated that temperature change was the leading cause of NDVI decreases, while precipitation played a minor role in high-latitude areas. Sun et al. [28] suggested that increasing precipitation led to an improvement in the vegetation cover, whereas temperature was not a limiting factor in arid and semi-arid regions of China. In cities, anthropogenic activities could impact vegetation cover in a positive or negative way [18,29]. On the one hand, large patches of fertile cropland and forest were transformed to urban constructed lands during the process of urbanization [30,31]; on the other hand, people have recognized the importance of vegetation in urban ecosystems, and have given more and more attention to urban greening plans [8,32]. However, it is difficult to balance between economic development and ecosystem protection, resulting in obvious constructed land growth and green area reduction [33,34]. In addition, the rate of local climate warming would increase if large scale vegetation was removed (e.g., urban heat island) during the process of urbanization [35], which contrarily altered vegetation growing period [36], as well as water and carbon cycles within the urban ecosystem [37]. Consequently, local climate change and urban land expansion are the two dominant factors for vegetation variations, even acting as obstructions to reasoned urban planning [8,36,38].

In ancient China, intensive human activities such as deforestation and cultivation have already evidently affected the vegetation and ecosystem [12]. In the context of global warming, China has been experiencing a particularly rapid climate change [39–41]. The impacts of climate change on vegetation cover variation have been frequently explored in previous studies [10,26,28,40], although they mainly focus on the regional scale rather than the city scale. China has experienced remarkably dramatic population growth and rapid urbanization in the past few decades, with the population size of near 1.4 billion in 2012 [29]. This implies that the environment (e.g., vegetation cover), particularly in large cities, has been largely impacted by urbanization [42]. For example, Li et al. [43] indicated that the rapid urbanization in 1988–2003 led to obvious vegetation cover degradation in the plain areas of Shenzhen city, China. Moreover, urban area in Beijing city nearly doubled during 1997–2002, whereas farmland decreased rapidly as a result of the rapid urbanization [44]. Although these studies indicated that urban land expansion significantly influenced the local vegetation variation at individual large cities, the quantitative impacts of urbanization were rarely estimated among different cities.

This study aims to explore the response of vegetation cover to environment change, and assess the impacts of urbanization on vegetation cover change in large cities of China. In this study, we first examine the spatiotemporal changes of climate factors, urban lands and vegetation cover in urban and suburban areas, respectively. Then the correlation analysis among urban land proportion (PU), growing-season NDVI (G-NDVI), temperature (G-T) and precipitation (G-P) is conducted to explain the diversities of G-NDVI change in cities. Finally, the impacts of urbanization on G-NDVI changes are quantified by a regression model. Generally, it would be better to analyse the long-term changes of vegetation in cities considering their histories, but we try to use recent observed data (1998–2012) to conduct study in an important "moment" or "window" of human history. It could also reflect long-term changes of the vegetation, landscape, and environment.

2. Materials and Methods

2.1. Data

City statistical data included urban built-up land area and population of municipal districts, which were often used to represent the size of cities [39]. The city statistical data for 1998–2012 were downloaded from China City Statistical Yearbook issued by the National Bureau of Statistics of China (http://www.stats.gov.cn).

NSL dataset from Defense Meteorological Satellite Program (DMSP) Operational Linescan System (OLS) was obtained from the NOAA's National Geophysical Data Center (NGDC) (https://ngdc.noaa.gov/eog/download.html). The spatial resolution and value range of NSL data were 1 km and 0 to 63, respectively. The background noise of the light images was represented by 0. The NSL dataset has been frequently used to extract urban lands with the empirical thresholding technique [30]. For example, Liu et al. [45] demonstrated that the threshold for the year 2000 was 49 in Northern Coastal China where many large cities showed prosperous economy (e.g., Beijing and Tianjin). In our study, The NSL data of 1998 and 2012 was adopted to extract urban lands.

SPOT-VGT NDVI datasets for 1998–2012 were obtained from the Image Processing Centre in Vlaamse Instelling voor Technologisch Onderzock (VITO) (Mol, Belgium) (http://www.vgt.vito.be). These data were 10-day composites of atmospherically corrected maximal values with the actual values ranging from −0.1 to 1. In order to eliminate the impact of large bare lands and water areas, grid cells with NDVI value greater than 0.15 were used. Spatial and temporal resolutions of the NDVI dataset are 1 km and 10 day respectively. Monthly NDVI value was synthesized by the maximum value composite (MVC) method based on the 10-day dataset. Given that the NDVI value is impacted by snow, we only focused on the mean NDVI in growing season (G-NDVI). Annual G-NDVI was generated based on the month mean NDVI for April to October in terms of Zhou et al. [27] and Piao et al. [26].

Surface climate data include monthly mean temperature and monthly total precipitation during 1998–2012, which were obtained from the China Meteorological Data Service Center (http://data.cma.cn/site/index.html). Annual growing-season temperature (G-T) and annual growing-season precipitation (G-P) were calculated based on the monthly climate data for April to October.

2.2. Methods

2.2.1. City Selection and Quantification of Urban Land Expansion

China has a vast territory with complicated geographical and climate conditions, as well as rich diversities in ecosystems. However, the rapid urbanization has profoundly impacted and changed the environmental, social and economic situations of China. In our study, the large cities of China and nearby meteorological stations were selected based on the following two criteria: The urban population in 1998 was larger than 0.5 million; the meteorological station close to city has a continuous climate data record covering the period of 1998–2012. Seventy one cities were eventually selected (Figure 1).

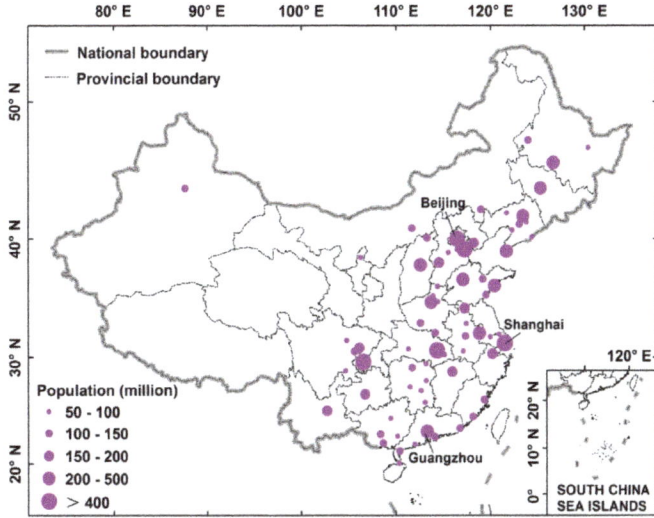

Figure 1. Locations of the selected 71 large cities in China and the corresponding urban populations in 1998.

In terms of Liu et al. [45], the different extraction thresholds were employed to extract urban lands of selected large cities in China based on the NSL data of 1998 and 2012 (the specific information of extraction thresholds was shown in Figure S1 of Supplementary Material). The geometric center of urban land in 2012 was determined for each city by the Feature to Point Tool in ArcGIS software, which was considered as the spatial city center. A radius was calculated based on the urban built-up land area of 2012 for each city. By means of the Edit Tool in ArcGIS software and based on the city center as well as the radius, a circle for each city was drawn centered on the city center and including the main city area—this was defined as the inner zone. A buffer area with a width of 10 km was generated outside this circle—this was defined as the outer zone. The inside and outside zones were named as urban area (Z1) and suburban area (Z2), respectively (Figure 2).

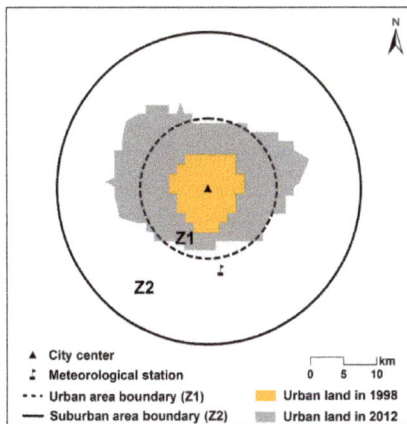

Figure 2. An example of the extracted urban lands of 1998 and 2012 in the urban area (Z1) and suburban area (Z2) of Nanchang city.

The areas of extracted urban lands in Z1 (AU_{Z1}) and Z2 (AU_{Z2}) for the 71 large cities were counted by ArcGIS software. The proportion of urban lands in Z1 (PU_{Z1}) and Z2 (PU_{Z2}) were calculated by the following formula:

$$PU_{Z1} = \frac{AU_{Z1}}{A_{Z1}} \times 100\% \tag{1}$$

where PU_{Z1} indicated the proportion of urban lands in Z1 (%); AU_{Z1} indicated the area of extracted urban lands in Z1 (km^2); A_{Z1} indicated the total area of Z1 (km^2). PU_{Z2} was calculated based on the area of extracted urban lands in Z2 (AU_{Z2}) and the total area of Z2 (A_{Z2}). The change of PU between 1998 and 2012 in Z1 (Z2) of each city, namely ΔPU_{Z1} (ΔPU_{Z2}), was calculated to represent the rate of urban land expansion during this period (Figure S1 of Supplementary Materials).

2.2.2. Analysis of Spatiotemporal Variations

The average G-NDVI values in Z1 (G-NDVI$_{Z1}$) and Z2 (G-NDVI$_{Z2}$) indicated the vegetation cover in Z1 and Z2 respectively. Because Z1 and Z2 are in the same city, their climate background was thought to be the same. The G-T and G-P were calculated using the meteorological data observed at a nearby meteorological station to represent the climate in growing-season for a given city (including Z1 and Z2).

This study firstly investigated the spatiotemporal variations of G-T, G-P, PU_{Z1} and PU_{Z2} among the selected 71 large cities between 1998 and 2012. Then, the spatiotemporal variations of G-NDVI$_{Z1}$ and G-NDVI$_{Z2}$ were also analysed with respect to G-T and G-P. The mean annual G-T, G-P and G-NDVI were calculated based on the datasets during 1998–2012. Linear trends in G-T, G-P and G-NDVI during 1998–2012 were examined using ordinary least-squares regression to analyze the inter-annual variations of G-T, G-P and G-NDVI. The trend rates of G-NDVI were used to reflect the directions of vegetation cover change [40].

2.2.3. Analyzing the Sensitivity of G-NDVI to Climate Change

Relationships between NDVI and other variables are generally examined via a correlation analysis [17,20,22,34,40]. In this study, the linear regression analysis was undertaken between mean annual G-NDVI and climate factors (G-P, G-T) to explain the spatial variabilities of G-NDVI. In order to assess the sensitivity of G-NDVI to climate change, the correlations between the inter-annual variations of G-NDVI and climate factors (1998–2012) and between the trends of G-NDVI and climate factors were examined by Pearson's correlation analysis.

2.2.4. Quantifying the Impact of Urbanization on G-NDVI Change

The differences of trend rates between G-NDVI$_{Z1}$ and G-NDVI$_{Z2}$ (trend rate of G-NDVI$_{Z1-Z2}$) were calculated. Because the climate changes in Z1 and Z2 are similar for a given city, the trend rate of G-NDVI$_{Z1-Z2}$ could be attributed to the different impacts of urban land expansion between Z1 and Z2. In addition, the differences between ΔPU_{Z1} and ΔPU_{Z2} (ΔPU_{Z1-Z2}) were calculated to show the different rates of urban land expansion between Z1 and Z2. The correlations between the trend rates of G-NDVI$_{Z1}$ (G-NDVI$_{Z2}$) and ΔPU_{Z1} (ΔPU_{Z2}) and between the trend rates of G-NDVI$_{Z1-Z2}$ and ΔPU_{Z1-Z2} were detected to analyze the impact of urbanization on vegetation cover change.

All the correlation analyses in this study were the single-variant analysis and performed using SPSS version 19.0. The significance of the correlation coefficients was estimated by t-tests at 0.1, 0.05 and 0.01 significance levels.

3. Results

3.1. Spatiotemporal Variation Analyses

3.1.1. Climate Factors

The climate condition of 71 cities varied from humid-hot climate to dry-cool climate with the latitude increasing (Figure 3a,b). Moreover, the inter-annual variations of G-T and G-P among the selected 71 cities were greatly different (Figure 3c,d). The trend rate of G-T and G-P ranged from −1.12 to 0.74 °C per decade and from −396 to 467 mm per decade respectively (Table 1). The G-T for 34 cities had ascending trends, which were mainly located in Middle Eastern and Middle Southern China (Figure 3c). More than a half of the cities (42 cities) experienced a descending G-P (Figure 3d). They are mainly located in Middle and Southern China.

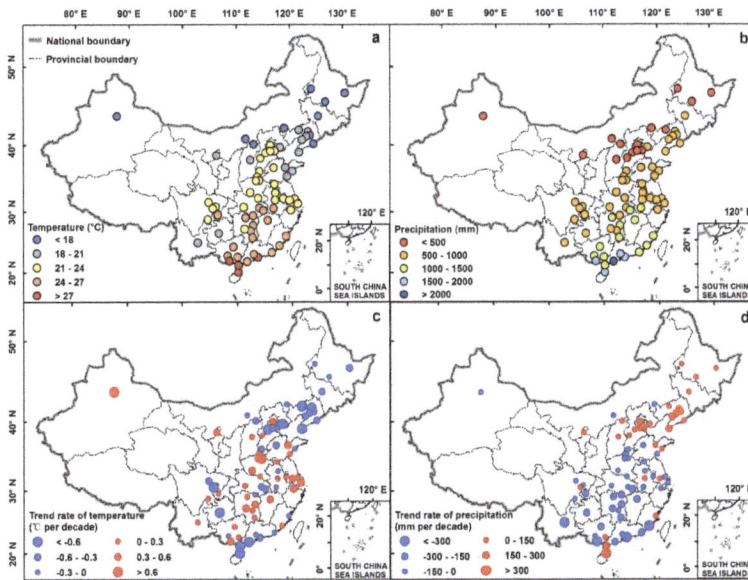

Figure 3. Spatial distribution of mean annual growing-season temperature (G-T) (**a**) and growing-season precipitation (G-P) (**b**) and inter-annual variations of G-T (**c**) and G-P (**d**) during 1998–2012 for the selected 71 large cities.

Table 1. Descriptive statistics of growing-season normalized difference vegetation index (G-NDVI), G-T, G-P and the change of urban land proportion (ΔPU) for the selected 71 large cities.

	Mean Annual Value				Change Trend				ΔPU_{Z1} [5]	ΔPU_{Z2} [6]
	$G\text{-}NDVI_{Z1}$ [1]	$G\text{-}NDVI_{Z2}$ [2]	G-T [3]	G-P [4]	$G\text{-}NDVI_{Z1}$	$G\text{-}NDVI_{Z2}$	G-T	G-P		
Mean	0.34	0.48	22.1	827	−0.007	0.016	−0.10	−37	37	22
Standard deviation	0.06	0.08	3.2	385	0.031	0.034	0.44	159	18	19
Range of value	0.29	0.35	12.3	1936	0.166	0.174	1.86	863	76	78
Minimum	0.23	0.29	15.3	178	−0.099	−0.082	−1.12	−396	1	0
Maximum	0.52	0.64	27.6	2114	0.067	0.092	0.74	467	77	78

[1] and [2] mean the growing-season NDVI of urban area and suburban area, respectively; [3] growing-season temperature, in °C; [4] growing-season precipitation, in mm; The unit of change trend is °C per decade for G-T and mm per decade for G-P; [5] and [6] the change of urban land proportion between 1998 and 2012 in urban and suburban area respectively, in %.

3.1.2. Urban Lands

PU_{Z1} and PU_{Z2} obviously increased from 1998 to 2012, indicating that the selected 71 cities have experienced rapid urbanization (Figure 4). But the ΔPU_{Z1} and ΔPU_{Z2} show great difference among the selected 71 cities. The ΔPU_{Z1} of nine cities is larger than 60%, while two cities have ΔPU_{Z2} of larger than 60%. The ΔPU_{Z2} of cities located in northern region and coast areas of China are larger than in southwestern China. Generally, the ΔPU_{Z1} of cities located in the inland of China were larger than ΔPU_{Z2} (i.e., $\Delta PU_{Z1-Z2} > 0$), while the ΔPU_{Z1} of some cities located in the coastal areas were smaller than ΔPU_{Z2} (i.e., $\Delta PU_{Z1-Z2} < 0$). Overall, the mean values of ΔPU_{Z1} and ΔPU_{Z2} are 37% and 22% respectively (Table 1), implying that urban areas of 71 cities have experienced faster urbanization than suburban areas.

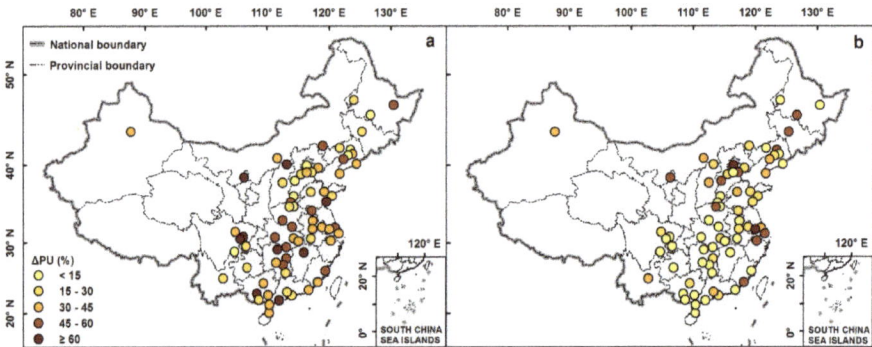

Figure 4. Spatial distribution of the ΔPU_{Z1} (**a**) and ΔPU_{Z2} (**b**) during 1998–2012 for the selected 71 large cities.

3.1.3. G-NDVI

Spatial distributions of mean annual growing-season NDVI (including $G\text{-}NDVI_{Z1}$ and $G\text{-}NDVI_{Z2}$) for the 71 cities were remarkably different (Figure 5a,b). $G\text{-}NDVI_{Z1}$ and $G\text{-}NDVI_{Z2}$ ranged from 0.23 to 0.52 and 0.29 to 0.64, with standard deviations of them being 0.06 and 0.08 respectively (Table 1). The $G\text{-}NDVI_{Z1}$ of 22 cities was lower than 0.3, while one cities had $G\text{-}NDVI_{Z1}$ of greater than 0.5. Two cities had $G\text{-}NDVI_{Z2}$ of lower than 0.3, while 34 cities had $G\text{-}NDVI_{Z2}$ of greater than 0.5. $G\text{-}NDVI_{Z2}$ was lower in northern region and coast areas of China than in southern China. For a given city, the mean annual $G\text{-}NDVI_{Z1}$ was generally lower than $G\text{-}NDVI_{Z2}$. The mean values of $G\text{-}NDVI_{Z1}$ and $G\text{-}NDVI_{Z2}$ were 0.34 and 0.48 respectively. Moreover, some adjacent cities with similar climate conditions had different G-NDVI. For example, $G\text{-}NDVI_{Z2}$ of Shanghai and Nantong were 0.36 and 0.51 respectively.

The inter-annual variations of G-NDVI during 1998–2012 also showed great difference among the 71 selected large cities (Figure 5c,d). The $G\text{-}NDVI_{Z1}$ for 40 cities experienced a decreasing trend (Figure 5c). The trend rates of $G\text{-}NDVI_{Z1}$ for 3 cities were greater than 0.05 per decade. Contrarily, the $G\text{-}NDVI_{Z2}$ for most of the cities (51 cities) experienced an increasing trend (Figure 5d). The trend rates of $G\text{-}NDVI_{Z2}$ for 13 cities were greater than 0.05 per decade. Cities with decreasing G-NDVI (particularly decreasing $G\text{-}NDVI_{Z2}$) were mainly located in the northern region and eastern coastal areas, where rapid urbanization and economic development occurred during 1998–2012. The trend rate of $G\text{-}NDVI_{Z1}$ and $G\text{-}NDVI_{Z2}$ ranged from −0.099 to 0.067 per decade and from −0.082 to 0.092 per decade, respectively (Table 1). Overall, the average trend rates of $G\text{-}NDVI_{Z1}$ and $G\text{-}NDVI_{Z2}$ were −0.007 and 0.016 per decade, respectively.

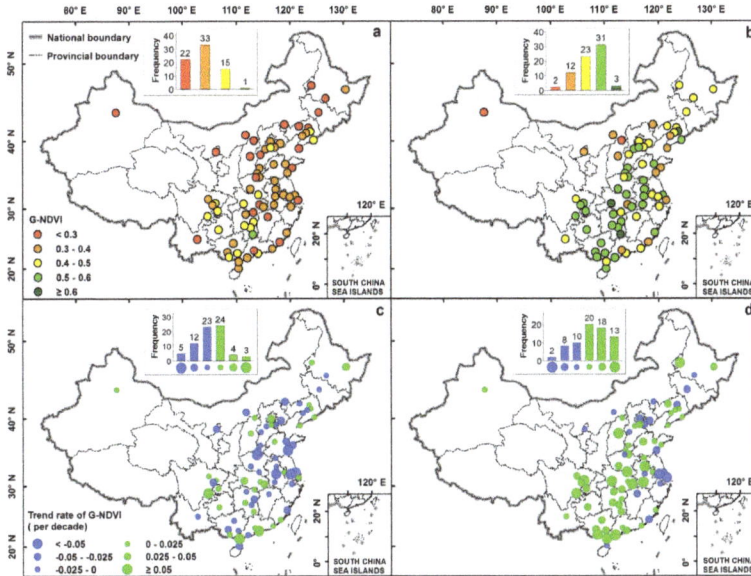

Figure 5. Spatial distribution of mean annual G-NDVI$_{Z1}$ (**a**) and G-NDVI$_{Z2}$ (**b**) and inter-annual variations of G-NDVI$_{Z1}$ (**c**) and G-NDVI$_{Z2}$ (**d**) during 1998–2012 in the selected 71 large cities.

A great difference in the changing trend of G-NDVI$_{Z1}$ and G-NDVI$_{Z2}$ was found for a given city (Figure 5c,d). For example, the trend rates of G-NDVI$_{Z1}$ and G-NDVI$_{Z2}$ for Beijing city were 0.026 and −0.027 per decade, respectively. Moreover, some adjacent cities with similar climate change conditions also showed great differences in the G-NDVI change. For instance, the trend rates of G-NDVI$_{Z1}$ for Shanghai and its adjacent city (Nantong) were 0.001 and −0.098 per decade respectively.

3.2. Sensitivity of G-NDVI to Climate Change

3.2.1. Relationship between Mean Annual G-NDVI and Climate

The G-NDVI was lower when the G-T was below 19 °C and the G-P was below 500 mm (Figure 6). G-NDVI generally increased with G-T and G-P. However, the G-NDVI value was relatively low when G-T exceeded 26 °C and G-P exceeded 1200 mm. High values of G-NDVI$_{Z1}$ occurred when G-T ranged between 23 and 24.5 °C and G-P ranged between 750 and 1000 mm (Figure 6a). High values of G-NDVI$_{Z2}$ occurred when G-T ranged between 23 and 25 °C and G-P ranged between 750 and 1200 mm (Figure 6b).

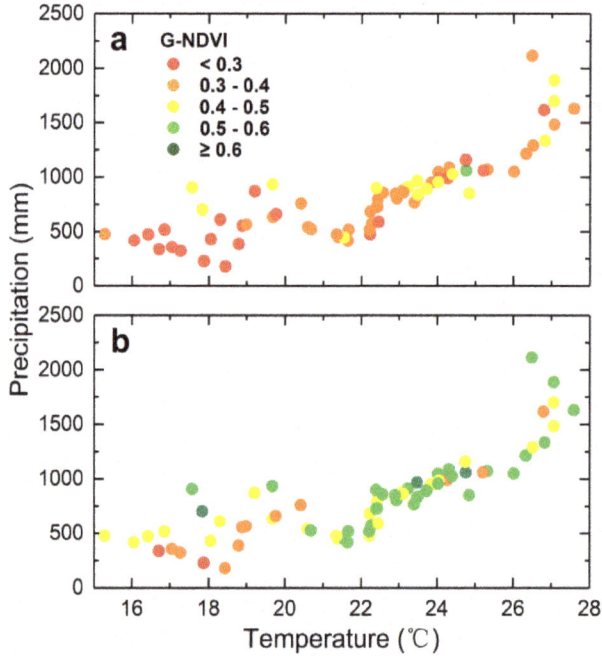

Figure 6. The relationship of mean annual G-T, G-P and G-NDVI$_{Z1}$ (**a**) G-NDVI$_{Z2}$ (**b**) of the selected 71 large cities. Each dot in the panel corresponds to three values (i.e., the mean annual G-NDVI, G-T and G-P).

The result of linear regression analysis showed that G-NDVI was positively correlated with climate variables (i.e., G-T and G-P) (Figure 7). The correlation coefficients between G-NDVI and climate variables were significant at the 0.01 significance level. Overall, the values of mean annual G-NDVI are closely related to climate factors. The mean annual G-NDVI of the selected 71 cities increased from areas with dry-cool climate to those with humid-hot climate.

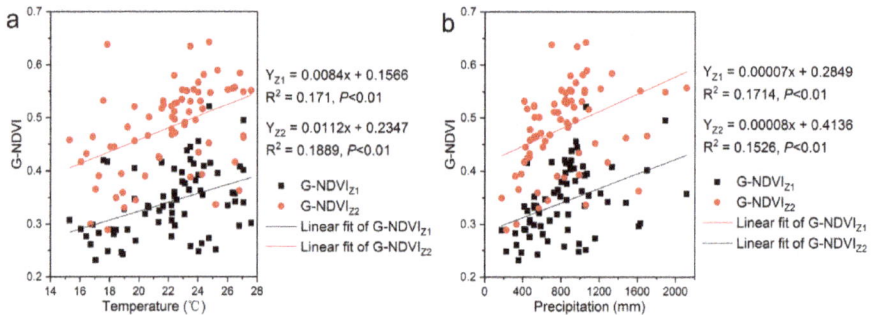

Figure 7. Correlations between the mean annual G-T and G-NDVI (**a**) and between the mean annual G-P and G-NDVI (**b**) of the selected 71 large cities.

3.2.2. Relationship between G-NDVI Variations and Climate Change

For the G-T, the correlation for the inter-annual variation between G-T and G-NDVI was insignificant for most of the selected cities (Figure 8a,c). For the G-P, cities with significantly positive relationships between G-NDVI and G-P were mainly located in Northern China where the mean annual precipitation was lower than 500 mm (Figure 8b,d). However, the correlation between the inter-annual variations of G-P and G-NDVI was insignificant for most of the cities (Figure 8). Moreover, the results of Pearson correlation test indicate that the correlations between the trend rates of G-NDVI and G-T and between the trend rates of G-NDVI and G-P are insignificant (Table 2). Overall, the inter-annual variations of G-NDVI were less sensitive to the changes of G-P and G-T.

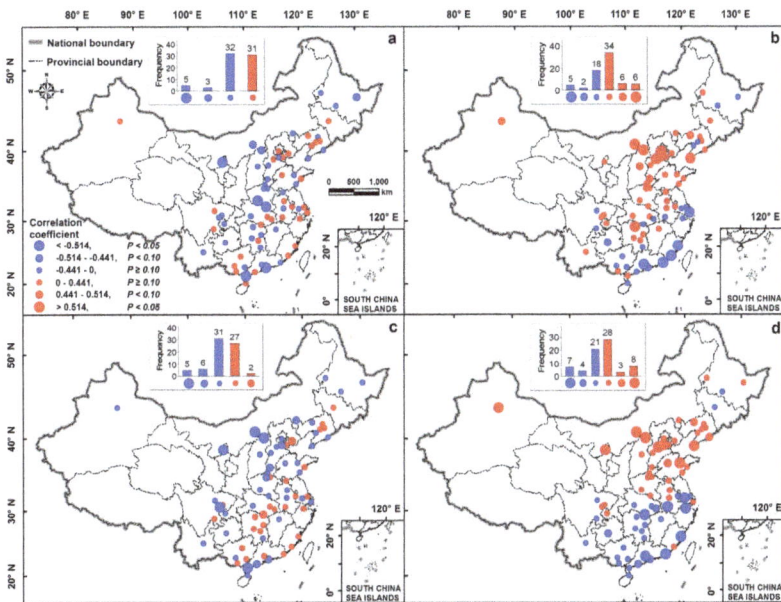

Figure 8. Distributions of the correction coefficients for the inter-annual variation between G-NDVI$_{Z1}$ and G-T (**a**), G-NDVI$_{Z1}$ and G-P (**b**), G-NDVI$_{Z2}$ and G-T (**c**), and G-NDVI$_{Z2}$ and G-P (**d**) in the selected 71 large cities during 1998–2012.

Table 2. The results of Pearson correlation test for the selected 71 large cities.

	ΔPU_{Z1} [1]	ΔPU_{Z2} [2]	Trend Rate of G-T [3]	Trend Rate of G-P [4]
Trend rate of G-NDVI$_{Z1}$ [5]	−0.37 **	−0.21	−0.17	−0.20
Trend rate of G-NDVI$_{Z2}$ [6]	0.04	−0.70 **	−0.10	−0.19

[1] and [2] indicate the change of urban land proportion between 1998 and 2012 in urban and suburban area respectively; [3] and [4] growing-season temperature and precipitation respectively; [5] and [6] change trends of growing-season NDVI in urban and suburban area respectively; ** means the correlation is significant at the 0.01 significance level.

3.3. Impact of Urbanization on G-NDVI Change

Table 2 shows that the negative correlations between ΔPU_{Z1} and the trend of G-NDVI$_{Z1}$ and between ΔPU_{Z2} and the trend of G-NDVI$_{Z2}$ are significant ($P < 0.01$), indicating that urban land expansion is closely associated with vegetation cover change in 71 large cities. In addition, the trends of G-NDVI$_{Z1-Z2}$ were negative for 54 cities mainly located in the inland of China, implying that the trend of G-NDVI in urban area was generally lower than that in suburban area for a given inland city (Figure 9). But the opposites were found in 17 cities mainly located in the coastal areas of China.

The smallest trend of G-NDVI$_{Z1-Z2}$ was found in Xinyang city with the rate being -0.093 per decade. The largest trend of G-NDVI$_{Z1-Z2}$ was found in Shanghai city with the rate being 0.083 per decade. Overall, vegetation cover in suburban areas has been greatly influenced by urbanization for cities located in the coastal area of China. But vegetation cover in urban areas has experienced more severe urbanization impact than in suburban areas for most of the inland cities.

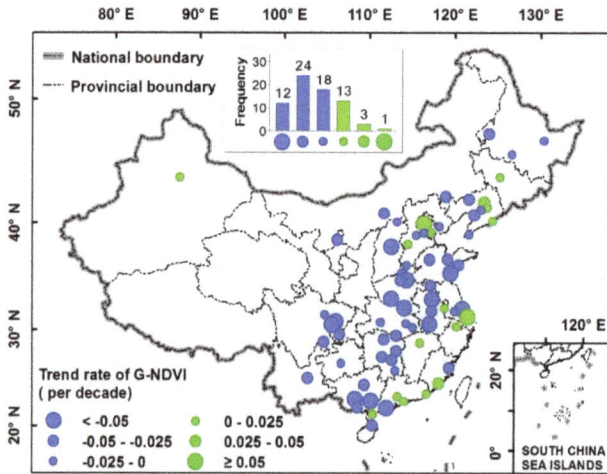

Figure 9. Trends of G-NDVI$_{Z1-Z2}$ for the selected 71 large cities.

Figure 10 shows that the trend of G-NDVI$_{Z1-Z2}$ is negatively correlated with ΔPU$_{Z1-Z2}$ ($R^2 = 0.3626$, $P < 0.01$). The regression equation between the trend of G-NDVI$_{Z1-Z2}$ and ΔPU$_{Z1-Z2}$ is $y = -0.0007x - 0.0012$. Through substituting the ΔPU$_{Z1}$ and ΔPU$_{Z2}$ of the 71 cities into this regression equation respectively (the constant being zero), the trends of G-NDVI induced by urbanization were calculated. We found that G-NDVI changes have been negatively impacted by urbanization in urban and suburban areas for most of the selected cities (Figure S1 of Supplementary Materials). On average, the impact of urbanization on G-NDVI change was estimated to be -0.026 per decade in Z1 and -0.015 per decade in Z2 during 1998–2012.

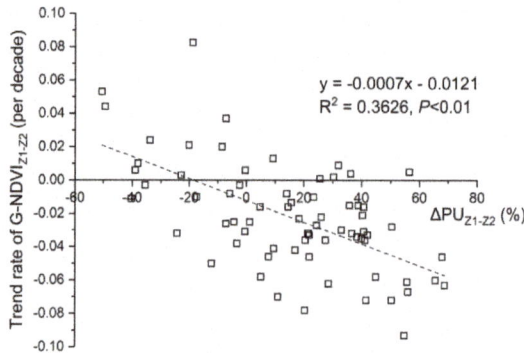

Figure 10. The relationship between trends of G-NDVI$_{Z1-Z2}$ and ΔPU$_{Z1-Z2}$ for the selected 71 large cities.

4. Discussion

4.1. Diversities in the Spatial Distribution of Mean Annual G-NDVI

The urbanization is an increasing active driving force of landscape change worldwide, and their long-term impacts on environmental change and sustainability could be described through concrete evidences within a relative short period. This study indicated that the mean annual G-NDVI were remarkably different among the selected 71 cities of China. This may be because the difference of NDVI values in China's cities is related to not only the climate conditions but also the development and management of cities (e.g., urbanization level) [34]. For larger cities, urban areas include a large proportion of constructed lands and a high density of building [46], resulting in relative lower NDVI in urban areas [34]. In addition, larger cities generally have more satellite towns and vaster intercity highway network around cities, implying that vegetation cover in suburban areas of larger city would be lower than that in smaller cities [47,48]. This is confirmed by the results presented in Figure 5, where the values of G-NDVI$_{Z2}$ in developed coast areas are generally lower than that in less developed inland areas. Despite this, the result of linear regression analysis showed a positive correlation between mean annual G-NDVI and climate variables (Figure 7). Consequently, the spatial difference of mean annual G-NDVI among the selected 71 cities is closely related to climate variabilities, but city development may play a certain role in local scales.

4.2. Driving Forces of the Temporal Variability of G-NDVI

The change of G-NDVI in the selected 71 cities was found to be less sensitive to climate change. Only few cities located in Northern China with the mean annual G-P being lower than 500 mm showed significant positive relationships between G-NDVI and G-P. Similar phenomenon was found by previous studies [25,26]. For example, there was a strong sensitivity of vegetation to precipitation in the regions where the annual rainfall ranged from 100 to 400 mm in Central Asia [23]. In general, the weak responses of NDVI to temperature change were attributed to increased water evaporation [26] and different vegetation types [10,24]. While the lower sensitivity of vegetation to precipitation in humid regions were ascribed to the limited water-use capacity of vegetation [49], the lower temperature and sunshine levels as precipitation increased [26], as well as the uncertainties of data adopted (e.g., cloud effects in the NDVI time-series) [50].

In cities, urbanization may be the crucial driver of effects on vegetation cover change and can be greater than climate factors [20,34] and hence mask the climate effects. For instance, urban land expansion, driven by urbanization in cities, could directly transform a large amount of agricultural lands to urban lands [43,44,47]. As a result, the effects of urban land expansion are able to disrupt the coupling between vegetation and precipitation [38]. The difference of inter-annual variation of G-NDVI between Z1 and Z2 for a given city also confirmed the impacts of urban land expansion in this study (Figure 9). The trend rate of G-NDVI change in some adjacent cities also showed a great difference. Climate conditions in a given city and among adjacent cities are similar, the difference was deemed to result from urban land expansion rather than climate variations particularly given that the selected 71 cities underwent rapid urbanization during the study period.

Our results indicated that for most of the selected cities changes in the G-NDVI were negatively impacted by urban land expansion in urban and suburban areas. Similar results were reported by previous studies based on different methods. For example, an analysis of NPP dynamics conducted by Peng et al. [18] showed that urbanization resulted in a lasting and observable loss of NPP over time and space. From the perspective of landscape patterns, Xia et al. [44] suggested that the urban area in Beijing city nearly doubled during 1997–2002, and farmland decreased rapidly as a result of urbanization. Using the correlation analysis, Sun et al. [34] found that NDVI change trends of China metropolises were negatively related to the change trends of urban area, population and GDP. While other studies qualitatively analyzed the impacts of urbanization on vegetation cover change, we quantified it based on a regression model between G-NDVI and PU. This method has excluded the impact of climate change on G-NDVI through calculating the difference of G-NDVI trend rate between Z1 and Z2 in a given city.

Certainly, the method presented here also has its limitations. One such example is that we have to rely on the assumption that the trend rate of G-NDVI$_{Z1-Z2}$ is mainly caused by urban land expansion rather than other anthropogenic activities (i.e., agricultural activities and greening measures) [18]. Because our sample size is big enough relative to the influences of these minor factors, we feel that this assumption is pretty reasonable. Another improvement based on the more specific land cover change will be conducted in further studies.

5. Conclusions

In this study, the spatiotemporal variations of vegetation cover and its sensitivity to climate change was investigated based on the growing-season NDVI, temperature and precipitation in 71 large cities of China during 1998–2012. The impact of urban expansion on vegetation cover change in urban and suburban areas was also quantitatively assessed. The main findings are as follows:

(1) The mean annual G-T, G-P and G-NDVI of the selected 71 cities are found to be greatly different. The spatial difference of G-NDVI is closely related to diverse climate conditions. Overall, the mean annual G-NDVI of 71 cities increases from dry-cool climate to humid-hot climate.

(2) The changes of G-T, G-P, PU and G-NDVI during 1998–2012 are different among the selected 71 cities. The mean values of ΔPU_{Z1} and ΔPU_{Z2} were 37% and 22% respectively, indicating that the selected 71 cities have experienced rapid urbanization during 1998–2012. The trend rates of G-NDVI$_{Z1}$ and G-NDVI$_{Z2}$ range from −0.099 to 0.067 per decade and −0.082 to 0.092 per decade respectively. G-NDVI changes are less sensitive to climate change, while closely related to urban land expansion. There is a negative correlation between G-NDVI trend and PU change, indicating vegetation cover in cities has been negatively impacted by urbanization.

(3) For most of the inland cities, vegetation cover in urban areas has experienced more severe urbanization impact than in suburban areas. But opposites occur in the 17 cities mainly located in the coastal areas of China. The average impacts of urbanization on G-NDVI change were estimated to be −0.026 per decade in Z1 and −0.015 per decade in Z2 during 1998–2012.

Cities generally have a long history and would keep on developing. However, many modern cities have encountered obstacles to maintain urban ecosystem sustainability which caused by intensive human activities (e.g., destroying vegetation). This study not only implies a significant long-term impact of human activities on the landscape, but also provides us information about human effects on the natural biotas. Therefore, we need to properly use these knowledges to plan the landscape management and ecological environment conservation for the future.

Supplementary Materials: The following are available online at http://www.mdpi.com/2071-1050/10/1/270/s1, Figure S1: Information of the selected 71 large cities in China.

Acknowledgments: The study was financially supported by the National Natural Science Foundation of China [41771558], the External Cooperation Program of BIC, Chinese Academy of Sciences [16146KYSB20150001], the European Commission Programme Horizon2020 project [635750] and the National Key Research and Development Program of China (2016YFC0501707).

Author Contributions: Kai Jin conceived and designed the study, completed the analyses, and drafted the manuscript. Fei Wang and Pengfei Li provided input to the study design, gave review suggestions for the manuscript on the whole writing process and polished the expression.

Conflicts of Interest: The authors declare no conflict of interest.

References

1. Grimm, N.B.; Faeth, S.H.; Golubiewski, N.E.; Redman, C.L.; Wu, J.; Bai, X.; Briggs, J.M. Global Change and the Ecology of Cities. *Science* **2008**, *319*, 756–760. [CrossRef] [PubMed]

2. Wu, J. Urban sustainability: An inevitable goal of landscape research. *Landsc. Ecol.* **2010**, *25*, 1–4. [CrossRef]

3.   Bolund, P.; Hunhammar, S. Ecosystem services in urban areas. *Ecol. Econ.* **1999**, *29*, 293–301. [CrossRef]
4.   Robinson, S.L.; Lundholm, J.T. Ecosystem services provided by urban spontaneous vegetation. *Urban Ecosyst.* **2012**, *15*, 545–557. [CrossRef]
5.   Gratani, L.; Varone, L.; Bonito, A. Carbon sequestration of four urban parks in Rome. *Urban For. Urban Green.* **2016**, *19*, 184–193. [CrossRef]
6.   Jonsson, P. Vegetation as an urban climate control in the subtropical city of Gaborone. *Botsw. Int. J. Climatol.* **2004**, *24*, 1307–1322. [CrossRef]
7.   Escobedo, F.J.; Nowak, D.J. Spatial heterogeneity and air pollution removal by an urban forest. *Landsc. Urban Plan.* **2009**, *90*, 102–110. [CrossRef]
8.   Dana, E.D.; Vivas, S.; Mota, J.F. Urban vegetation of Almería City—A contribution to urban ecology in Spain. *Landsc. Urban Plan.* **2002**, *59*, 203–216. [CrossRef]
9.   Jenerette, G.D.; Harlan, S.L.; Stefanov, W.L.; Martin, C.A. Ecosystem services and urban heat riskscape moderation: Water, green spaces, and social inequality in Phoenix, USA. *Ecol. Appl.* **2011**, *21*, 2637–2651. [CrossRef] [PubMed]
10.   Marignani, M.; Chiarucci, A.; Sadori, L.; Mercuri, A.M. Natural and human impact in Mediterranean landscapes: An intriguing puzzle or only a question of time? *Plant Biosyst.* **2016**, *151*, 900–905. [CrossRef]
11.   Mercuri, A.M.; Marignani, M.; Sadori, L. Palaeoecology and long-term human impact in plant biology. *G. Bot. Ital.* **2015**, *149*, 136–143. [CrossRef]
12.   Li, Y.; Wu, J.; Hou, S.; Shi, C.; Mo, D.; Liu, B.; Zhou, L. Palaeoecological records of environmental change and cultural development from the Liangzhu and Qujialing archaeological sites in the middle and lower reaches of the Yangtze River. *Quat. Int.* **2010**, *227*, 29–37. [CrossRef]
13.   Wang, J.; Meng, J.J.; Cai, Y.L. Assessing vegetation dynamics impacted by climate change in the southwestern karst region of China with AVHRR NDVI and AVHRR NPP time-series. *Environ. Geol.* **2008**, *54*, 1185–1195. [CrossRef]
14.   Ruimy, A.; Saugier, B.; Dedieu, G. Methodology for the estimation of terrestrial net primary production from remotely sensed data. *J. Geophys. Res. Atmos.* **1994**, *99*, 5263–5283. [CrossRef]
15.   Fu, Y.; Lu, X.; Zhao, Y.; Zeng, X.; Xia, L. Assessment Impacts of Weather and Land Use/Land Cover (LULC) Change on Urban Vegetation Net Primary Productivity (NPP): A Case Study in Guangzhou, China. *Remote Sens.* **2013**, *5*, 4125–4144. [CrossRef]
16.   Zeng, X.; Dickinson, R.E.; Walker, A.; Shaikh, M. Derivation and Evaluation of Global 1-km Fractional Vegetation Cover Data for Land Modeling. *J. Appl. Meteorol.* **1999**, *39*, 826–839. [CrossRef]
17.   Suzuki, R.; Nomaki, T.; Yasunari, T. Spatial distribution and its seasonality of satellite-derived vegetation index (NDVI) and climate in Siberia. *Int. J. Climatol.* **2001**, *21*, 1321–1335. [CrossRef]
18.   Peng, J.; Shen, H.; Wu, W.; Liu, Y.; Wang, Y. Net primary productivity (NPP) dynamics and associated urbanization driving forces in metropolitan areas: A case study in Beijing City, China. *Landsc. Ecol.* **2016**, *31*, 1077–1092. [CrossRef]
19.   Eastman, J.R.; Sangermano, F.; Machado, E.A.; Rogan, J.; Anyamba, A. Global Trends in Seasonality of Normalized Difference Vegetation Index (NDVI), 1982–2011. *Remote Sens.* **2013**, *5*, 4799–4818. [CrossRef]
20.   Li, H.; Li, Y.; Gao, Y.; Zou, C.; Yan, S.; Gao, J. Human Impact on Vegetation Dynamics around Lhasa, Southern Tibetan Plateau, China. *Sustainability* **2016**, *8*, 1146. [CrossRef]
21.   Wen, Z.F.; Wu, S.J.; Chen, J.L.; Lv, M.Q. NDVI indicated long-term interannual changes in vegetation activities and their responses to climatic and anthropogenic factors in the Three Gorges Reservoir Region, China. *Sci. Total Environ.* **2017**, *574*, 947–959. [CrossRef] [PubMed]
22.   Blok, D.; Schaepman-Strub, G.; Bartholomeus, H.; Heijmans, M.M.P.D.; Maximov, T.C.; Berendse, F. The response of Arctic vegetation to the summer climate: Relation between shrub cover, NDVI, surface albedo and temperature. *Environ. Res. Lett.* **2011**, *6*, 035502. [CrossRef]
23.   Gessner, U.; Naeimi, V.; Klein, I.; Kuenzer, C.; Klein, D.; Dech, S. The relationship between precipitation anomalies and satellite-derived vegetation activity in Central Asia. *Glob. Planet. Chang.* **2013**, *110*, 74–87. [CrossRef]
24.   Bai, J.J.; Bai, J.T.; Wang, L. Spatio-temporal change of vegetation NDVI and its relations with regional climate in Northern Shaanxi Province since implementation of returning farmland to forests project. In Proceedings of the 2013 Second International Conference on Agro-Geoinformatics (Agro-Geoinformatics), Fairfax, VA, USA, 12–16 August 2013. [CrossRef]

25. Fuller, D.O.; Prince, S.D. Rainfall and foliar dynamics in tropical southern Africa: Potential impacts of global climate change on savanna vegetation. *Clim. Chang.* **1996**, *33*, 69–96. [CrossRef]

26. Piao, S.; Mohammat, A.; Fang, J.; Cai, Q.; Feng, J. NDVI-based increase in growth of temperate grasslands and its responses to climate changes in China. *Glob. Environ. Chang.* **2006**, *16*, 340–348. [CrossRef]

27. Zhou, L.; Tucker, C.J.; Kaufmann, R.K.; Slayback, D.; Shabanov, N.V.; Myneni, R.B. Variations in northern vegetation activity inferred from satellite data of vegetation index during 1981 to 1999. *J. Geophys. Res. Atmos.* **2001**, *106*, 20069–20084. [CrossRef]

28. Sun, Y.; Zhao, S.; Qu, W. Quantifying spatiotemporal patterns of urban expansion in three capital cities in Northeast China over the past three decades using satellite data sets. *Environ. Earth Sci.* **2015**, *73*, 7221–7235. [CrossRef]

29. Wu, J.; Xiang, W.N.; Zhao, J. Urban ecology in china: Historical developments and future directions. *Landsc. Urban Plan.* **2014**, *125*, 222–233. [CrossRef]

30. Imhoff, M.L.; Lawrence, W.T.; Elvidge, C.D.; Paul, T.; Levine, E.; Privalsky, M.V.; Brown, V. Using nighttime DMSP/OLS images of city lights to estimate the impact of urban land use on soil resources in the United States. *Remote Sens. Environ.* **1997**, *59*, 105–117. [CrossRef]

31. Lin, T.; Grimm, N.B. Comparative study of urban ecology development in the U.S. and China: Opportunity and Challenge. *Urban Ecosyst.* **2015**, *18*, 599–611. [CrossRef]

32. Li, F.; Wang, R.; Paulussen, J.; Liu, X. Comprehensive concept planning of urban greening based on ecological principles: A case study in Beijing, China. *Landsc. Urban Plan.* **2005**, *72*, 325–336. [CrossRef]

33. Jim, C.Y.; Chen, S.S. Comprehensive greenspace planning based on landscape ecology principles in compact Nanjing city, China. *Landsc. Urban Plan.* **2003**, *65*, 95–116. [CrossRef]

34. Sun, J.; Wang, X.; Chen, A.; Ma, Y.; Cui, M.; Piao, S. NDVI indicated characteristics of vegetation cover change in China's metropolises over the last three decades. *Environ. Monit. Assess.* **2011**, *179*, 1–14. [CrossRef] [PubMed]

35. Memon, R.A.; Leung, D.Y.; Chunho, L. A review on the generation, determination and mitigation of urban heat island. *J. Environ. Sci.* **2008**, *20*, 120–128.

36. Lu, P.; Qiang, Y.; Liu, J.; Lee, X. Advance of tree-flowering dates in response to urban climate change. *Agric. For. Meteorol.* **2006**, *138*, 120–131. [CrossRef]

37. Buyantuyev, A.; Wu, J. Urbanization diversifies land surface phenology in arid environments: Interactions among vegetation, climatic variation, and land use pattern in the Phoenix metropolitan region, USA. *Landsc. Urban Plan.* **2012**, *105*, 149–159. [CrossRef]

38. Buyantuyev, A.; Wu, J. Urbanization alters spatiotemporal patterns of ecosystem primary production: A case study of the Phoenix metropolitan region, USA. *J. Arid Environ.* **2009**, *73*, 512–520. [CrossRef]

39. Jin, K.; Wang, F.; Chen, D.; Jiao, Q.; Xia, L.; Fleskens, L. Assessment of urban effect on observed warming trends during 1955–2012 over China: A case of 45 cities. *Clim. Chang.* **2015**, *132*, 631–643. [CrossRef]

40. Wang, J.; Wang, K.; Zhang, M.; Zhang, C. Impacts of climate change and human activities on vegetation cover in hilly southern China. *Ecol. Eng.* **2015**, *81*, 451–461. [CrossRef]

41. Cao, L.; Zhu, Y.; Tang, G.; Yuan, F.; Yan, Z. Climatic warming in China according to a homogenized data set from 2419 stations. *Int. J. Climatol.* **2016**, *36*, 4384–4392. [CrossRef]

42. Cai, D.; Fraedrich, K.; Guan, Y.; Guo, S.; Zhang, C. Urbanization and the thermal environment of Chinese and US-American cities. *Sci. Total Environ.* **2017**, *589*, 200–211. [CrossRef] [PubMed]

43. Li, Y.J.; Zeng, H.; Wei, J.B. Vegetation change in Shenzhen City based on NDVI change classification. *Chin. J. Appl. Ecol.* **2008**, *19*, 1064–1070. (In Chinese)

44. Xia, B.; Yu, X.X.; Ning, J.K.; Wang, X.P.; Qin, Y.S.; Chen, J.Q. Landscape pattern evolution of Beijing in recent 20 years. *J. Beijing For. Univ.* **2008**, *30*, 60–66. (In Chinese)

45. Liu, Z.; He, C.; Zhang, Q.; Huang, Q.; Yang, Y. Extracting the dynamics of urban expansion in China using DMSP-OLS nighttime light data from 1992 to 2008. *Landsc. Urban Plan.* **2012**, *106*, 62–72. [CrossRef]

46. Han, G.F.; Xu, J.H. Vegetation change trajectory and the reasons in Shanghai City. *Acta Ecol. Sin.* **2009**, *29*, 1793–1803. (In Chinese)

47. Tian, G.; Jiang, J.; Yang, Z.; Zhang, Y. The urban growth, size distribution and spatio-temporal dynamic pattern of the Yangtze River Delta megalopolitan region, China. *Ecol. Model.* **2011**, *222*, 865–878. [CrossRef]

48. Liu, C.L.; Yu, R.L. Spatial Accessibility of Road Network in Wuhan Metropolitan Area Based on Spatial Syntax. *Acta Geogr. Sin.* **2012**, *4*, 128–135. [CrossRef]

49. Angert, A.L.; Huxman, T.E.; Barron-Gafford, G.A.; Gerst, K.L.; Venablem, D.L. Linking growth strategies to long-term population dynamics in a guild of desert annuals. *J. Ecol.* **2007**, *95*, 321–331. [CrossRef]

50. Julien, Y.; Sobrino, J. Global land surface phenology trends from GIMMS database. *Int. J. Remote Sens.* **2009**, *30*, 3495–3513. [CrossRef]

sustainability

MDPI

Article

Historical Arable Land Change in an Eco-Fragile Area: A Case Study in Zhenlai County, Northeastern China

Yuanyuan Yang [1,2,*] **and Shuwen Zhang** [3]

1   Institute of Geographic Sciences and Natural Resources Research, Chinese Academy of Sciences,
    Beijing 100101, China
2   Faculty of Geographical Science, Beijing Normal University, Beijing 100875, China
3   Northeast Institute of Geography and Agroecology, Chinese Academy of Sciences,
    Changchun 130102, China; zhangshuwen@neigae.ac.cn
*   Correspondence: yangyy@igsnrr.ac.cn

Received: 21 August 2018; Accepted: 26 October 2018; Published: 30 October 2018

Abstract: Long-term land changes are cumulatively a major driver of global environmental change. Historical land-cover/use change is important for assessing present landscape conditions and researching ecological environment issues, especially in eco-fragile areas. Arable land is one of the land types influenced by human agricultural activity, reflecting human effects on land-use and land-cover change. This paper selected Zhenlai County, which is part of the farming–pastoral zone of northern China, as the research region. As agricultural land transformation goes with the establishment of settlements, in this research, the historical progress of land transformation in agricultural areas was analyzed from the perspective of settlement evolution, and the historical reconstruction of arable land was established using settlement as the proxy between their inner relationships, which could be reflected by the farming radius. The results show the following. (1) There was little land transformation from nonagricultural areas into agricultural areas until the Qing government lifted the ban on cultivation and mass migration accelerated the process, which was most significant during 1907–1912; (2) The overall trend of land transformation in this region is from northeast to southwest; (3) Taking the topographic maps as references, the spatial distribution of the reconstructed arable land accounts for 47.79% of the maps. When this proxy-based reconstruction method is applied to other regions, its limitations should be noticed. It is important to explore the research of farming radius calculations based on regional characteristics. To achieve land-system sustainability, long-term historical land change trajectories and characteristics should be applied to future policy making.

Keywords: historical land-cover/use change; land reconstruction; arable land; farming radius; eco-fragile area; Northeast China

1. Introduction

Land-use/cover change (LUCC) is a fundamental component of global environmental change [1]. Human activities over the last centuries have significantly transformed the Earth's environment, primarily through the conversion of natural ecosystems to agriculture [2]. Both human-induced and natural land-cover/use changes interact with the global carbon cycle, biodiversity, climate change, landscape, and ecology [3–6]. The impact of changes in land cover over historical periods should be included in the process of building models that accurately simulate global environmental change [7]. Moreover, land-use history is very important for assessing present landscape conditions, because land-use legacies influence climatic and ecological processes that occur today [8,9]. This often requires the reconstruction of land-use/cover history over the time slice.

Yet, the availability of historical land-cover/use data is often limited. Historical land-cover/use data tend to be fragmented and difficult to obtain due to unfavorable restraints, including copyright, accessibility, language barriers, and secrecy status. The currently available global-scale reconstructions are based on a combination of data and modeling, which incorporates the dynamics of long-term human–environment relationships [2,4,10–13]. Most studies are restricted to local and regional levels using natural archives, historical documents, statistical datasets, old maps, pictures, and model simulations, which potentially filled the data gap before the advent of satellite archives [14–20]. Especially, this would be much more easily achieved when historical data are combined with GIS (Geographic Information Systems) tools [21–23]. Land cover/use often needs to be reconstructed in both quantity and spatial distribution. After selecting valid data, they must be implemented in a GIS tool to permit their spatial evaluation. Historical documents contain qualitative or semiquantitative information about past land cover, and are the main data sources for researching historical changes in land cover from regional to global levels. They are often recorded in administrative units such as towns, counties, cities, and provinces. Historical records provide land managers with information that can be used to understand LUCC trajectories, and a more complete picture of historical land use/cover could be obtained by integrating multisource data [24]. According to the data characteristics from historical documents, there are generally two reconstruction methods. First, useful information about historical land cover could be extracted directly from the old documents after a qualitative or semiquantitative analysis, and then uniform tabular data could be formed. However, it must be noted that large differences often occur between historical data and present statistics due to the impact of policies, economies, wars, and other factors, such as numerical characteristics and statistical standards. Therefore, it is necessary to revise and calibrate these historical data prior to use [25]. In this context, the second method is proposed: under some assumptions, circumstantial evidence is obtained from historical documents, and then, a quantitative land-cover analysis is performed based on the relationships between variables, such as between cropland areas and rates of population change, speed of urban sprawl and increase in population, or pastoral areas and numbers of animals.

Arable land is a land type that is influenced by agricultural activity, reflecting humanity's impact on LUCC. In farming regions, transforming land from nonagricultural areas into agricultural areas often occurs with the establishment of residential settlements to satisfy the food demands of local residents. Arable land has a close relationship with settlements, which could comprehensively reflect the interaction between human activities and the natural environment. The site selection of settlements integrates such basic elements as human survival mode and environmental selection. Generally, human beings decide to locate or relocate to a certain site after considering water sources, the history of cultivation, traffic conditions, and other natural and social factors to form settlements. Then, natural land cover changes to meet agricultural land, resulting in significant LUCC. Therefore, in consideration of the law that the establishment of settlements is synchronized with the transformation from nonagricultural areas to agricultural areas, the development and evolution of settlement patterns may reflect the history of land transformation in a local region [26]. In a traditional agricultural production pattern, farmland is usually distributed around settlements and influenced by human activities. Thus, applying the mutual relationship between the location of settlements (residential land uses) and arable land potentially provides a way to reconstruct historical areas of arable land, which could be reflected through the farming radius.

The farming–pastoral ecotone of Northern China, as a sensitive region of terrestrial ecosystems, is very vulnerable to global change and human disturbance [27]. Human activities, such as excessive reclamation, grazing, excavation, and abandonment, have generated enormous negative environmental impacts in the region, especially regarding the destruction of natural vegetation, which has constantly drained the service functions of the local ecosystems [27]. Changes in land use are likely the most ancient of all human-induced environmental impacts [28]; therefore, large areas of grassland have been converted into cropland due to climate warming, increased population, and food demand, causing the pattern of land cover in the farming–pastoral ecotone of Northern China to change quickly and continually. With the encroachment of cropland in traditional pastoral zones, the picture of resilient and balanced rangeland ecology has been turned upside down.

Bearing all this in mind, this paper researches arable land change in an eco-fragile area after reconstructing historical arable land using available documents. Especially, this research tries to explore arable land reconstruction based on information about settlements from the local gazetteer according to their relationship, which could be reflected by the farming radius. Zhenlai County, a part of the farming–pastoral ecotone of Northern China, is located in northwestern Jilin Province, where the eco-environment is very fragile and vulnerable to disturbance. Over the past century, northeastern China has experienced dramatic land-cover/use changes. In this context, the study area was selected as the target for this research. The objectives of the research are as follows: (1) to analyze historical settlement changes and land transformation processes from gazetteer records through a historical lens; and (2) to reconstruct spatial arable land data based on the distribution of settlements and the farming radius in the 1930s.

2. Materials and Methods

2.1. Study Area

Zhenlai County (45°28′ N–46°18′ N, 122°47′ E–124°04′ E) (Figure 1) lies in Baicheng City of northwestern Jilin Province, with a high-lying northwest and low southeast in its complex and diverse topography. Adjacent to the Great Khingan region in the northwest, the county's central area is mostly fluctuant hilly land, while its south and east are surrounded by the Tao'er River and the Nenjiang River, respectively, shaping fertile alluvial plains lining the riverbanks. Located in inland areas of mid-latitude, the region has distinct seasons and a temperate continental monsoon climate. Its mean annual temperature is around 4.9 °C. The mean annual rainfall is 402 mm, which is unevenly distributed over time [29,30]. The low amount of precipitation and the high amount of evaporation mainly result in a drought-prone climate, especially in spring. The study area is located in the transition zone from the first step to the second step, and in the climate transition belt from the humid East Asian monsoon to arid inland. As part of the farming–pastoral region, its natural geographical conditions show marginal and transitional characteristics. The region, with a fragile eco-environment, is a sensitive area responding to global change [31]. In the past century, increasing human agricultural activities on land conversion mainly for grain crops has significantly influenced land-use and land-cover change.

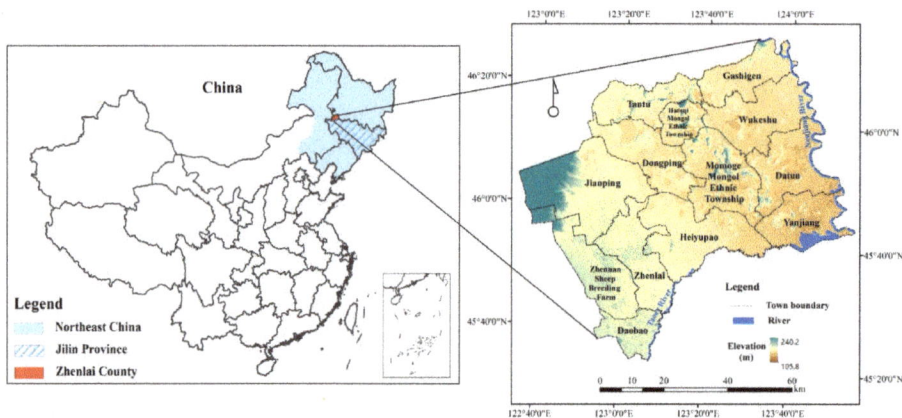

Figure 1. Location of the study area, Zhenlai County.

The region was the nomadic land for Mongol princes, and inhabitants were not allowed to reclaim it until the enactment of the ban on transforming land from nonagricultural areas into agricultural areas in the late Qing Dynasty (1902). The county was established in 1910 as Zhendong County, and Laibei County was merged into this region in 1947, with the new name Zhenlai County. As Zhenlai County's administrative border changed over the past century, this research used the border from *Zhenlai Gazetteer* (1985) as the study range, thereby ensuring comparability across different periods of time. According to the gazetteer, Zhenlai County includes 11 towns/townships and one farm, which are: Zhenlai (the county seat), Heiyupao, Yanjiang, Tantu, Datun, Wukeshu, Dongping, Jianping, Ganshigen Hatuqi Mongol Ethnic Township, Momoge Mongol Ethnic Township, and Zhennan Sheep Breeding Farm (Figure 1).

2.2. Data Collection and Processing

Considering the advantages and constraints regarding land use reconstruction through the historical documents [14] illustrated in Figure 2, in this study, historical documents were collected, including *Zhenlai Gazetteer (1985)* [32] and *Geographic Names of Baicheng Region (1984)*, while other documents were also referred to, such as the *Zhenlai County Annals, Dalai County Annals*, and *Zhendong County Annals* in *Integrated Local Records in China—Album of Jilin County Annals (1931)*.

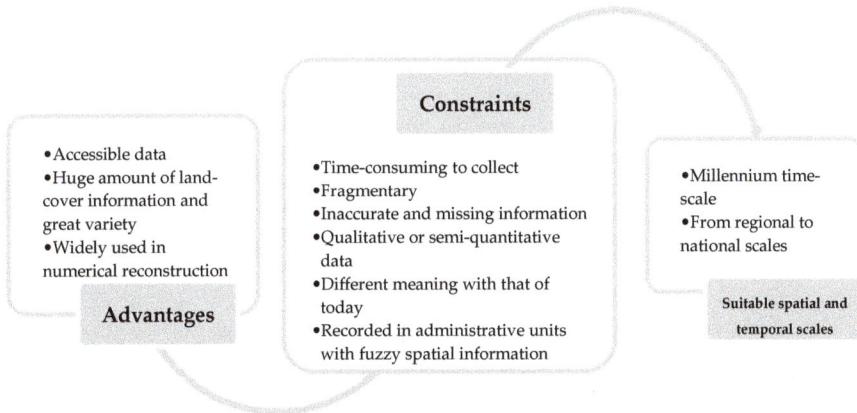

Figure 2. Advantages and constraints of reconstruction using historical documents.

Among those documents, the *Zhenlai Gazetteer (1985)* is very important for this research. The gazetteer records include detailed geographic names, definitions, history, land area, population, geographical entities, artificial construction, etc., and also describes and accurately investigates the regional scale, the size and number of villages or towns, and the main production modes, which could reflect the evolution of how people change natural land cover and how nature is influenced by human activities. These factors provide valuable information for research of the man–land relationship and the future development and utilization of the environment [26,33]. The *Zhenlai Gazetteer (1985)* mainly consists of the following information: village names and changes, year of founding, type of agriculture (e.g., cultivation or grazing), the ethnic groups that established the villages, the method of building villages, the locations, administrative subordination relationships, population, and number of households. For example, it is recorded that the Sibe people lived a nomadic life and built Wulinsibe village in Zhenlai County, while Ma's family began to cultivate and build Yuanbaotu village in 1853 during the years of the Qing Emperor Kangxi. Data about settlements can be extracted, and an attribute table can be formed mainly from the *Zhenlai Gazetteer* (Table 1). So, making full use of these documents, detailed settlement information was obtained according to the flow path (Figure 3). In general, the establishment of settlements is synchronized with land transformation. Thus, the development and evolution history of settlement patterns may reflect the history of agricultural land transformation, and the reconstruction of arable land based on the distribution of settlements and farming radius is available. It should be noted that not all of the settlement data are from the gazetteer, and the missing data were collected from the other documents mentioned above. Then, the attribute table generated from the *Zhenlai Gazetteer* (Table 1) was joined to the spatial location of settlements with *x* and *y* axes using ArcGIS.

Table 1. Attribute table generated from the *Zhenlai Gazetteer*.

Village Name	Year of Founding	Type of Agriculture	Ethnic Group	Number of Households in the 1980s	Population in the 1980s	Administrative Subordination
Baxizhao	1875	Nomadism	Mongol	74	374	Baxizhao village
Dongmoshihai	1912	Cultivation and grazing	Mongol	105	521	Ximingga village
Erlongsuokou	1898	Cultivation	Mongol	112	589	Changfa village
Shuangyushu	1909	Cultivation	Mongol	51	277	Teli village
Houwulantu	1899	Cultivation	Han	142	795	Lixin village
... ...	... ...	... ...	... ...	... ...	... ...	... ...

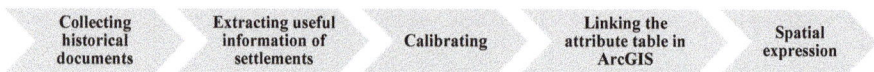

Collecting historical documents	Extracting useful information of settlements	Calibrating	Linking the attribute table in ArcGIS	Spatial expression

Figure 3. Flow path of settlement data capture from the *Zhenlai Gazetteer*.

In addition, the study area is covered by a series of topographic maps dating back to the 1930s at a scale of 1:50,000, which were obtained from a project plan for the production of digital images of the Chinese mainland by the Japanese Academy of Sciences. In this study, the old maps were scanned at high resolution due to the high requirements for interpretation and the quality of the maps. Then, a minimum of four ground control points were chosen from the scanned maps to rectify them using polynomial equations in ArcGIS software. When the result proved to be unsatisfactory, new control points were chosen or uncertain control points were omitted, and the transformation was repeated. Manual geo-referencing is necessary when using the historical data. Map coordinates and unique landmarks such as churches, crossroads, and coastal shapes were used for geo-referencing [34]. Considering the lack of latitude and longitude information in the 1:50,000 topographic maps, geo-referencing was undertaken by identifying common objects such as settlements, unchanged road junctions, and other similar land features with the same position in the 1970s and 1930s maps. Although there were some name changes with the settlements, the majority of names were transliterated based on their pronunciation. The historical maps were used to calculate the farming radius, assuming that they could reflect the same real land-use/cover distribution as the reference maps.

For the research of historical arable land change, we also collected data of land-use and land cover from the Data Center for Resource and Environment Science of the Chinese Academy of Sciences. In this study, we adopted the current land-use classification system, including six categories and 25 subclasses [35], where arable land refers to agricultural land, including paddy fields and rainfed cropland.

2.3. Methodology

2.3.1. Gravity-Center Model

A gravity-center model, which has been extensively applied in the fields of land-use planning, economic geography, and land-use science, is a modeling approach that identifies the movement direction and distance to the center of gravity for targeted objects [36,37]. These can reflect changes in quantity and change trends of the targeted object over time. Gravity-center models are spatially explicit; the degree of spatial representation that they offer depends on the number of zones in which the study region is subdivided. The models are static or quasi-static (or comparatively static) at best, which means that they do not account for the dynamics underlying the observed interactions. In terms of the level of detail of the land uses considered and the spatial behavior modeled, the most common forms of gravity model concern two main types of land use, e.g., residential and commercial, residential and employment, or residential and recreational [38]. In this paper, the gravity-center model was applied to obtain the spatial changes of historical settlements (residential areas), and is not expected to explain driving forces of change. The land development stages were reclassified, and the gravity-center coordinates of each stage were calculated. As the attribute of settlements in this research is a point (not a polygon with area), the equation for the center of gravity of settlements was applied as follows:

$$X_t = \sum_{i=1}^{n} X_{ti}/n \tag{1}$$

$$Y_t = \sum_{i=1}^{n} Y_{ti}/n \tag{2}$$

where X_t and Y_t are the X and Y gravity center coordinates, respectively, of settlements for period t; X_{ti} and Y_{ti} are the x and y coordinates, respectively, of point i; and n is the total number of settlements in period t.

2.3.2. Proxy-Based Reconstruction Method

Historical documents, maps, models, and proxy-based reconstruction studies of LUCC and agricultural land cover have been widely used for reproducing past land-use information, not only for the quantity of land cover/use in a historical period, but also for the spatial distribution. In particular, proxy-based reconstruction is a convenient and effective method to reconstruct one type of land cover/use by using deductions from the relationship between the land type and other variables. The proxy method means that data recorded indirectly could be replaced by similar data according to different statistical standards (Equation (3)) [27]. For example, Goldewijk (2001) assumed that there should always be some agricultural activities where people are living (especially in the past), and therefore, he used population density as an acceptable proxy for the allocation of cropland and pasture when estimating global land-use change over the past 300 years [39]. The name of a settlement intuitively records conditions of new land transformation when humans arrive; therefore, it is important in understanding the land development process and the historical course of reconstructing land-use changes:

$$y = f(x) \tag{3}$$

where x refers to the acceptable proxy index, which is easy to obtain from historical documents, and y refers to the targeted reconstructed land.

2.3.3. GIS-Based Analysis Methodology

This research tried to reconstruct arable land based on its relationship with settlements, which could be reflected by the farming radius. So, how to calculate the farming radius is the key in this proxy-based reconstruction. In general, there are two ways to achieve it: an equalization-based calculation, and a cultivation–settlement ratio-based calculation. In this context, the use of a GIS tool that could capture, store, manipulate, analyze, manage and represent spatial data could be very helpful [21].

(1) Equalization-Based Calculation Method

Influenced by the actual conditions of farming and agricultural production levels, it is assumed that the farming radius of settlements is limited, with a fixed value in the region. Theoretically, when the buffer area, determined by taking the settlement as the center with a certain radius, is equal to the area of cultivated land in the study area, this radius represents the farming radius of the settlement [40,41].

(2)   Cultivation–Settlement Ratio-Based Calculation Method

Jin (1988) [42] proposed a formula that could reflect the relationships among farming radius, number of population, and per capita cultivated farmland (Equation (4)). For a certain settlement, its total area of cultivated land could be calculated by the product of per capita cultivated land area and population number. The ratio of cultivated land and rural settlement represents the regional agricultural production level [41]. Combining Equations (4) and (5), the farming radius of a settlement can be deduced as Equation (6):

$$NM = K\pi R^2 \tag{4}$$

$$G = \frac{NM}{J} \tag{5}$$

$$R = \sqrt{\frac{GJ}{K\pi}} \tag{6}$$

where N is the population; M is the per capita cultivated land area; K is the ratio of cultivated land according to the area of the administrative region, and is also named the land transformation coefficient; G is the ratio of cultivated land to rural settlement; J is the area of settlement; and R is the farming radius.

(3)   Comparison of the Two Methods

The equalization-based calculation method, as a regular method, assumes that the farming radius is a fixed value, and all of the settlements in the region cultivate land in the range of circles with the same radii. However, the cultivation–settlement ratio-based calculation method considers the sizes of settlements. When the environments are similar and the per capita rural residential land is basically the same in the study region, the settlements with large populations tend to have a relatively large farming radius to maintain a certain per capita amount of production land (Figure 4).

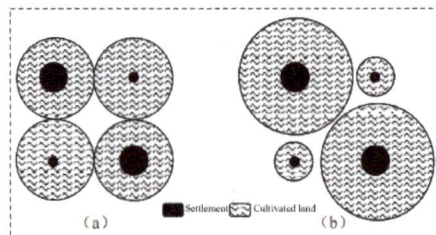

Figure 4. Farming radius calculated by (**a**) equalization-based method and (**b**) cultivation–settlement ratio-based method.

In this research, the digitized data from the topographic maps of the 1930s were considered as the actual spatial distribution of historical land use/cover. Here, land-use types mainly include settlements and arable land. GIS-based buffer analysis was used to calculate the farming radius. In the equalization-based calculation method, a buffer area map was generated with 100-m step lengths by taking each settlement from the 1930s maps as a center. After establishing topological relationships, the buffer area was revised according to the administration range of the study area, and its size was obtained after repeated calculations. Then, a distribution map of all of the settlements was produced. Next, comparing the produced buffer area with the actual farmland distribution map from the 1930s maps, the cultivation radius of Zhenlai County in the 1930s could be worked out. When the buffer area within the research region was closest to the value of arable land area determined from the topographic maps, the buffer distance could be considered as the farming radius. In the cultivation–settlement ratio-based calculation method, the farming radius varies according to different regions. The town/township administrative map of Zhenlai County was combined with the farmland area extracted from topographic maps from the 1930s to calculate the farming radius based on Equation (6).

3. Results

3.1. Settlement Change According to Zhenlai Gazetteer

According to the records of the *Zhenlai Gazetteer*, two villages were built in 1853 that reflect and represent the early land conversion after people settled in the study area, but these were small-scale land transformations due to the low technology level at that time, which had little influence on the overall land-use and land-cover change there. Large-scale development began in 1875 after the ban on cultivation was lifted, and the most recent settlement was founded in 1975.

Human activities have a profound impact on LUCC, and the evolution of settlements could reflect the transformation process of cultivated land. Combined with the history of Northeast China and the evolution characteristics of new villages, the process of land development in the study region is divided into the following five stages (Figure 5): (1) at the end of the Qing Dynasty (1875–1906), 198 new villages; (2) from Xuantong of the Qing Dynasty to the first year of the Republic of China (1907–1912), 159 new villages; (3) during the period of the Republic of China (1913–1931), 83 new villages; (4) during Manchoukuo (1932–1945), 34 new villages; and (5) after liberation (1946–1985), 47 new villages. Figure 6 shows the spatial distributions of villages, the years of founding, and the spatial development trend in the study area during 1875–1985.

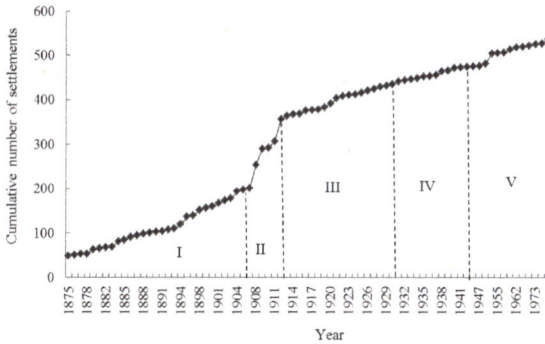

Figure 5. Cumulative curve of the number of villages.

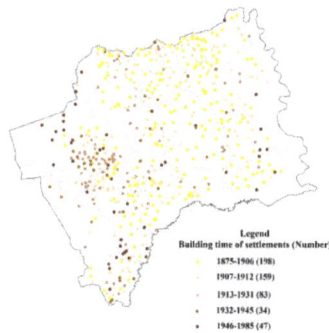

Figure 6. Spatial distribution of settlements during 1875–1985.

3.2. Land Development and Transformation Process

Based on the above classification, the corresponding development process of land-cover/use change in each stage was as follows. (1) Before 1875, there was little land transformation, and the main land cover was natural land, including grassland and wetland, influenced by the Qing government's prohibition policy; (2) After 1875, an influx of refugees migrated to the northeast, and the Qing government began to lift the ban on cultivation. The legal land-change policy, huge population pressure, and years of famine in the north in Shanhaiguan Pass prompted large-scale immigration. From 1875 to 1906, mass migration accelerated the natural land-cover change to meet agricultural land; (3) The period 1907–1912 had the greatest increase in villages, and was also the most prominent stage for land-use development in Zhenlai County; (4) Due to natural disasters and the warlord dogfight in Shanhaiguan Pass, many immigrants migrated to the northeast during 1913–1931. Land conversion extended to the west of Zhenlai based on the existing land development; (5) During 1932–1945, the western flat area in the research region was gradually reclaimed and developed based on the previously cultivated land; (6) There was mainly an expansion of existing cultivated land or preferential development in the undeveloped area after 1945. Laibei County was merged with Zhendong County in 1947 to create Zhenlai County, indicating that this region had been transformed from animal husbandry-dominated to agriculture-dominated. The cumulative curve of the number of villages that were built each year (Figure 5) shows that approximately 84% of land development and utilization occurred between 1875–1931, and the number of settlements increased most rapidly from 1907 to 1912. The Qing government announced the comprehensive opening of the area in 1905, and immigration and land transformation in Jilin Province experienced the most active period shortly thereafter. Zhenlai County is located in the northwest of Jilin Province; thus, its entry into the active period was a little later. After several years of rapid development, land transformation in the study region slowed down, and stepped into the normal development stage.

The elevation of Zhenlai County ranges from 105.8 m to 240.2 m, and most of the region is located between 130–140 m, with small fluctuations. The altitude of the northwest area is between 180–232.5 m. In the middle part, there are undulating sand dunes and scattered swamps, with altitudes of 130 m to 179.2 m. In the east, it is low-lying and flat, with altitudes between 129–149 m. Figure 5 reflects the general trend of land development in Zhenlai County from northeast to southwest. Zhenlai County has sloping terrain that is higher in the northwest and lower in the southeast. To analyze the influence of elevation on village distribution, equal intervals for elevation were not adopted, while elevation was divided into intervals that are suitable for the microslope in the study region according to the statistical analysis of actual regional elevation. The region is divided into six intervals of elevation: 105.8–130 m, 130–135 m, 135–140 m, 140–145 m, 145–180 m, and 180–240.2 m, which can be produced by the reclassification of a digital elevation model (DEM) raster. By overlapping the reclassified elevation and settlement files, a sketch map of the movement of settlement gravity on various elevations can be obtained (Figure 7, Table 2) to analyze the influence of elevation on the spatial location of villages. Table 2 reveals that most of the settlements were located at 130–140 m during the 1875–1985 period, and approximately half were at 130–135 m. After 1913, it is very obvious that settlements were concentrated at 135–140 m. During 1932–1945 and 1946–1985, most of the settlements were located at elevations of 140–145 m. Gradually, the southeastern area with lower elevation became a key area for people to build villages from 1946 to 1985. Overall, land development expanded from flat areas to higher terrain, as seen in Figure 7. That is, the land development and transformation of Zhenlai County started in the eastern and middle parts, the western part followed, and the high-elevation area in the northwest corner was developed last. By further referring to the river system map and disaster information in the historical records, it can be learned that few people at earlier stages lived or reclaimed land around rivers, such as the Nenjiang and Tao'er Rivers, to avoid floods. However, with the reinforcement of embankments, some places near rivers started to be exploited. Therefore, the eastern and southeastern edge of Zhenlai County was developed later.

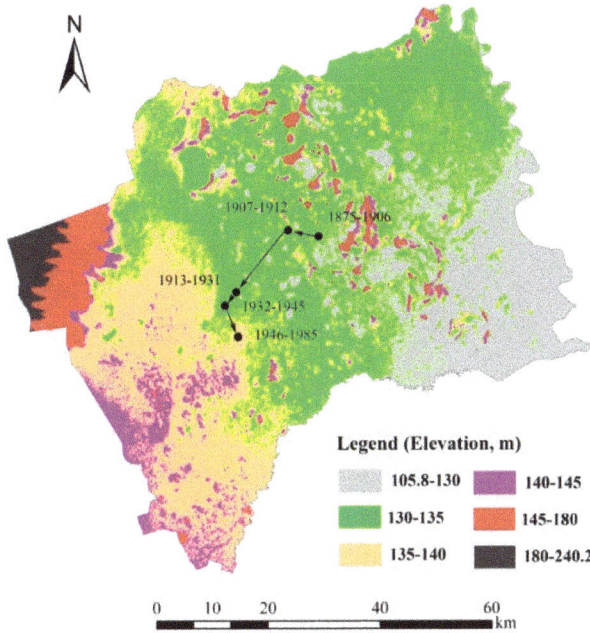

Figure 7. Sketch map of movement of settlement gravity during 1875–1985.

Table 2. Distribution of settlements at different elevations.

Elevation (m)	1875–1906		1907–1912		1913–1931		1932–1945		1946–1985	
	N	P (%)	N	P (%)	N	P (%)	N	P (%)	N	P (%)
105.8–130	2	1.01	1	0.63	0	0.00	0	0.00	5	10.64
130–135	96	48.48	79	49.69	19	22.89	10	29.41	4	8.51
135–140	90	45.45	66	41.51	58	69.88	17	50.00	25	53.19
140–145	10	5.05	13	8.18	6	7.23	6	17.65	9	19.15
145–180	0	0.00	0	0.00	0	0.00	1	2.94	4	8.51
180–240.2	0	0.00	0	0.00	0	0.00	0	0.00	0	0.00

N, number; P, percentage.

After juxtaposing the soil type and settlement maps, the influence of soil type on settlement locations can also be analyzed. Table 3 shows that the settlements were mainly distributed on chernozem in each period. In addition, people liked to build their settlements on meadow soil and alkaline soil during 1875–1906, 1907–1912, 1913–1931, and 1932–1945. As for 1946–1985, most of the villages were located on dark chestnut and meadow soil, followed by alluvial and meadow soil. Overall, settlements were mainly built on fertile land, such as chernozem and meadow soil; however, the central alkaline soil area was also an important location where people constructed villages due to the limited land resources. During 1946–1985, alluvial soil near the eastern Nenjiang River and south of the Taoerhe River and dark chestnut calcium soil of the northwest also gradually became key areas where villages were founded.

Table 3. Distribution of settlements in different soil types.

Soil Type	1875–1906		1907–1912		1913–1931		1932–1945		1946–1985	
	N	**P (%)**	**N**	**P (%)**	**N**	**P (%)**	**N**	**P (%)**	**N**	**P (%)**
Alluvial soil	7	3.54	10	6.29	0	0.00	3	8.82	8	17.02
Dark chestnut soil	2	1.01	0	0.00	0	0.00	0	0.00	11	23.40
Marsh soil	0	0.00	0	0.00	0	0.00	0	0.00	0	0.00
Peat soil	0	0.00	0	0.00	0	0.00	0	0.00	0	0.00
Alkaline soil	39	19.70	34	21.38	18	21.69	5	14.71	5	10.64
Meadow soil	57	28.79	50	31.45	14	16.87	7	20.59	10	21.28
Aeolian sandy soil	18	9.09	9	5.66	1	1.20	0	0.00	1	2.13
Chernozem	75	37.88	56	35.22	50	60.24	19	55.88	12	25.53
Total	198	100.00	159	100.00	83	100.00	34	100.00	47	100.00

N, number; P, percentage.

3.3. Settlement as a Proxy for Historical Arable Land Reconstruction

In general, the circular buffer area obtained on the basis of a farming radius and using the settlement as the center of the circle becomes the maximum spatial scope of agricultural activity within the settlement. However, this type of settlement–farming distribution pattern is under ideal conditions. In reality, the practical distribution of farmland is affected by terrain, landform, gradient, soil, moisture, and other factors, and is usually irregular. The gross area of farmland in reality is often larger than it is within the buffer area. The larger the proportion of farmland area distributed in the buffer of settlements in total farmland area, the closer the degree of agricultural development. Farming radius is one of the important factors affecting the distribution of rural residential settlements. In turn, the distribution of settlements may indirectly reflect the farming radius of the study area.

3.3.1. Calculation of Farming Radius Adopting Two Methods Based on Data from Topographic Maps

(1)　Farming Radius Using the Equalization-Based Calculation

When the buffer distance is 1092 m (Figure 8), the buffer area in the research region reaches 143,749 hectares, and there is only a difference of 4.24 hectares (0.003%) compared to the total area of farmland determined from topographic maps (Figure 9). To analyze the spatial difference, the buffer area map is juxtaposed with the topographic maps. The farmland within the buffer area is 72,301.12 hectares, accounting for 50.30% of the practical farmland area (Figure 10). The result reveals that since the buffer distance constantly increases, the buffer area may gradually cover the entire county, and the farmland area within the buffer area will be closer to the practical farmland area.

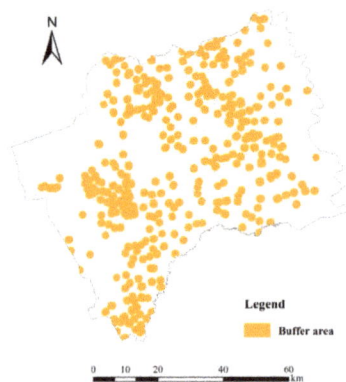

Figure 8. Buffer area map with a 1092-m cultivation radius.

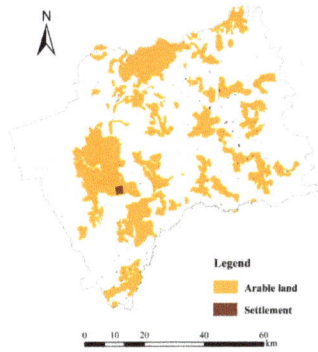

Figure 9. Distribution of arable land and settlements from topographic maps from the 1930s.

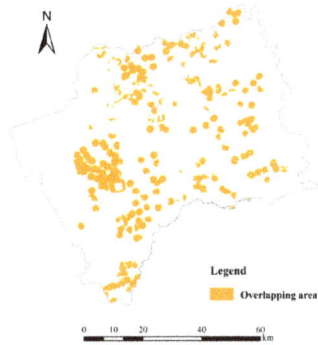

Figure 10. Overlapping arable land between buffer area and topographic map.

(2)   Farming Radius Based on Cultivation–Settlement Ratio

Considering that Zhenlai County has sloping terrain, tilting from northwest to southeast, the farming radius varies in different regions. The township administrative map of the county was combined with farmland area extracted from topographic maps from the 1930s to calculate the land transformation coefficients and cultivation–settlement ratios of all of the towns (Table 4).

Table 4. Farming radius (*R*) in the administrative districts of Zhenlai County.

Towns	Cultivated Area (ha) [1]	Total Area of Administrative Region (ha)	*K* (%) [2]	Settlement Area (ha) [1]	*G* [2]	*R* (m) [2]
Tantu	16,793	32,332	0.519	106	158.70	1014.98
Ganshigen	9670	31,958	0.303	92	105.59	1009.12
Daobao	6960	19,478	0.357	185	37.60	787.79
Zhenlai	19,344	39,496	0.490	862	22.45	1121.80
Hatuqi Mongol Ethnic Township	4203	15,329	0.274	34	122.03	698.86
Momoge Mongol Ethnic Township	10,485	48,408	0.217	143	73.33	1241.96
Jianping	18,853	76,884	0.245	129	146.48	1529.55
Dongping	8299	40,509	0.205	129	64.21	1136.10
Heiyupao	19,707	58,550	0.337	147	134.11	1365.86
Wukeshu	9906	48,547	0.204	178	55.71	1243.76
Datun	8355	45,107	0.185	257	32.54	1198.90
Yanjiang	7319	35,854	0.204	90	81.69	1068.88
Zhennan Sheep Breeding Farm	2818	32,332	0.519	33	86.34	1014.98

[1] Cultivated area and settlement area are calculated from the topographic maps from the 1930s. [2] *K* is the ratio of cultivated land according to the area of administrative region; *G* is the ratio of cultivated land and rural settlement; *R* is the farming radius.

According to the land-use map from the 1930s topographic maps, the buffer area is established based on the cultivation radius of rural residential areas in various towns, as shown in Table 4. After integrating the overlapping areas, a buffer area of 157,199 hectares is obtained (Figure 11). The area of cultivated land within the buffer area is 77,380.37 hectares, accounting for 53.83% of the actual cultivated land (Figure 12).

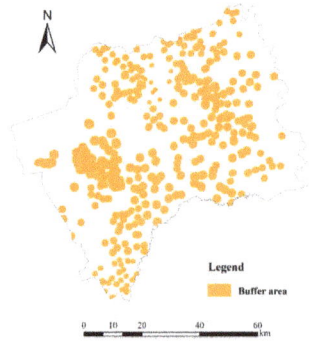

Figure 11. Cultivation and settlement ratio-based buffer area map.

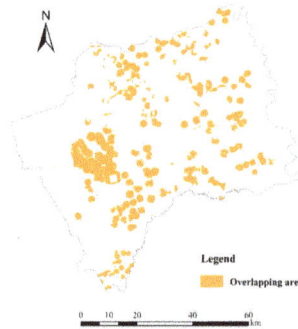

Figure 12. Overlapping areas of buffer region and topographic maps.

(3) Differences of Farming Radius from Different Methods

Through a comparative analysis of the results obtained from the two methods (Table 5), it could be learned that according to the equalization-based computing method, the farming radius is 1092 m, and the area of cultivated land within the buffer area accounts for 50.30% of the actual area of cultivated land, which is lower than the result (53.83%) obtained by the cultivation–settlement ratio computing method. Therefore, this research adopts the latter method, and uses the data extraction of historical settlements to reconstruct historical cultivated land based on the settlement distribution and farming radius.

Table 5. Comparison of buffer distance and farmland area based on equalization and cultivation–settlement ratio calculations.

Method	Actual Farmland Area (ha)	Buffer Radius (m)	Buffer Area (ha)	Farmland Area within Buffer Area (ha)	Ratio between Farmland Area in Buffer Area and Actual Farmland Area (%)
Equalization-based	143,744.76	1092	143,749	72,301.12	50.30%
Cultivation–settlement ratio-based		$R = \sqrt{\frac{GI}{K\pi}}$	157,199	77,380.37	53.83%

3.3.2. Historical Arable Land Reconstruction Based on Settlements and Farming Radius

Based on the above analysis, this study adopts the cultivation–settlement ratio-based method to reconstruct historical arable land based on spatial settlement distribution and farming radius. According to the distribution of settlements during 1875–1931 extracted from the *Zhenlai Gazetteer*, an analysis of the buffer areas of various villages (towns) is carried out by taking the buffer radii in Table 4 calculated using topographic data. After integrating the overlapped pattern spots, the spatial distribution of arable land within the study area in the 1930s can be obtained. The reconstruction results (Figure 13) indicate that Zhenlai County had 155,358 hectares of arable land in 1932.

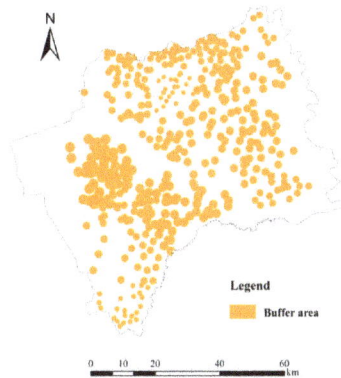

Figure 13. Arable land reconstruction in 1932 based on settlement distribution and farming radius.

By overlaying the arable land reconstruction map with the digitized arable land data from the topographic maps of the 1930s, we can see that the overlapping area amounts to 68,700.01 hectares, accounting for 47.79% of the arable land area in the topographic maps. The spatial distribution of overlapping sections is shown in Figure 14.

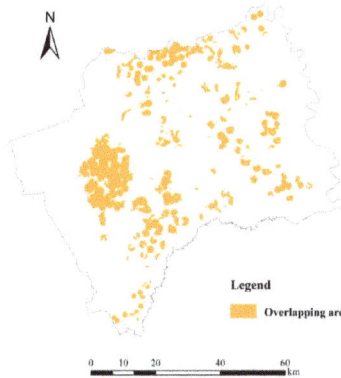

Figure 14. Overlap analysis map of buffer area and arable land area in the topographic maps.

4. Discussion

Historical changes in arable land on decadal to centennial time scales could reflect the effects of human activities on environmental change, which are significant drivers of many environmental issues. Historical land-use/cover reconstruction is the key and a source of difficulty in the research of historical land-use/cover change. Before the advent of remote sensing technology, historical records provided land managers with information that could be used to understand LUCC trajectories, and a more complete picture of historical land use/cover could be obtained by integrating multisource data [24]. Agricultural expansion can also be conceived as a spatial diffusion process driven by a set of decisions by agents to migrate into frontier areas and clear land for crop production [43]. In this context, this paper uses the in-depth research and analysis of the historical progress of land transformation from nonagricultural areas into agricultural areas from the perspective of settlement evolution by referring to the *Zhenlai Gazetteer* and other historical documents.

The above results show that before 1875, there was little land transformation from nonagricultural to agricultural, and the main land cover was grassland and wetland. After that time, mass migration accelerated the process of land transformation, which was most significant during 1907–1912, revealing that settlement establishment was often accompanied by land transformation. It is easy to deduce from the aforementioned development process that policy has played an important role in the land development and utilization of Zhenlai County [44]. Over the last century, northeastern China also has experienced a large-scale population influx from Chuang Guandong migration, and a major land transformation process influenced by the ban on the Qing government's prohibition policy being lifted [44,45]. This was mainly caused by the following. (1) North China had a large population and limited land at the time of the research, and frequent man-made and natural disasters forced local people to leave their homes and migrate to the northeast; (2) Northeast China is geographically vast, thinly populated, and has fertile soil. Regional development and construction require much labor, and the extensive railway network in northeastern areas provided convenient transportation for immigrants; (3) Relevant policies, such as lifting the ban on the Qing government's prohibition policy, the immigration encouragement policy carried out by the Central Government of the Republic of China and local governments in the northeast, fare waivers, etc. facilitated the large-scale immigration of Chuang Guandong [46]. The overall trend of transformation to agricultural land in this region is from northeast to southwest. Settlements were mainly built on fertile land, such as chernozem and meadow soil, and gradually turned to other soil types due to the limited land resources. These characteristics reveal the agricultural change order in the spatial distribution. People prefer to first develop agricultural land that is easily reclaimed at low cost.

In addition, this paper also provides a basis for understanding land-use changes in eco-fragile areas, and serves as an attempt to reconstruct historical land transformation patterns on the basis of settlement evolution history. It innovates and establishes a new methodological framework for the historical reconstruction of arable land not only in quantity, but also in spatial distribution. By applying the mutual relationship between the location of settlements and arable land, which can be reflected by the farming radius, historical arable land can be built, and analyzing historical changes in land cover/use is feasible. It is more common to obtain agricultural activity using population as a proxy, and most of the current studies focus on the relationship between cropland area and rate of population change [47]. This method enriches the current historical methods of arable land reconstruction by using settlement as a proxy based on other proxies that integrate multisource data. As global agricultural land is impacted by human activities, it is also workable when this method is applied to other regions as long as the distribution of settlement and farming radius are available. Sometimes, the farming radius might be difficult to calculate in some regions due to data availability; if so, its value could be replaceable by the data in similar regions. Results indicate that farming radii calculated using the cultivation–settlement ratio method tends to have higher accuracy than those calculated using the equalization-based method, because the former fully considers the different influences of the sizes of settlements on the farming radius. In this research, spatial arable land data could be obtained from the topographical maps, but to offer a new thought for researchers when they face the difficulties in collecting historical data, a proxy-based reconstruction method is innovated based on the relationships between arable land and settlements, which could be reflected by the farming radius. Taking the topographic maps as the references, the spatial distribution of the reconstructed arable land accounts for 47.79% (almost half) of that in the topographic maps, which seems good, but still has space to advance. Yet, it is possible that any precise measurement of simulation accuracy is unattainable due to time point inconsistencies concerning land category definitions [48]. This inaccuracy could be explained as follows. (1) The majority of the topographic maps that were used in this research were produced between 1932–1935, and are limited primarily in that they were created with a specific military purpose. As paddy fields were rare in this study area in the 1930s, arable land mainly refers to rainfed land. Rainfed lands and blank areas lacking a symbol were often difficult to decipher from grassland and wildland, making it challenging to extract and digitize grassland data. Inevitably, the area of arable land is larger on topographic maps than in the proxy-based model due to the difficulties in deducing mixed grassland in the process of digitization; (2) The farming radius was calculated in each town or township, as the population data could be easily collected according to the administrative division. The spatial distribution of arable land deduced from the farming radius depends on the spatial zones into which the study region is subdivided. To improve the accuracy of historical arable land reconstruction, it will be necessary to explore the research of the farming radius calculation method based on regional characteristics such as terrain, water supplies, and population density in the future.

Human agricultural activity has greatly influenced the patterns of local land use/cover and landscape. Land-use transitions could be associated with negative feedback that arises from a depletion of key resources or a decline in the provision of important ecosystem goods and services [43]. From the current literature [27], human-induced land changes, constituting 26.89% of Zhenlai County during 1954–2005, played an important role in environmental change. Besides, demographic change often results in major land transformation and the expansion of agricultural areas to meet the food requirements of local people, which is the main cause of pressure on historical grassland and wetland in the research area. In this research, 521 settlements were established during 1875–1985 in the study area, reflecting intensive human activities. It is also said that the increasing annual average temperature and decreasing annual rainfall in Zhenlai County over the past century resulted in high evaporation and frequent droughts. These brought serious damage to natural ecosystems such as grassland and wetland [48]. Eventually, national political decisions that led to the agricultural land transition were also influenced by a perception of ecological degradation. Yet, a transition is not a fixed pattern, nor is it deterministic [43]. Economic development has created enough nonfarm jobs to pull farmers off the land, thereby inducing the conversion of fields into grasslands or other lands. It should be noted that a lot of arable land has been abandoned with the acceleration of urbanization. This is because the labor force is driven from agriculture to other economic sectors, and from rural to urban areas for better living and earnings. So, long-term historical land change trajectories and characteristics should be applied to future land policy making to achieve sustainable land development. Facing increasing land degradation, it is urgent to adjust and optimize the land-use structure to promote the integrated development of production–living–ecology.

In addition, the constraints of using historical documents should be noticed, as they are limited by the history of writing records and maps. In this context, paleoenvironmental research can provide information for reconstructing past vegetation in the millennial time-scale or longer [14]. It has its unique possibilities for the study of time periods where historical records are lacking [49]. Reconstructions based on natural archives, such as pollen, plant macrofossils, animal deposits, and annual plant and animal growth cycles were widely used to reconstruct paleoclimate and paleoecology, while also gradually being used to analyze historical land-use/cover change [49–51]. Especially, the modeling of pollen–vegetation relationships has made such spatial vegetation/landscape reconstruction feasible [52,53]. The pollen assemblages have close relationships with human-induced vegetation related to various traditional human activities. This offers a basis for the interpretation of past human impact on vegetation and the quantitative reconstruction of human-induced landscapes from fossil pollen data.

5. Conclusions

In this research, the historical progress of land transformation for agricultural areas was analyzed from the perspective of settlement evolution by referring to the *Zhenlai Gazetteer* and other historical documents. In addition, the historical reconstruction of arable land was established using settlement as a proxy between inner relationships, which could be reflected by farming radius. The farming radius was calculated at the town/township level using the digitized land-cover/use data from topographic maps, while settlement information was mainly obtained from gazetteer records within the research region. Then, we produced and reconstructed the spatial arable land data based on the distribution of settlements and the farming radius in Zhenlai County, Northeast China. The results show the following. (1) Before 1875, there was little land transformation from nonagricultural areas into agricultural areas, and the main land cover was grassland and wetland until mass migration accelerated the process of land transformation, which was most significant during 1907–1912, revealing that settlement establishment was often accompanied by land transformation; (2) The overall trend of land transformation for agricultural land in this region is from northeast to southwest. Land development and transformation in Zhenlai County started in flat areas in the eastern and middle parts; the western part followed, and the high-elevation areas in the northwest corner were developed last; (3) Settlements were mainly built on fertile land, such as chernozem and meadow soil, and gradually turned to other soil types due to the limited land resources; (4) Taking the topographic maps as references, the spatial distribution of the reconstructed arable land accounts for 47.79% of that in those maps. As for the proxy-based method reconstruction, most of the current studies have focused on the relationship between cropland area and rate of population change, and it is still necessary to explore the relationships and reconstruct arable land based on other proxies that integrate multisource data. This research innovates and establishes a new methodological framework for the historical reconstruction of arable land in both quantity and spatial distribution based on the distribution of settlements and the farming radius, making a certain contribution to the current literature. Meanwhile, the limitations of this proxy-based reconstruction method should be noted, and it is necessary to explore the research of the farming radius calculation method based on regional characteristics. Moreover, paleoenvironmental research can provide historical land-cover/use information for longer time series. By applying historical land change trajectories and characteristics to land use policy making, a better way could be realized to achieve sustainable development.

Author Contributions: Y.Y. and S.Z. conceived and designed the experiments; Y.Y. analyzed the data and wrote the paper. All authors have read and approved the final manuscript.

Funding: This research was supported by the National Natural Science Foundation for Young Scientists of China (Grant No: 41601173) and the China Postdoctoral Science Foundation (Grant No: 2016M600954). We thank Yan Lv from the second topographic survey team of the National Administration of Surveying, Mapping and Geo information for her data collection.

Conflicts of Interest: The authors declare no conflict of interest.

References

1.   International Geosphere-Biosphere Programme (IGBP). *Global Change and Earth System: A Planet Under Pressure*; IGBP Science Series; IGBP: Stockholm, Swedish, 2001; p. 4.
2.   Ramankutty, N.; Foley, J.A. Estimating historical changes in global land cover: Croplands from 1700 to 1992. *Glob. Biogeochem. Cycles* **1999**, *13*, 997–1027. [CrossRef]
3.   Goldewijk, K.K.; Verburg, P.H. Uncertainties in global-scale reconstructions of historical land use: An illustration using the HYDE data set. *Landsc. Ecol.* **2013**, *28*, 861–877. [CrossRef]
4.   Fuchs, R.; Herold, M.; Verburg, P.H.; Clevers, J.G.P.W.; Eberle, J. Gross changes in reconstructions of historic land cover/use for Europe between 1900 and 2010. *Glob. Chang. Boil.* **2015**, *21*, 299–313. [CrossRef] [PubMed]
5.   Fuchs, R. *A Data–Driven Reconstruction of Historic Land Cover/Use Change of Europe for the Period 1900 to 2010*; Wageningen University: Wageningen, The Netherlands, 2015.

6. Petit, C.C.; Lambin, E.F. Impact of data integration technique on historical land-use/land-cover change: Comparing historical maps with remote sensing data in the Belgian Ardennes. *Landsc. Ecol.* **2002**, *17*, 117–132. [CrossRef]

7. Turner, B.L.; Skole, D.L.; Sanderson, S.; Fischer, G.; Fresco, L.; Leemans, R. *Land-Use and Land-Cover Change: Science/Research Plan*; Report 35; International Geosphere-Biosphere Programme: Stockholm, Swedish, 1995.

8. Leite, C.C.; Costa, M.H.; Soares-Filho, B.S.; de Barros Viana Hissa, L.C. Historical land use change and associated carbon emissions in Brazil from 1940 to 1995. *Glob. Biogeochem. Cycles* **2012**, *26*. [CrossRef]

9. Chang-Martínez, L.A.; Mas, J.F.; Valle, N.T.; Torres, P.S.U.; Folan, W.J. Modeling Historical Land Cover and Land Use: A Review from Contemporary Modeling. *ISPRS Int. J. Geo-Inf.* **2015**, *4*, 1791–1812. [CrossRef]

10. Goldewijk, K.K.; Beusen, A.; Janssen, P. Long-term dynamic modeling of global population and built-up area in a spatially explicit way: HYDE 3.1. *Holocene* **2010**, *20*, 565–573. [CrossRef]

11. Goldewijk, K.K.; Beusen, A.; Drecht, G.V.; De Vos, M. The HYDE 3.1 spatially explicit database of human–induced global land-use change over the past 12,000 years. *Glob. Ecol. Biogeogr.* **2011**, *20*, 73–86. [CrossRef]

12. Hurtt, G.C.; Chini, L.P.; Frolking, S.; Betts, R.A.; Feddema, J.; Fischer, G.; Fisk, J.P.; Hibbard, K.; Houghton, R.A.; Janetos, A.; et al. Harmonization of land–use scenarios for the period 1500–2100: 600 years of global gridded annual land–use transitions, wood harvest, and resulting secondary lands. *Clim. Chang.* **2011**, *109*, 117–161. [CrossRef]

13. Olofsson, J.; Hickler, T. Effects of human land-use on the global carbon cycle during the last 6000 years. *Veget. Hist. Archaeobot.* **2008**, *17*, 605–615. [CrossRef]

14. Yang, Y.Y.; Zhang, S.W.; Yang, J.C.; Bu, K.; Xing, X.X. A review of historical reconstruction methods of land use/land cover. *J. Geogr. Sci.* **2014**, *24*, 746–766. [CrossRef]

15. Boucher, Y.; Arseneault, D.; Sirois, L.; Blais, L. Logging pattern and landscape changes over the last century at the boreal and deciduous forest transition in Eastern Canada. *Landsc. Ecol.* **2009**, *24*, 171–184. [CrossRef]

16. Hamre, L.N.; Domaas, S.T.; Austad, I.; Rydgren, K. Land–cover and structural changes in a western Norwegian cultural landscape since 1865, based on an old cadastral map and a field survey. *Landsc. Ecol.* **2007**, *22*, 1563–1574. [CrossRef]

17. Hurtt, G.C.; Frolking, S.; Fearon, M.G.; Moore, B.; Shevliakova, E.; Malyshev, S.; Pacala, S.w.; Houghton, R.A. The underpinnings of land–use history: Three centuries of global gridded land-use transitions, wood–harvest activity, and resulting secondary lands. *Glob. Chang. Biol.* **2006**, *12*, 1208–1229. [CrossRef]

18. Kumar, S.; Merwade, V.; Rao, P.S.C.; Pijanowski, B.C. Characterizing Long–Term Land Use/Cover Change in the United States from 1850 to 2000 Using a Nonlinear Bi–analytical Model. *AMBIO* **2013**, *42*, 285–297. [CrossRef] [PubMed]

19. Pongratz, J.; Reick, C.; Raddatz, T.; Claussen, M. A reconstruction of global agricultural areas and land cover for the last millennium. *Glob. Biogeochem. Cycles* **2008**, *22*. [CrossRef]

20. Wulf, M.; Sommer, M.; Schmidt, R. Forest cover changes in the Prignitz region (NE Germany) between 1790 and 1960 in relation to soils and other driving forces. *Landsc. Ecol.* **2010**, *25*, 299–313. [CrossRef]

21. García-Ayllón, S. Retro-diagnosis methodology for land consumption analysis towards sustainable future scenarios: Application to a mediterranean coastal area. *J. Clean. Prod.* **2018**, *195*, 1408–1421. [CrossRef]

22. García-Ayllón, S. Predictive Diagnosis of Agricultural Periurban Areas Based on Territorial Indicators: Comparative Landscape Trends of the So-Called "Orchard of Europe". *Sustainability* **2018**, *10*, 1820. [CrossRef]

23. Wulf, M.; Rujner, H. A GIS–based method for the reconstruction of the late eighteenth century forest vegetation in the Prignitz region (NE Germany). *Landsc. Ecol.* **2011**, *26*, 153–168. [CrossRef]

24. Bürgi, M.; Hersperger, A.M.; Hall, M.; Southgate, E.W.; Schneeberger, N. Using the past to understand the present land use and land cover. In *A Changing World*; Kienast, F., Wildi, O., Ghosh, S., Eds.; Springer: Cham, The Netherlands, 2007; pp. 133–144.

25. He, F.N.; Li, S.C.; Zhang, X.Z. The reconstruction of cropland area and its spatial distribution pattern in the mid–northern Song dynasty. *Acta Geogr. Sin.* **2011**, *66*, 1531–1539. (In Chinese)

26. Zeng, Z.Z.; Fang, X.Q.; Ye, Q. The Process of Land Cultivation Based on Settlement Names in Jilin Province in the Past 300 Years. *Acta Geogr. Sin.* **2011**, *66*, 985–993. (In Chinese)

27. Yang, Y.Y.; Zhang, S.W.; Wang, D.Y.; Yang, J.C.; Xing, X.S. Spatiotemporal changes of farming-pastoral ecotone in Northern China, 1954–2005: A case study in Zhenlai County, Jilin Province. *Sustainability* **2014**, *7*, 1–22. [CrossRef]

28. de Sherbinin, A. *A Guide to Land-Use and Land-Use Cover Change (LUCC)*; A Collaborative Effort of SEDAC and the IGBP/IHDP LUCC Project; Columbia University: New York, NY, USA, 2002.

29. Xie, R.Q.; Liu, F.M. Study on changes in landscape pattern of land use based on fractal theory-a case study of Zhenlai Town of Zhenlai County. *Res. Soil Water Conserv.* **2013**, *20*, 217–222. (In Chinese)

30. Zhang, G.K.; Song, K.S.; Zhang, S.Q.; Liang, Y.H. The change of landscape pattern in Zhenlai Xian, Jilin Province in recent ten years. *Acta Ecol. Sin.* **2012**, *32*, 3958–3965. (In Chinese) [CrossRef]

31. Cui, H.S.; Zhang, B.; Liu, X.N. Forecast of land desertification in northern Jilin province of China. *J. Desert Res.* **2004**, *24*, 235–239. (In Chinese)

32. Geographical Names Committee Office of Zhenlai County. *Zhenlai Gazetteer, Zhenlai County People's Government*; Geographical Names Committee Office of Zhenlai County: Baicheng, China. 1985. (In Chinese)

33. He, F.; Li, M.; Li, S. Reconstruction of Lu-level cropland areas in the Northern Song Dynasty (AD976–1078). *J. Geogr. Sci.* **2017**, *27*, 606–618. [CrossRef]

34. Fuchs, R.; Verburg, P.H.; Clevers, J.G.P.W.; Herold, M. The potential of old maps and encyclopaedias for reconstructing historic European land cover/use change. *Appl. Geogr.* **2015**, *59*, 43–55. [CrossRef]

35. Liu, J.Y.; Liu, M.L.; Zhuang, D.F.; Zhang, Z.X.; Deng, X.Z. Study on spatial pattern of land-use change in China during 1995–2000. *Sci. China* **2003**, *46*, 374–384.

36. He, Y.B.; Chen, Y.Q.; Tang, H.J.; Yao, Y.M.; Yang, P.; Chen, Z.X. Exploring spatial change and gravity center movement for ecosystem services value using a spatially explicit ecosystem services value index and gravity model. *Environ. Monit. Assess.* **2011**, *175*, 563–571. [CrossRef] [PubMed]

37. Yang, Y.Y.; Liu, Y.S.; Li, Y.R.; Du, G.M. Quantifying spatio-temporal patterns of urban expansion in Beijing during 1985–2013 with rural-urban development transformation. *Land Use Policy* **2018**, *74*, 220–230. [CrossRef]

38. Briassoulis, H. *Analysis of Land Use Change: Theoretical and Modeling Approaches*; West Verginia University: Morgantown, WV, USA, 2000.

39. Goldewijk, K.K. Estimating global land use change over the past 300 years: The HYDE database. *Glob. Biogeochem. Cycles* **2001**, *15*, 417–433. [CrossRef]

40. Ye, Q.L.; Wang, C.; Jiang, F.X.; Zhao, S.H. A farming radius-based research of clustering scale of pure agricultural rural households in hilly areas—A case study of Bailin Village, Shapingba district, Chongqing. *J. Southwest Univ. Nat. Sci. Ed.* **2013**, *35*, 1–8. (In Chinese)

41. Qiao, W.F.; Wu, J.G.; Zhang, X.L.; Ji, Y.Z.; Li, H.B.; Wang, Y.H. Optimization of spatial distribution of rural settlements at county scale based on analysis of farming radius —Case study of Yongqiao district in Anhui province. *Resour. Environ. Yangtze Basin* **2013**, *22*, 1557–1563. (In Chinese)

42. Jin, Q.M. *Rural Settlement Geography*; Science Press: Beijing, China, 1988. (In Chinese)

43. Lambin, E.F.; Meyfroidt, P. Land use transitions: Socio-ecological feedback versus socio-economic change. *Land Use Policy* **2010**, *27*, 108–118. [CrossRef]

44. Lv, Y.; Zhang, S.W.; Yang, J.C. Application of toponymy to the historic LUCC researches in Northeast China: Taking Zhenlai county of Jilin province as an example. *J. Geo-Inf. Sci.* **2010**, *12*, 174–179. (In Chinese)

45. Lv, Y.; Zhang, S.W.; Yang, J.C. Reconstruct the spatial distribution of land use/land cover in the early reclaimed time of the western Jilin Province—Based on the GEOMOD model. *J. Anhui Agric. Sci.* **2015**, *43*, 304–308. (In Chinese)

46. Fan, L.J. *Immigration and Social Change in Modern Northeast China*; Zhejiang University: Hangzhou, China, 2005. (In Chinese)

47. Yang, Y.Y.; Zhang, S.W.; Yang, J.C.; Xing, X.S.; Wang, D.Y. Using a Cellular Automata–Markov Model to Reconstruct Spatial Land–Use Patterns in Zhenlai County, Northeast China. *Energies* **2015**, *8*, 3882–3902. [CrossRef]

48. Yang, Y.Y.; Zhang, S.W.; Liu, Y.S.; Xing, X.S.; de Sherbinin, A. Analyzing historical land use changes using a Historical Land Use Reconstruction Model: A case study in Zhenlai County, northeastern China. *Sci. Rep.* **2017**, *7*, 41275. [CrossRef] [PubMed]

49. Jie, D.M.; Wang, S.Z.; Guo, J.X.; Lv, J.F.; Li, J. Pollen combination and paleo-environment of Momoge Lake since 1500 years before. *J. Appl. Ecol.* **2004**, *15*, 575–578.

50. Jackson, S.T.; Overpeck, J.T.; Webb-III, T.; Keattch, S.E.; Anderson, K.H. Mapped plant-macrofossil and pollen records of late Quaternary vegetation change in eastern North America. *Quat. Sci. Rev.* **1997**, *16*, 1–70. [CrossRef]

51. Fritts, H.C.; Swetnam, T.W. Dendroecology: A tool for evaluating variations in past and present forest environments. *Adv. Ecol. Res.* **1989**, *19*, 111–189.
52. Dahlström, A. Grazing dynamics at different spatial and temporal scales: Examples from the Swedish historical record A.D.1620–1850. *Veg. Hist. Archaeobot.* **2008**, *17*, 563–572. [CrossRef]
53. Sugita, S. Pollen representation of vegetation in Quaternary sediments: Theory and method in patchy vegetation. *J. Ecol.* **1994**, *82*, 881–897. [CrossRef]

![sustainability logo] *sustainability*

MDPI

Article

Spatio-Temporal Variation of Land-Use Intensity from a Multi-Perspective—Taking the Middle and Lower Reaches of Shule River Basin in China as an Example

Libang Ma *, Wenjuan Cheng, Jie Bo, Xiaoyang Li and Yuan Gu

College of Geography and Environmental Science, Northwest Normal University, Lanzhou 730000, China; chengwenjuannw@163.com (W.C.); bojie17325125716@163.com (J.B.); leexy1999@163.com (X.L.); apprentice0717@126.com (Y.G.)
* Correspondence: malb0613@nwnu.edu.cn; Tel.:+86-931-797-1754

Received: 31 December 2017; Accepted: 9 March 2018; Published: 11 March 2018

Abstract: The long-term human activities could influence land use/cover change and sustainability. As the global climate changes, humans are using more land resources to develop economy and create material wealth, which causes a tremendous influence on the structure of natural resources, ecology, and environment. Interference from human activities has facilitated land utilization and land coverage change, resulting in changes in land-use intensity. Land-use intensity can indicate the degree of the interference of human activities on lands, and is an important indicator of the sustainability of land use. Taking the middle and lower reaches of Shule River Basin as study region, this paper used "land-use degree (LUD)" and "human activity intensity (HAI)" models for land-use intensity, and analyzed the spatio-temporal variation of land-use intensity in this region from a multi-perspective. The results were as follows: (1) From 1987 to 2015, the land use structure in the study region changed little. Natural land was always the main land type, followed by semi-natural land and then artificial land. (2) The LUD in the study region increased by 35.36 over the 29 years. It increased the most rapidly from 1996 to 2007, and after 2007, it still increased, but more slowly. A spatial distribution pattern of "low land-use degree in east and west regions and high land-use degree in middle region" changed to "high land-use degree in east and middle regions and low land-use degree in west region". (3) The human activity intensity of artificial lands (HAI-AL) in the study region decreased from 1987 to 1996, and then increased from 1996 to 2015. The human activity intensity of semi-artificial lands (HAL-SAL) in the study region increased over the 29 years, and more rapidly after 1996. (4) 1996–2007 was a transition period for the land-use intensity in the study region. This was related to the implementation of the socio-economy, policies such as "Integrated Development of Agricultural Irrigation and Immigrant Settlement in Shule River Basin (1996–2006)", and technologies.

Keywords: land-use intensity; spatio-temporal pattern; land-use degree; human activity intensity; sustainability; middle and lower reaches of Shule River Basin

1. Introduction

The unprecedented growth in the human population in the last centuries translates to escalated resource consumption, as manifested in relatively high rates of agriculture and food production, industrial development, energy production, and urbanization. These human enterprises lead to local land-use and land-cover changes [1]. Land use is one of the main ways through which human activities act on natural environment [2]. It directly reflects the relationship between human activities and environmental changes [1–3]. However, palaeoecological and palaeoenvironmental records show that land use changes are the main driving force of long-term changes in biodiversity, and in the stability and function of ecosystem services [4–7]. They include land-cover change and land-use intensity change [8]. Current research on land-use changes mainly focuses on land-cover change, such as its pattern and process, driving mechanism, spatio-temporal prediction, dynamic simulation, and ecological and environmental effects, and many valuable results have been obtained [7–10]. As a comparison, land-use intensity and its spatial variation have not yet become a mainstream topic for research during the past several decades [8]. Land-use intensity can indicate the degree of the interference of human activities on lands, and its variation has important influence on biodiversity and ecosystem service function [9–14].

During the past decades, developed countries have experienced unprecedented industrialization and urbanization. Populations in these countries increase rapidly, and urban land area continues to expand. In order to meet the growing needs of humans for products and services, ways of land investment become more diversified, leading to increased spatial difference in land-use intensity [15–17]. An agreement on the importance of land-use intensity and its changes has been reached. From the perspective of interaction between humans and natural environment, some researchers selected land cover, social economy, and ecological environment as indexes, and built an index system for land-use intensity [18–20], on which basis they studied the land-use intensity and its spatial variation on different scales [21–23], regions [15,24], and objects [25–27]. Since 2000, China has experienced rapid industrial and urban transformation, and its economic development has been greatly promoted. However, problems also occur: the shortage of cultivated land, along with the expansion of urban land area, as well as unbalanced regional development and food safety issues caused by unreasonable utilization of lands [28–30]. At the same time, under limited land resources, population and consumption of food have been continuously increased, and led to an increase in demands for products and services. Thus, many methods have been proposed to improve the output of limited land resources and its efficiency. For example, by increasing investment in production factors, such as fertilizer, farm chemicals, and labor force, and increasing the density and volume of buildings, the utilization of land resources can be intensified and exhibit certain spatial variation. In fact, reasonably improving current land-use intensity is the main way to guarantee balanced regional development and national food safety [6]. Research on land-use intensity mainly focuses on grading evaluation, as well as intensive use of cultivated lands, construction lands, and other lands where human activities are concentrated [31–33]. However, most research only explores comprehensive, large-scale land use from a single perspective. Although they can reflect land-use intensity on a macro scale, there is a lack of analysis of internal spatial heterogeneity and the use intensity of single type of land. Moreover, quantitative descriptions of land use intensity and its spatial pattern in most regions still lack a clear understanding [11,34–36]. Thus, developing an evaluation system for regional land-use intensity and analyzing its spatio-temporal variation, as well as the use intensity of different types of land, are very necessary.

The middle and lower reaches of Shule River are located at the westernmost part of Hexi Corridor, where population and social economic activities are the most concentrated. We did not find relevant research on the land (or land-use) transformations over the millennia in the Shule River Basin; indeed, we know that there were no large-scale human activities before the last 30 years in the area, and therefore, we assume that the observed changes in the ecological environment may be related to both natural factors and human activities [37–39]. The development of artificial oasis in Han and Tang dynasties led to the lake drying up, and the river receded of lower reaches, but the river diversion caused by the war and man-made water conservancy facilities made the ancient artificial oasis become desertified and barren [37]. Humans did not learn from history. After 1949, around the development of oil and agricultural land, the population was continuously moved into Guazhou County and Yumen City. The influence of human activities after the 1980s gradually increased. In 1983 and 1996, "Agricultural Construction and Immigrant Settlement Project in Two West Regions" and "Integrated Development of Agricultural Irrigation and Immigrant Settlement in Shule River Basin" were implemented in Shule River Basin, respectively. After 1997, Chinese government implemented a series of policies emphasizing regional development and environmental protection, such as "Reconstruction of Hexi Corridor", "Western Development", "Three-North Shelterbelt Project",

"Grain for Green Project", etc., which directly or indirectly influenced land use in Shule River Basin. Recently, a lot of research on Shule River Basin has been carried out from perspectives of land use/cover changes, quality of ecosystem, ecological and environmental effects and so on [37–40]. Nonetheless, these studies do not consider spatio-temporal variation of land-use intensity and variation caused by immigration, national policies, technologies, and other human activities.

Taking Guazhou County and Yumen City at the middle and lower reaches of Shule River Basin as examples, the main objective of this article is to use "land-use degree (LUD)" and "human activity intensity (HAI)" models for land-use intensity on the basis of land type classification indexes, then analyze the spatio-temporal variation of land-use intensity from a multi-perspective, reveal the impact of human activities, especially socio-economic, national policies, and technologies on the change of land-use intensity, and clarify how the human activities influence land use/cover change. Generally, it is suggested to analyze the long-term variation of land-use intensity in the middle and lower reaches of Shule River Basin considering its long history. However, using remote sensing imagery, we mainly focused on the period (1987–2015), when human activities had the most significant influence on the variation of land-use intensity. The results could also reflect the long-term changes of land and environment. On the one hand, this study provides a new perspective for the study on the change of land-use intensity in arid and semi-arid areas; on the other hand, it has effectively revealed the spatio-temporal heterogeneity of land use intensity in arid and semi-arid areas, it helps to deeply understand the interactive pattern and eco-environmental effects of human beings and natural ecosystems and provides a scientific basis for the formulation of future sustainable land use policy in arid and semi-arid areas.

2. Overview of the Study Region

2.1. Natural Geography

Shule River is located between 92°11′ E and 98°30′ E and between 38°0′ N and 42°48′ N, at the westernmost of Hexi Corridor, a transition zone between Qinghai–Tibet Plateau and Alashan Plateau of Inner Mongolia. It is one of the three inland rivers in Hexi Corridor, and occupies an area of 100,000 km^2. Shule River Basin has a temperate arid climate, with sufficient and strong solar radiation. Since this inland region is located far away from sea areas, it has only a little rainfall, with large temperature difference and strong evaporation. The annual average temperature is 7–9 °C. The annual average rainfall is less than 60 mm, but the evaporation amount reaches 1500–3000 mm. The rainfall is often concentrated in June–September, accounting for 80% of the rainfall in a whole year. The annual temperature difference ranges from 31.5 to 34.1 °C. The highest temperature exceeds 40.0 °C, and the lowest temperature is lower than −30.0 °C. Changma Valley and Shuangta Reservoir divide Shule River into upstream portion, midstream portion, and downstream portion. Specifically, the upstream portion is between the source and Changma Valley, the midstream portion is between Changma Valley and Shuangta Reservoir, and the downstream portion is portion at the downstream of Shuangta Reservoir. In the upstream region, Shule River has abundant water and flows rapidly through mountains that are steep. There are typical polar continental glacier and a large permafrost region. Thus, the upstream region is as the water source area. In the midstream and downstream regions of Shule River, the terrain is very flat. Oasis and desert coexists. Since Han Dynasty, there has been agricultural development and water conservancy construction in the middle and lower reaches of Shule River. Irrigated agriculture is well developed in Yumen City, Guazhou County, and Dunhuang City. The branches of Shule River include Yulin River, Shiyou River, etc. Four reservoirs, including Changma reservoir, Shuangta reservoir, Chijin reservoir, and Yulin reservoir, are built around Shule River, and they provide water for agricultural irrigation in three irrigation areas, including Changma irrigation district, Shuangta irrigation district, and Huahai irrigation district. In this paper, the region at the south of Yumen City and Guazhou County (94.81–98.25° E; 39.63–41.47° N) was studied. The region is adjacent to Jinta County at its east and to Dunhuang City at its west, with an area of 27,160.2 km^2 (Figure 1).

Sustainability **2018**, *10*, 771

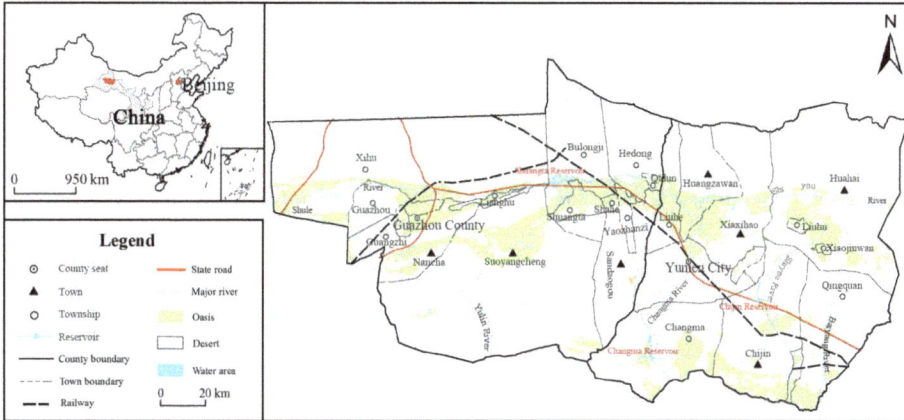

Figure 1. Survey map of the study region.

2.2. Social Economy

In 2015, the total population in Yumen City and Guazhou County was 289,800. Among them, agricultural population was 197,600, accounting for 73.76% of the total population. The GDP of this region was 17.747 billion Yuan. Gross agricultural output was 19.47 billion Yuan, accounting for 10.97% of the GDP. The ratio of primary, secondary, and tertiary sectors was 11.54:49.61:38.85. Per capita GPD was 66,254 Yuan (Per capita GDP of Gansu Province was 26,165 Yuan). Rural per capita income was 13,351 Yuan (Rural per capita income of Gansu Province was 6936 Yuan). Clearly, this region develops relatively well in Gansu Province. Guazhou County and Yumen City have jurisdiction over 15 towns in 1987, 16 towns in 1996, 21 towns in 2007, and 23 towns in 2015.

3. Data Sources and Research Methods

3.1. Data Sources

The data used in this paper came from four sources: (1) Remote sensing data. Landsat5 TM remote sensing images for 1987, 1996, 2007, and 2015 were obtained from United States Geological Survey (USGS) website and Institute of Remote Sensing and Digital Earth (RADI), Chinese Academy of Sciences. The path/row of the images are 135/32, 136/31, 137/31, and 137/32, respectively. A total of 20 images were obtained for the four years. The images were taken in July or August when the vegetation coverage rate is the highest. The average cloud cover of the images was no higher than 10%, so the quality of the data can meet the requirements. (2) Basic maps. Topographic map for Guazhou County and Yumen City (1:250,000), and vector administrative boundary (1:250 000) were obtained from Gansu Province Surveying and Mapping Bureau. Land use survey data (1: 1,000,000) in 1998 and 2008 were obtained from Gansu Province Land and Resources Department. Land use change data (vector format) in 2015 were also obtained from Gansu Province Land and Resources Department. (3) Social economy statistics. Statistics about population, social economy, labor force, agricultural machinery, fertilizer, and agricultural film from 1987 to 2015 were obtained from Yumen Statistics Bureau and Guazhou Statistics Bureau. The basic data of Immigrant Development in Shule River Basin Project were obtained from Gansu Province Shule River Construction Administration Bureau (Project Completion Report of the World Bank Loan Project for Hexi Corridor (Shule River) in Gansu Province of China). (4) Field survey data. From 2007 and 2015, we carried out field survey for seven times in the study region and interviewed the staff in local related departments, such as agriculture, forestry, water conservancy, and nature reserves, etc.

3.2. Data Processing

Geometric correction and visual interpretation of the remote sensing images for 1987, 1996, 2007, and 2015 were performed. The classification results were checked according to the field survey data and land use survey data. Interpretation precisions (Kappa coefficient) were 0.752, 0.761, 0.861, and 0.865, respectively, all higher than the minimum allowable precision (0.7). ArcGIS10.2 (dissolve tool) was used to perform spatial overlap and field fusion of interpreted remote sensing data and vector administrative boundary of Guazhou County and Yumen City. Then, land use type data for Guazhou County and Yumen City were obtained. According to the requirements of this study, the interference of human activities on the lands, and the degree of difficulty of restoring natural lands, the lands were classified into four types: (1) Artificial lands. They are characterized by water impervious surface and can hardly restore a natural state. Typical example is urban construction land. (2) Semi-artificial lands. The topsoil is frequently disturbed, and they can be restored to natural state. Typical example is cultivated land. (3) Semi-natural lands. The topsoil is sometimes disturbed, and the vegetation is frequently disturbed. Typical examples include woodland, grassland, wetlands, etc. (4) Natural lands. The topsoil and vegetation are seldom disturbed. Typical examples include Gobi Desert and sandy land.

3.3. Research Methods

Land-use intensity indicates the degree of the interference of human activities on lands. The spatio-temporal variation of land-use intensity in the middle and lower reaches of Shule River Basin was analyzed from the perspective of "land-use degree (LUD)" and "human activity intensity (HAI)". LUD mainly reflects the breadth and depth of land use. It not only reflects the natural attributes of land, but also reflects the combined effect of human factors and natural environment factors. HAI reflects the impact of human economic and social activities on a certain geographical natural environment. First, land use process was analyzed. The land use structure and it changes were used to characterize LUD and its variation. Second, human activities were considered. The land-use intensity under the influence of human activities was analyzed. HAI and its changes were used to characterize the use intensity of single type of land and its variation.

3.3.1. Land-Use Degree (LUD)

According to the natural balance of land under the influence of social factors, LUD was classified into three grades, and a value was assigned for each grade (Table 1). On this basis, LUD was calculated and used to indicate land-use intensity. The detailed method was explained as follows [41].

$$L_a = 100 \times \sum_{i=1}^{n} A_i \times C_i \, L_a \in (100, 400) \tag{1}$$

where L_a is the comprehensive LUD index of the study region, A_i is the grading index of land type i in the study region, C_i is the percentage of the area of land type i in the study region, and n is the number of land types. L_a values can be classified into five groups: low land-use degree ($L_a < 125$), relatively low land-use degree ($125 < L_a < 150$), medium land-use degree ($150 < L_a < 175$), relatively high land-use degree ($175 < L_a < 200$), and high land-use degree ($L_a > 200$).

Table 1. The classification values of land use degree.

Land Use Type	Self-Use Land	Reclaimed Land	Non-Renewable Land
Land type	Semi-natural land (woodland, grassland, wetland)	Semi-artificial land (cultivated land)	Artificial land (urban construction land)
Grading index	1	2	3

3.3.2. Human Activity Intensity (HAI)

In the middle and lower reaches of Shule River Basin, human activities are mainly concentrated in artificial lands and semi-artificial lands. Thus, human activity intensity of artificial land and semi-artificial land were mainly analyzed.

(1) Human activity intensity of artificial lands (HAI-AL)

HAL-AL includes breadth, depth, and frequency. In order to ensure the comparability of the three indicators, they were standardized and turned into dimensionless values. A combination of arithmetic mean method and geometric mean method were used to calculate HAL-AL.

$$B_i = \frac{1}{2}\left(\frac{B_g + B_s + B_p}{3} + \sqrt[3]{B_g B_s B_p}\right) \tag{2}$$

where B_i is HAL-AL, B_g is the breadth of artificial land use, B_s is the depth of artificial land use, and B_p is the frequency of artificial land use. HAL-AL can be classified into five groups: low land-use intensity ($B_i < 0.075$), relatively low land-use intensity ($0.075 < B_i < 0.15$), medium land-use intensity ($0.15 < B_i < 0.225$), relatively high land-use intensity ($0.225 < B_i < 0.3$), and high land-use intensity ($B_i > 0.3$).

$$B_g = S_B/S * 100\% \tag{3}$$

$$B_s = S_B/G_{23} \tag{4}$$

$$B_p = P/S_B \tag{5}$$

where S_B is the total construction area (km^2), S is the total land area (km^2), G_{23} is the added value of secondary and tertiary industries (10,000 Yuan), and P is the population.

(2) Human activity intensity of semi-artificial lands (HAI-SAL)

HAL-SAL was calculated according to investment per unit area of cultivated land. Investment includes labor force, agricultural machinery, fertilizer, and agricultural film. These indicators have different dimensions, which makes it difficult for comparison. Thus, the concept of "emergy" was introduced which is proposed by Odum [42]. Emergy is defined as available energy consumed by both direct and indirect manufacturing services and products, and it is usually quantified as the equivalent of solar energy and expressed as solar joules (sej) [42].

$$P_i = (L_E + Q_E + F_E + B_E)/S_P \tag{6}$$

where P_i is cultivated land-use intensity (sej/km^2), S_p is total cultivated area (km^2), L_E is the emergy of labor force (sej), Q_E is the emergy of agricultural machinery (sej), F_E is the emergy of fertilizer (sej), and B_E is the emergy of agricultural film (sej). The emergy values were obtained according to Odum [42] and Lan [43]. By using natural breakpoint method, HAL-SAL can be classified into five groups: low land-use intensity ($P_i < 50$), relatively low land-use intensity ($50 < P_i < 100$), medium land-use intensity ($100 < P_i < 200$), relatively high land-use intensity ($200 < P_i < 300$), and high land-use intensity ($P_i > 300$).

(a) Emergy of labor force:

$$L_E = T_l \cdot C_l \cdot N_l \tag{7}$$

where T_l is the conversion rate of the emergy of labor force (taken as 3.8×10^5 sej/J in this paper), C_l is energy transformation coefficient for labor force (taken as 1.26×10^7 J/person), and N_l is the total amount of labor force (persons).

(b) Emergy of agricultural machinery

$$Q_E = T_q \cdot C_q \cdot N_q \cdot Q \cdot 0.1 \tag{8}$$

where T_q is the conversion rate of the emergy of agricultural machinery (7.5×10^3 sej/J), C_q is energy transformation coefficient for agricultural machinery (taken as 2.1×10^8 J/kg), N_q is the total amount of input power of agricultural machinery (KW), Q is the mass-to-energy conversion coefficient (taken as 04.72 kg/KW), and 0.1 is depreciation coefficient.

(c) Emergy of fertilizer

$$F_E = R_N \cdot T_N + R_P \cdot T_P + R_K \cdot T_K + R_F \cdot T_F \tag{9}$$

where R_N, R_P, R_K, and R_F are the total use amounts of nitrogen fertilizer, phosphate fertilizer, potash fertilizer, and compound fertilizer (t), respectively; T_N, T_P, T_K, and T_F are the emergy conversion rates of nitrogen fertilizer, phosphate fertilizer, potash fertilizer, and compound fertilizer, respectively (taken as 3.8×10^{15} sej/t, 3.9×10^{15} sej/t, 1.1×10^{15} sej/t and 2.8×10^{15} sej/t, respectively).

(d) Emergy of agricultural film

$$B_E = T_B \cdot N_B \tag{10}$$

where T_B is the emergy conversion rate of agricultural film (taken as 3.8×10^{14} sej/t) and N_B is the total use amount of agricultural film (t).

4. Results

4.1. The Overall Characteristics of Land Use in the Middle and Lower Reaches of Shule River Basin

From 1987 to 2015, land use structure in the middle and lower reaches of Shule River Basin changed little (Figure 2 and Table 2). Natural land was always the main land type, whose area accounted for more than 80% of the total land area, followed by semi-natural land, and then artificial land (area less than 1%). From 1987 to 2015, the area of oasis consisting of semi-natural land, semi-artificial land, and artificial land increased by 549.16 km², with an average annual increase of 18.94 km². From 1996 to 2007, oasis area increased the most, by 398.75 km², accounting for 72.61% of the total increase in oasis area, with an average annual increase of 36.25 km². After 2007, oasis area still increased, but more slowly, and the average annual increase decreased by 36.88% compared with that from 1996 to 2007. From 1987 to 2015, the artificial and semi-artificial land areas increased with increase in oasis area, whereas the semi-natural land area decreased. The semi-artificial land area increased more rapidly than artificial land area, by 977.96 km² over the 29 years with an average annual increase of 33.72 km². Especially from 1996 to 2007, the semi-artificial land area increased by 593.61 km², accounting for 60.70% of the total increase, with an average annual increase of 53.96 km².

After 2007, the artificial land area increased rapidly, by 46.40 km² from 2007 to 2015, accounting for 51.69% of the total increase, with an average annual increase of 5.80 km².

Table 2. Area and proportion of land type in the middle and lower reaches of the Shule River Basin (1987–2015).

Land Types		1987	1996	2007	2015
Natural land	Area (km²)	22768.38	22801.22	22402.24	22219.2
	Proportion (%)	83.83	83.95	82.48	81.81
Semi-natural land	Area (km²)	3429.69	3287.94	3060.24	2911.12
	Proportion (%)	12.63	12.11	11.27	10.72
Semi-artificial land	Area (km²)	902.5	1001.11	1594.71	1880.47
	Proportion (%)	3.32	3.69	5.87	6.92
Artificial land	Area (km²)	59.67	70.18	103.03	149.43
	Proportion (%)	0.22	0.26	0.38	0.55

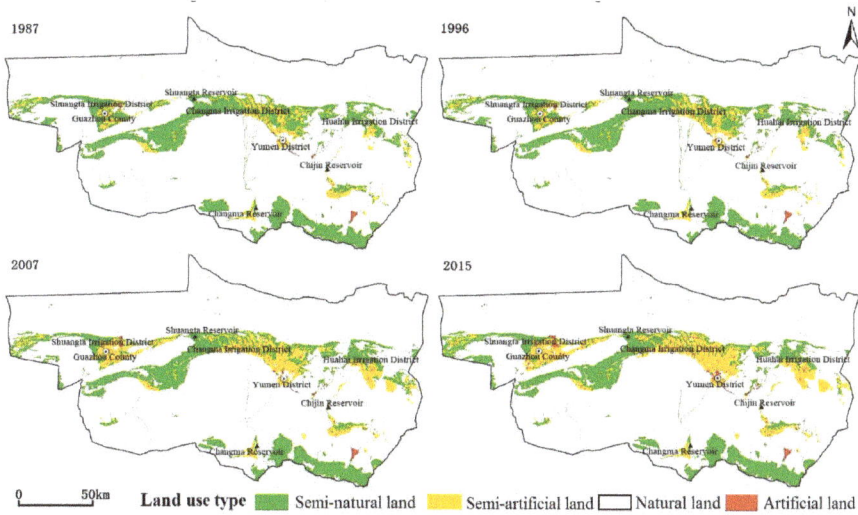

Figure 2. Spatial distribution of different land types in the middle and lower reaches of the Shule River Basin (1987–2015).

4.2. Spatio-Temporal Variation of LUD in the Middle and Lower Reaches of Shule River Basin

4.2.1. Temporal Variation of LUD

From 1987 to 2015, LUD of the middle and lower reaches of Shule River Basin increased by 35.36, with an average annual increase of 1.22 (Figure 3). There was a transition of LUD from relatively low level to medium level. Especially, LUD increased the most rapidly from 1996 to 2007, by 24.17, accounting for 68.35% of the total increase. After 2007, LUD still increased, but more slowly.

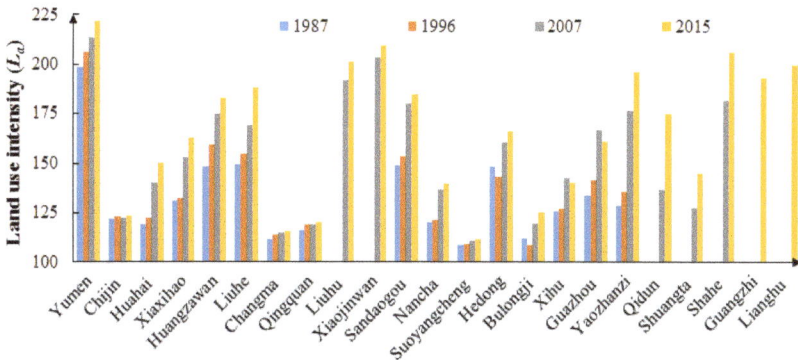

Figure 3. Temporal variation of LUD in the middle and lower reaches of the Shule River (1987–2015).

From 1987 to 2015, the LUD of the towns in the study region also increased. In the four years, the LUD of Yumen Town was always the highest, with an average intensity of 209.73, indicating high land-use intensity. The LUD of Suoyang Town was always the lowest, and was 111.64 in 2015, indicating low land-use intensity. Over the 29 years, the LUD of Yaozhanzi Town increased the most, by 67.42, with an average annual increase of 2.32. Following Yaozhanzi Town, Liuhe Town, Sandaogou Town, Huangzhawan Town, and Huahai Town also experienced significant increases in LUD, which were 38.33, 35.91, 34.40, and 31.05, respectively. The increase in LUD of Suoyang Town was the smallest (2.88). The LUD of 86.36% of the towns increased more rapidly from 1996 to 2007 than in other periods. From 1996 to 2007, Yaozhanzi Town experienced the greatest increase in LUD, which was 40.82, accounting for 60.55% of the total increase (newly constructed towns were not included). In the same period, the increase in the LUD of Hedong Town accounted for 91.5% of its total increase.

4.2.2. Spatial Variation of LUD

In order to study the spatial distribution of LUD on town scale, the study region was divided into three irrigation areas: Huahai irrigation district, Changma irrigation district, and Shuangta irrigation district. On this basis, the difference in LUD among three irrigation areas can be observed (Figure 4). In 1987, the development of land was slow. Except Yumen Town, with relatively high land-use degree, other towns had low or relatively low land-use degree. Among the three irrigation areas, Changma irrigation district in the middle of the study region had relatively high land-use degree, whereas the other two irrigation areas had relatively low levels of land development. In 1996, Huahai and Shuangta irrigation districts experienced obvious changes in LUD. Yumen Town in Changma irrigation district experienced a transition from relatively high land-use degree to high land-use degree. Sandaogou Town, Liuhe Town, and Huangzhawan Town experienced a transition from relatively low land-use degree to medium land-use degree. For the study region, there was a distribution pattern of "low land-use degree in east and west regions, and high land-use degree in the middle region". From 1996 to 2007, the LUD increased significantly and the spatial distribution of LUD changed the most evidently than in other periods. In 2007, the LUD of Changma irrigation district was the highest, with concentrated regions of medium or high land-use degree. Huahai irrigation district experienced a transition from low land-use degree to relatively low land-use degree, and the LUD of some towns even reached relatively high or high levels. In Shuangta irrigation district, all the towns except Suoyang Town, had low or medium land-use degree. From 2007 to 2015, the LUD of the study region further increased. Changma irrigation district and the region of Huahai irrigation district at the north of Chijin reservoir experienced significant changes in LUD. The LUD of 95% of the towns reached medium or high levels, showing a spatial distribution pattern of "high land-use degree in east and middle regions and low land-use degree in west region".

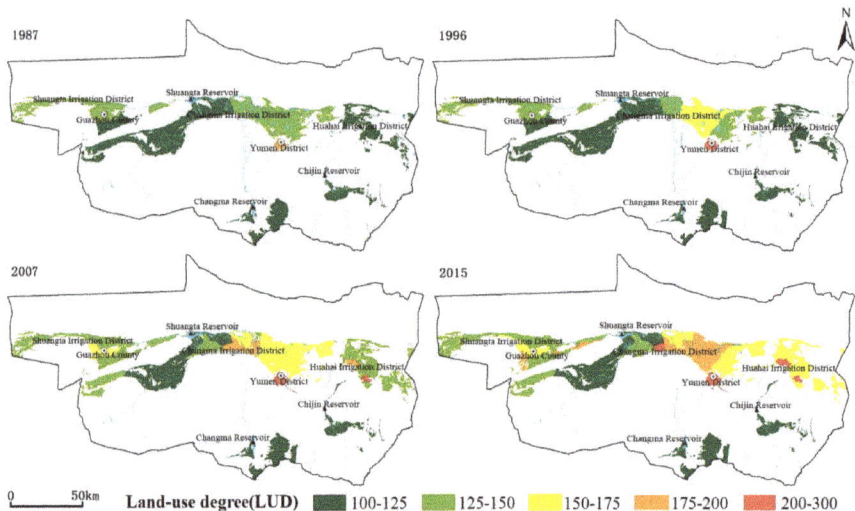

Figure 4. Spatial distribution of LUD in the middle and lower reaches of the Shule River Basin (1987–2015).

4.3. Spatio-Temporal Variation of HAL-AL and HAL-SAL in the Middle and Lower Reaches of Shule River Basin

In the middle and lower reaches of Shule River Basin, human activities are mainly concentrated in artificial and semi-artificial lands. This paper also investigated the HAL-AL and HAL-SAL under the influence of human activities.

4.3.1. Spatio-Temporal Variation of HAL-AL

(1)   Temporal variation of HAL-AL

HAL-AL decreased from 1987 to 1996, and then increased from 1996 to 2015 (Table 3 and Figure 5). In 1987, HAL-AL was low. There were no towns with high HAL-AL. Eleven towns had low or medium HAL-AL, accounting for 73.33% of all towns. From 1987 to 1996, HAL-AL of all the towns decreased. Fourteen towns had low or medium HAL-AL, accounting for 87.5% of all towns. The newly constructed Yaozhanzi Town had the highest HAL-AL, which was 0.376. From 1996 to 2007, five new towns were constructed, and the HAL-AL of all towns increased. In 2007, towns with low or medium HAL-AL accounted for only 38.1%, a decrease of 49.4% compared with that in 1996. The number of towns with relatively high or high HAL-AL reached 13, accounting for 61.9% of all towns. From 2007 to 2015, HAL-AL of all the towns still increased, but very slowly. In 2015, there were still eight towns with low or medium HAL-AL. However, two new towns were constructed. Thus, the proportion of towns with low or medium HAL-AL decreased to 34.78%. The number of towns with relatively high or high HAL-AL reached 15, accounting for 65.22% of all towns.

Table 3. Temporal variation of HAL-AL and HAL-SAL in the middle and lower reaches of the Shule River Basin (1987–2015).

Town	HAI-AL (B_i)				HAI-SAL (P_i- $\times$ 10^{15})			
	1987	1996	2007	2015	1987	1996	2007	2015
Yumen Town	0.250	0.250	0.250	0.250	92.53	123.23	172.61	151.80
Chijin Town	0.179	0.123	0.214	0.222	50.50	73.42	135.42	221.07
Huahai Town	0.207	0.096	0.185	0.212	66.69	89.43	83.74	165.16
Liuhe Town	0.164	0.134	0.282	0.232	57.56	96.86	93.23	80.53
Xiaxihao Town	0.133	0.089	0.133	0.129	58.01	80.60	52.65	53.02
Huangzhawan Town	0.165	0.122	0.237	0.237	62.06	108.20	90.56	79.72
Changma Town	0.250	0.121	0.250	0.227	62.42	113.11	91.36	133.40
Qingquan Town	0.120	0.071	0.122	0.119	48.96	90.64	97.43	95.84
Liuhu Town			0.461	0.472			237.81	631.04
Xiaojinwan Town			0.558	0.562			333.90	348.14
Sandaogou Town	0.253	0.175	0.314	0.380	98.45	228.04	296.85	343.84
Nancha Town	0.131	0.084	0.112	0.127	101.45	170.69	134.58	188.53
Suoyang Town	0.066	0.057	0.104	0.147	51.45	88.94	261.74	148.41
Guazhou Town	0.153	0.119	0.255	0.286	110.70	145.61	110.83	120.43
Xihu Town	0.074	0.056	0.062	0.068	55.89	217.23	112.79	122.41
Yaozhanzi Town		0.376	0.394	0.412		85.53	282.01	204.81
Hedong Town	0.236	0.148	0.283	0.303	83.95	161.25	153.99	170.25
Bulongji Town	0.129	0.088	0.149	0.141	68.78	139.41	108.4	107.21
Shuangta Town			0.377	0.300				235.54
Shahe Town			0.512	0.493				99.07
Guangzhi Town				0.416				535.84
Qidun Town			0.542	0.325				191.45
Lianghu Town				0.374				437.26

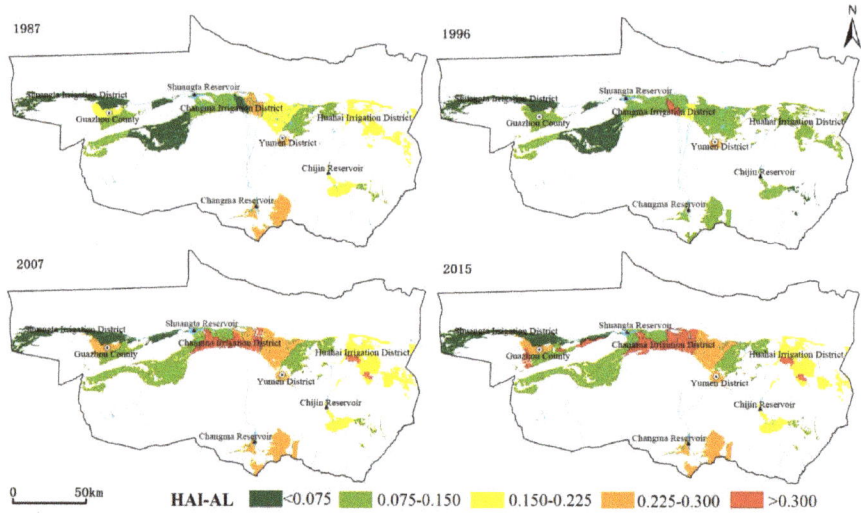

Figure 5. Spatial distribution of HAL-AL in the middle and lower reaches of the Shule River Basin (1987–2015).

(2)   Spatial variation of HAL-AL

From 1987 to 2015, the difference in HAL-AL between towns and between irrigation areas became more and more evident (Figure 5). In 1987, Huahai and Changma irrigation districts mainly had high HAL-AL, whereas Shuangta irrigation district had low or relatively low HAL-AL. From 1987 to 1996, the HAL-AL decreased in general. HAL-AL in Huahai irrigation district and Changma irrigation district decreased from low or medium levels, to low or relatively low levels. In Shuangta irrigation district, HAL-AL changed little, except that HAL-AL of Guazhou Town decreased from medium level to relatively low level. From 1996 to 2007, HAL-AL increased. Changma irrigation experienced the greatest change in HAL-AL. HAL-AL in most towns in Changma irrigation district increased from relatively low levels to relatively high or high levels. The HAL-AL in Huahai irrigation district also increased, from relatively low level to medium level. As a comparison, Shuangta irrigation district experienced the smallest changes in HAL-AL. From 2007 to 2015, the spatial distribution of HAL-AL tended to be stable, reflecting a pattern of "high HAL-AL in east and middle regions and low HAL-AL in west region".

4.3.2. Spatio-Temporal Variation of HAL-SAL

(1)   Temporal variation of HAL-SAL

From 1987 to 2015, the HAL-SAL in the study region increased with fluctuation (Table 3). The HAL-SAL and their changes were different between towns (Figure 6). In 1987, the HAL-SAL was low in all towns. The HAL-SAL was the highest in Guazhou Town (110.70×10^{15} sej/km^2), and the lowest in Qingquan Town (48.96×10^{15} sej/km^2). Except Nancha Town and Guazhou Town, all other towns had HAL-SAL lower than 100×10^{15} sej/km^2. The average intensity of semi-artificial land use was only 71.29×10^{15} sej/km^2. From 1987 to 1996, the HAL-SAL increased rapidly, and the average values increased to 125.76×10^{15} sej/ km^2, by 76.41%. The HAL-SAL of Sandaogou Town was 228.04×10^{15} sej/km^2, far higher than the average value and 3.11 times of that of Chijin Town (which was the lowest). From 1996 to 2007, the HAL-SAL of towns (except Yumen Town, Chijin Town, Qingquan Town, Sandaogou Town, Suoyang Town, and Yaozhanzi Town) decreased. In 2007, the average value was 158.33×10^{15} sej/km^2, increased by 25.89% compared with that in

1996. Only five towns had HAL-SAL higher than the average value. In 2015, the average value was 211.51×10^{15} sej/km^2, increased by 33.59% than that in 2007. The HAL-SAL was the highest in Liuhu Town (631.04×10^{15} sej/km^2) and the lowest in Xiaxihao Town (53.02×10^{15} sej/km^2). From 2007 to 2015, the total investment in cultivated lands in most towns increased. Liuhu town experienced the greatest increase in the HAL-SAL, which was 392.23×10^{15} sej/km^2. The HAL-SAL in some towns decreased, and decreased the most (113.33×10^{15} sej/km^2) in Suoyang Town.

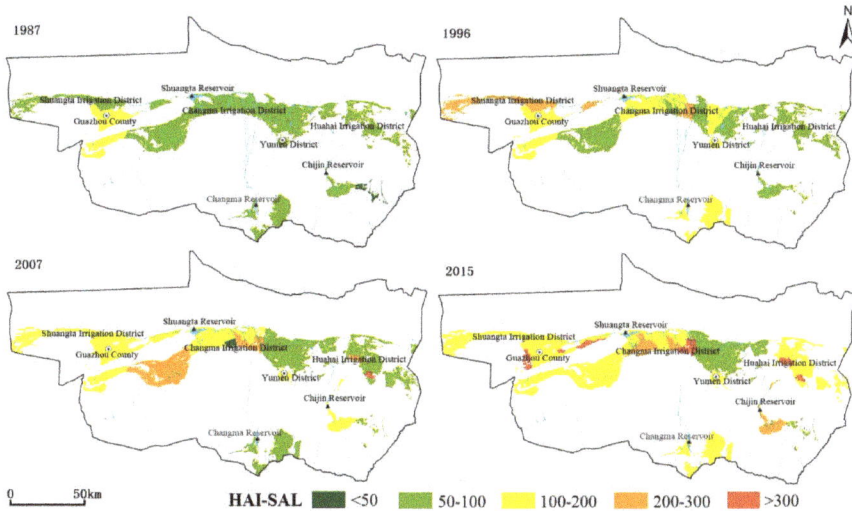

Figure 6. Spatial distribution of HAL-SAL in the middle and lower reaches of the Shule River Basin (1987–2015).

(2) Spatial variation of the HAL-SAL

The HAL-SAL in the middle and lower reaches of Shule River Basin increased over the 29 years, but there was difference in the intensity between irrigation areas (Figure 6), which was due to natural and social economic factors. In 1987, all areas, except Nancha Town and Guazhou Town in Shuangta irrigation district, showed relatively low intensity of semi-artificial land use. From 1987 to 1996, Xihu Town in Shuangta irrigation district experienced a transition from low intensity to relatively high intensity of semi-artificial land use. The west and south parts of Changma irrigation district experienced a transition from low to medium intensity of semi-artificial land use. The HAL-SAL in Huahai irrigation district changed little. From 1996 to 2007, the HAL-SAL in three irrigation areas fluctuated, and the intensity basically followed the order: Shuangta irrigation district > Changma irrigation district > Huahai irrigation district. In Shuangta irrigation district, the HAL-SAL in all towns reached medium or higher levels. Notably, the HAL-SAL in some towns decreased. For example, Xihu town in Shuangta irrigation district experienced a transition from relatively high intensity to medium intensity of semi-artificial land use. The HAL-SAL in Huangzhawan Town and Changma Town in Changma irrigation district decreased from medium level to relatively low level. Liuhu Town and Xiaojinwan Town were newly constructed, and they showed relatively high intensity of semi-artificial land use. From 2007 to 2015, the HAL-SAL in three irrigation areas increased. This was especially true for Shuangta irrigation district, Huahai irrigation district, and the west part of Changma irrigation district, accounting for 71.1% of the total area of semi-artificial land. The HAL-SAL in the east part of Changma irrigation district was relatively low. The overall distribution pattern was of "high HAL-SAL in east and west regions and low HAL-SAL in middle region".

5. Discussion and Conclusions

5.1. Discussion

Land use intensity reveals the extent to which human activities affect natural ecosystems. Revealing the spatio-temporal variation of land-use intensity is the key to monitoring the influences of land use changes on society and the environment, and to identifying the driving factors of land use changes. The driving factors interact with and restrict each other. They function together in the process of exploitation and utilization of land resources, leading to the spatial differentiation of land use intensity [6]. Shule River Basin is an oasis-desert ecosystem in arid area, where scarce water resources and low vegetation coverage are the main characteristics influencing its sustainable development. Meanwhile, because of the double pressure of population growth and rapid economic development, the ecosystem is very fragile, and the land utilization structure can reflect the ecological environmental problems in arid area [44,45]. With the development of society and advancement of technology, the land utilization has been transformed from natural exploitation to complex management model. In other words, driven by economic benefits, more investment has been put into machinery, fertilizer, and technology for agricultural land, new crop types are planted, and new management measures are adopted. Besides, a large number of buildings and water conservancy facilities have been constructed. Trees and grass have also been planted to improve the property of land that can hardly be utilized. All of these lead to increase in the degree and complexity of land utilization, which will inevitably affect regional ecological environment and its response to global changes [46].

5.1.1. Impacts of Socio-Economic Factors on Land-Use Intensity

Population plays an important role in the process of land use, and it is the most dynamic factor influencing land use. Population growth directly leads to land use changes. Specifically, it influences the way and intensity of land use by influencing the demand for food, housing, and products [18,30,47]. First, population growth will inevitably lead to increase in the demand for food. One way to solve this problem is to expand cultivated land area. Meanwhile, population growth will also result in growing demand for housing and living environment. In order to meet these demands, humans will expand land for living, transportation, public service facilities, and exploit unutilized land, leading to the transformation of land use types [48,49].

From 1987 to 2015, the population has increased sharply in the middle and lower reaches of Shule River Basin, from 232.213 to 289.813 thousand, by 24.81%. Among them, the immigrant population accounted for a large proportion. With the increase of population and the change of lifestyle, some families are gradually changed from large ones into smaller ones, and the total number of households has increased from 56,550 in 1987 to 87,436 in 2015. In terms of population composition, the agricultural population was 129.531 thousand in 1987, accounting for 55.78% of the total population, and 148.926 thousand in 2015, accounting for 51.39%. Although the proportion has decreased, the population structure mainly composed of agricultural population has not changed yet. In the background of population growth, household increasing, and dominated agricultural population, the demand for food, clothing, housing, and transportation will inevitably increase [37]. First, in order to meet the demand for food, the increased agricultural production caused by the cultivation of virgin land is the most significant, in addition to that brought by trade input and the advancement of technology. This explains the continuous expansion of arable land in the middle and lower reaches of Shule River Basin. Second, the increasing population and households have led to the expansion of construction land in both rural and urban areas. Especially in rural areas, the new comers often build new homes to meet their housing demand, thus increasing the area of construction land. Third, the transfer of population from villages to towns has driven the development of tourism and other industries in the region, promoted the improvement of infrastructure and the progression of urbanization, accelerated the transformation of land use mode, and increased the land use intensity [50].

Additionally, socio-economic activity is the basis for the existence and development of regions, and is a very important human activity. Economic growth, industrial restructuring, urbanization, and industrialization are all factors influencing the land-use intensity in Shule River Basin, and their influences have increased significantly in recent 30 years. Economic growth causes the adjustment of regional industrial structure and the change of land input, further influencing the total amount and structure of land demanded. In fact, different industrial structures require different lands [51]. The economic development and life quality improvement promote urbanization and industrialization, which accompany the socio-economic development and promotes non-agricultural land use by means of population and industry concentration, as well as the expansion of urban industrial and mining land [15,49,51–53]. At the same time, urbanization and industrialization cause the change of lifestyle and value concepts, which leads to changes of the original land use structure, and then affects the land-use intensity [54].

5.1.2. Impacts of Policy Factors on Land-Use Intensity

Land use intensity has a significant relationship with the policy. However, this impact is often overlooked [55,56]. In the past 50 years, the Chinese government has repeatedly put forward the policies concerning local development and protection in the Shule River Basin to realize local development and meet the social demands [37,57,58]. As a macroscopic and external force, policy directly or indirectly influences human behavior in the process of implementation, affecting land use [59–63].

The inland river of arid area not only nourishes watershed creatures, and maintains ecological stability, but also provides an important guarantee for social and economic development [29,37]. As the global climate changes, interference from human activities has exacerbated the regional land use and land coverage, which influences the intensity of land use. However, human activities are mainly affected by the policies as the result of natural background superimposition and artificial influence [38]. From the investigation of three inland river basins in the Hexi Corridor of Gansu Province, development intensity of land resources is different from east to west (Shiyang River development and utilization up to 171%, Hei River development and utilization rate to 92%, Shule River development and utilization rate to 75%), the contradiction between "man and land" also gradually gets slow from the east to the west [39,40,60]. A comprehensive survey of human development process is made in three inland rivers in Gansu Province, we found a large-scale migration development in Shiyang River and Hei River except historical period (especially Han, Tang, Ming and Qing). Meanwhile, no large-scale immigrant development in the two major basins is constructed, and land resources are in a state of incremental development after the founding of the People's Republic of China. On the contrary, large-scale migration development in the middle and lower reaches of the Shule River basin has been carried out under the promotion of national policies [38,39,44,45]. "The comprehensive development project for agricultural irrigation and resettlement in the Shule River" lasted 10 years, and was initiated in 1996, which was the most recent and largest organized migration to the basin [37]. During the implementation of this 10-year project, 1.848 billion Yuan were invested, 62,000 people immigrated to and settled down in Shule River Basin, forest area increased by 42.19 km^2, and cultivated land area increased by 142.12 km^2. Additionally, the areas of infrastructure, public service facilities, and residential land for immigrant housing are also increasing, with expected increases of 5.62 hm^2 in ten years due to a large number of immigrants. Large-scale immigration and land development changed the ways of land use [37–40]. Artificial and semi-artificial lands gradually replaced some part of the natural and semi-natural lands. The original water circulation and the original hydrological ecosystem of oasis were disturbed. Moreover, almost all surface water was channeled to farming area for agricultural production. This changed the natural distribution of water resources and the ecological environment which relies on water circulation.

From 1996 to 2007, the development of all the towns was promoted. Changma irrigation district and Huahai irrigation district were the main regions that underwent development. This period was a transition period for the land development and utilization in the middle and lower reaches of Shule River Basin. It can be attributed to the implementation of the "Integrated Development of Agricultural Irrigation and Immigrant Settlement in Shule River Basin (1996–2006)". Until 31 December 2006, 6 new towns and 46 new villages were built. A total of 62,000 immigrant people settled down in the study region, mainly in Changma irrigation district and Huahai irrigation district. The population in these two irrigation areas accounted for 61.14% and 23.67%, respectively, of the total population in the study region. Total investment reached 1.971 billion Yuan, aiming to support the construction of Changma Reservoir and immigrant settlement. About 93.68% of the investment was spent on agricultural development. After 1996, the implementation of policies such as "Reconstruction of Hexi", "Western Development", "Construction of New Countryside" led to further land development and utilization in the study region. National policies can help solve some practical problems, but they might lead to changes in land use, vegetation coverage, and ecosystem [64].

Above results and our field investigation indicate that 1996 was the key year for artificial and semi-artificial land use in the study region [44,45]. After 1996, HAI increased more rapidly. Before 1996 (1987–1996), the development of social economy in Shule River Basin was slow. Urbanization and industrialization lagged. Population mainly increased in rural regions, so the urban construction area did not expand too much [37]. After 1996, because of the implementation of policies such as "Integrated Development of Agricultural Irrigation and Immigrant Settlement in Shule River Basin (1996–2006)", "Reconstruction of Hexi (1997)", and "Western Development (2000)", as well as the progression of industrialization and urbanization, the population increased rapidly and gradually transferred to urban areas. This led to the expansion of the urban construction area and increase in HAL-AL [37–39,44]. Meanwhile, because of the immigrant project, people's needs for food and profits increased. In order to obtain more agricultural output, farmers increased agricultural input and improved food production per unit area of cultivated land. Also, farmers continuously cultivated virgin soils and increased income by expanding the cultivated land area. It can be seen that national policies have indirect, but the most significant influences on the semi-artificial land use intensity.

5.1.3. Impacts of Technological Factors on Land-Use Intensity

Technological progress is a fundamental factor that determines land-use intensity and environmental change in a region. Particularly, the progression of agricultural science and technology has an important influence on land utilization and ecological environment evolution, as it can effectively improve the efficiency of agricultural production and promote the change of land use type [30,44]. This is mainly shown in the following aspects:

In 1987, the amount of fertilizer used in Yumen City and Guazhou County was only 5831.05 t and was 39,408.08 t in 2015, with a total increase by 5.76 times, and an increase in the amount of fertilizer used per unit area by 35.12 t. The amount of agricultural film used was 1017.5 t in 1996 and 4228.8 t in 2015, which was 4.16 times of the former, with the amount of agricultural film used per unit area increasing by 2.62 t. The increasing amount of fertilizer and agricultural film used per unit area has promoted the growth of crops, improved crop yield and land output, promoted agricultural development, and increased the income level of farmers. Increased investment in fertilizer, agricultural film, and other aspects has led to the continuous development of agriculture, which can provide strong economic and technical support for the further development and utilization of land resources, thus promoting the full utilization of land resources [65].

In 1987, the total power of agricultural machinery used in Yumen City and Guazhou County was only 67.7×10^4 KW, which increased by 10.16 times to 755×10^4 KW in 2015. Technological advances have driven the modernization of agricultural machinery, effectively improved the efficiency of agricultural production, and saved a lot of human and financial resources. On this basis, the potential of land resources can be fully exploited, and the transformation of non-agricultural land to agricultural land is accelerated, leading to increase in cultivated land area [66]. Additionally, due to the wide application of machinery in agricultural production, the surplus labor force in rural areas will increase, and the transfer of surplus labor force will affect the use intensity of artificial land to a certain extent. Some research demonstrates that the transfer of rural labors can lead to an "N" shape change of the utilization intensity of cultivated land [26]. The transfer of surplus labor force can result in higher income level of farmers, more intensive utilization of artificial and semi-artificial land, and changes of land-use intensity.

5.2. Conclusions

Taking Guazhou County and Yumen City at the middle and lower reaches of Shule River Basin as the study region, this paper analyzed the spatio-temporal variation of LUD and HAI. The results might provide insights into the theory of land use changes in inland river basin in Hexi Corridor. The results were as follows:

(1) From 1987 to 2015, the land use structure in the middle and lower reaches of Shule River Basin changed little. Natural land was always the main land type, accounting for more than 80% of the total land area, followed by semi-natural land, and then artificial land (less than 1%). The semi-artificial and artificial land areas increased with increase in oasis area, while the semi-natural land area decreased gradually.

(2) The LUD in the study region increased by 35.36 over the 29 years, with an average annual increase of 1.22. From 1996 to 2007, LUD increased the most rapidly than in other periods, by 24.17, accounting for 68.35% of the total increase. After 2007, the LUD still increased, but more slowly. A spatial distribution pattern of "low land-use degree in east and west regions, and high land-use degree in middle region" changed to "high land-use degree in east and middle regions and low land-use degree in west region".

(3) The HAL-AL in the study region decreased from 1987 to 1996, and then increased from 1996 to 2015. In 1987, 73.33% of the towns showed medium or relatively low HAL-AL. In 1996, this percentage increased to 87.5%. From 1996 to 2007, HAL-AL in all towns increased. Towns with relatively high or high HAL-AL accounted for 61.9%. From 2007 to 2015, HAL-AL still increased but only a little. In 2015, towns with relatively high or high HAL-AL accounted for 65.22%. A spatial distribution pattern of "high HAL-AL in east and middle regions and low HAL-AL in west region" was formed.

(4) The HAL-SAL in the study region increased from 1987 to 2015. In 1987, the average intensity of semi-artificial land use was 71.29 sej/km^2. Except Nancha Town and Guazhou Town in Shuangta irrigation district, all other areas showed relatively low HAL-SAL. From 1987 to 1996, the HAL-SAL increased very rapidly. The average intensity of semi-artificial land use increased to 125.76 sej/km^2 (76.41%). From 1996 to 2007, the HAL-SAL still increased, but more slowly than in the previous period. Some towns even showed decreased intensity of semi-artificial land use. From 2007 to 2015, the HAL-SAL in three irrigation areas increased. A spatial distribution pattern of "high HAL-SAL in east and west regions and low HAL-SAL in middle region" was formed.

(5) 1996–2007 was a transition period for the land use intensity in the study region. A spatial distribution pattern of "high land use intensity in two of the three irrigation areas and low land use intensity in one irrigation area" was formed. The LUD and the HAL-AL increased significantly, while the HAL-SAL decreased. These trends were related to immigrant development and other national policies.

It should be pointed out that the spatio-temporal variation of land-use intensity is caused by socio-economic, policy, and technological factors. It is a multi-scale and multi-dimensional process with interaction between humans and land. Therefore, the selection of a multi-perspective in the construction of land-use intensity evaluation system reflects the multi dimensions and characteristics of land-use intensity. In addition, all land use types were considered, on which basis a more systematic assessment of the impact of different land use types and intensities on the ecological environment can be ensured. Still, much attention was paid to the use intensity of artificial land and semi-artificial land, so that we could analyze, in depth, the causes of land use changes and the influences of human activities.

Acknowledgments: This work was supported by the National Natural Science Foundation of China (Grant No. 41661105).

Author Contributions: Libang Ma and Wenjuan Cheng designed the study and processed the data. Jie Bo, Xiaoyang Li and Yuan Gu gave comments on the manuscript. All authors Contributed to the results, related discussions and manuscript writing.

Conflicts of Interest: The authors declare no conflict of interest.

References

1. Braimoh, A.K.; Osaki, M. Land-use change and environmental sustainability. *Sustain. Sci.* **2010**, *5*, 5–7. [CrossRef]
2. Blüthgen, N.; Dormann, C.F.; Prati, D.; Klaus, V.H.; Kleinebecker, T.; Hölzel, N.; Alt, F.; Boch, S.; Gockel, S.; Hemp, A.; et al. A quantitative index of land-use intensity in grasslands: Integrating mowing, grazing and fertilization. *Basic. Appl. Ecol.* **2012**, *13*, 207–220. [CrossRef]
3. Lioubimtseva, E.; Henebry, G.M. Climate and environmental change in arid Central Asia: Impacts, vulnerability, and adaptations. *J. Arid Environ.* **2009**, *73*, 963–977. [CrossRef]
4. Nd, T.B.; Lambin, E.F.; Reenberg, A. The emergence of land change science for global environmental change and sustainability. *Proc. Natl. Acad. Sci. USA* **2007**, *104*, 20666–20671.
5. Reyers, B.; O'Farrell, P.J.; Cowling, R.M.; Egoh, B.N.; Le Maitre, D.C.; Vlok, J.D.C. Ecosystem services, land-cover change, and stakeholders: Finding a sustainable foothold for a semiarid biodiversity hotspot. *Ecol. Soc.* **2009**, *14*, 1698–1707. [CrossRef]
6. Ruddiman, W.F. The anthropogenic greenhouse era began thousands of years ago. *Clim Chang.* **2003**, *61*, 261–293. [CrossRef]
7. Vitousek, P.M.; Mooney, H.A.; Lubchenco, J.; Melillo, J.M. Human domination of Earth's ecosystems. *Science* **1997**, *277*, 494–499. [CrossRef]
8. Erb, K.; Niedertscheider, M.; Dietrich, J.P.; Schmitz, C.; Verburg, P.H.; Jepsen, M.R.; Haberl, H. Conceptual and empirical approaches to mapping and quantifying land use intensity. In *Ester Boserup's Legacy on Sustainability: Orientations for Contemporary Research*; Springer: Dordrecht, The Netherlands, 2014; pp. 61–86.
9. Sala, O.E.; Wall, D.H. Global biodiversity scenarios for the year 2100. *Science* **2000**, *287*, 1770–1774. [CrossRef] [PubMed]
10. Burney, J.A.; Davis, S.J.; Lobell, D.B. Greenhouse gas mitigation by agricultural intensification. *Proc. Natl. Acad. Sci. USA* **2010**, *107*, 12052–12057. [CrossRef] [PubMed]
11. Wang, G.J.; Liao, S.G. Spatial heterogeneity of land use intensity. *Chin. J. Appl. Ecol.* **2006**, *17*, 611–614. (In Chinese)
12. Liiri, M.; Häsä, M.; Haimi, J.; Setälä, H. History of land-use intensity can modify the relationship between functional complexity of the soil fauna and soil ecosystem services—A microcosm study. *Appl. Soil. Ecol.* **2012**, *55*, 53–61. [CrossRef]
13. Tardy, V.; Spor, E.; Mathieu, O.; Eque, J.L.E.; Terrat, S.E.; Plassart, P.; Regnier, T.; Bardgett, R.; Putten, W.H.V.D.; Roggero, P.P.; et al. Shifts in microbial diversity through land use intensity as drivers of carbon mineralization in soil. *Soil. Biol. Biochem.* **2015**, *90*, 204–213. [CrossRef]
14. Li, Q.; Zhang, X.F.; Liu, Q.F.; Liu, Y.; Ding, Y.; Zhang, Q. Impact of land use intensity on ecosystem services: An example from the agro-pastoral ecotone of central inner mongolia. *Sustainability* **2017**, *9*, 1030. [CrossRef]
15. Jiang, L.; Deng, X.Z.; Seto, K.C. The impact of urban expansion on agricultural land use intensity in China. *Land Use Policy* **2013**, *35*, 33–39. [CrossRef]
16. Stone, G.D. Theory of the square chicken: Advances in agricultural intensification theory. *Asia. Pac. Viewp.* **2010**, *42*, 163–180. [CrossRef]
17. Shriar, A.J. Determinants of agricultural intensity index "scores" in a frontier region: An analysis of data from northern Guatemala. *Agr. Hum. Values* **2005**, *22*, 395–410. [CrossRef]
18. Václavík, T.; Lautenbach, S.; Kuemmerle, T.; Seppelt, R. Mapping global land system archetypes. *Glob. Environ. Chang.* **2013**, *23*, 1637–1647. [CrossRef]

19. Kühling, I.; Broll, G.; Trautz, D. Spatio-temporal analysis of agricultural land-use intensity across the Western Siberian grain belt. *Sci. Total Environ.* **2016**, *544*, 271–280. [CrossRef] [PubMed]

20. Petz, K.; Alkemade, R.; Bakkenes, M.; Schulp, C.J.E.; Velde, M.; Leemans, R. Mapping and modelling trade-offs and synergies between grazing intensity and ecosystem services in rangelands using global-scale datasets and models. *Glob. Environ. Chang.* **2014**, *29*, 223–234. [CrossRef]

21. Wellmann, T.; Haase, D.; Knapp, S.; Salbach, C.; Selsam, P.; Lausch, A. Urban land use intensity assessment: The potential of spatio-temporal spectral traits with remote sensing. *Ecol. Indic.* **2018**, *85*, 190–203. [CrossRef]

22. Sluis, T.V.D.; Pedroli, B.; Kristensen, S.P.; Cosor, G.L.; Pavlis, E. Changing land use intensity in Europe—Recent processes in selected case studies. *Land Use Policy* **2016**, *57*, 777–785. [CrossRef]

23. Wang, F.H.; Antipova, A.; Porta, S. Street centrality and land use intensity in Baton Rouge, Louisiana. *J. Transp. Geogr.* **2011**, *19*, 285–293. [CrossRef]

24. Pfestorf, H.; Weiß, L.; Müller, J.; Boch, S.; Socher, S.A.; Prati, D.; Schöning, I.; Weisser, W.; Fischer, M.; Jeltsch, F. Community mean traits as additional indicators to monitor effects of land-use intensity on grassland plant diversity. *Perspect. Plant Ecol.* **2013**, *15*, 1–11. [CrossRef]

25. Ge, X.D.; Dong, K.K.; Luloff, A.E.; Wang, L.Y.; Xiao, J. Impact of land use intensity on sandy desertification: An evidence from Horqin Sandy Land, China. *Ecol. Indic.* **2016**, *61*, 346–358. [CrossRef]

26. Dietrich, J.P.; Schmitz, C.; Müller, C.; Fader, M.; Lotze-Campen, H.; Popp, A. Measuring agricultural land-use intensity-a global analysis using a model-assisted approach. *Ecol. Model.* **2012**, *232*, 109–118. [CrossRef]

27. Berner, D.; Marhan, S.; Keil, D.; Poll, C.; Schützenmeister, A.; Piepho, H.P.; Kandeler, E. Land-use intensity modifies spatial distribution and function of soil microorganisms in grasslands. *Pedobiologia* **2011**, *54*, 341–351. [CrossRef]

28. Zhao, X.G.; Pan, Y.J.; Zhao, B.; He, R.F.; Liu, S.F.; Yang, X.Y.; Li, H.X. Temporal-spatial evolution of the relationship between resource-environment and economic development in China: A method based on decoupling. *Prog. Geogr.* **2011**, *30*, 456–463. (In Chinese)

29. Yang, Y.T.; Shi, P.J.; Pan, J.H. Analysis of land use difference degree in arid inland river basin—Taking Zhangye, Ganzhou District as an example. *J. Arid. Land. Resour. Environ.* **2012**, *26*, 102–107. (In Chinese)

30. Liu, G.S.; Wang, H.M.; Cheng, Y.X.; Zheng, B.; Lu, Z.L. The impact of rural out-migration on arable land use intensity: Evidence from mountain areas in Guangdong, China. *Land Use Policy* **2016**, *59*, 569–579. [CrossRef]

31. Liu, J.Y.; Zhang, Z.X.; Xu, X.L.; Kuang, W.H.; Zhang, S.W.; Li, R.D.; Yan, C.Z.; Yu, D.S.; Wu, S.X.; Jiang, N. Spatial patterns and driving factors of land use change in China during the early 21st century. *J. Geogr. Sci.* **2010**, *20*, 483–494. [CrossRef]

32. Wu, L.N.; Yang, S.T.; Liu, X.Y.; Luo, Y.; Zhou, X.; Zhao, H.G. Response analysis of land use change to the degree of human activities in Beiluo River basin since 1976. *Acta. Geogr. Sin.* **2014**, *69*, 54–63. (In Chinese)

33. Liu, H.; Zhang, R.Q.; Hao, J.M.; Ai, D. Tupu analysis of land use intensity using semi-variance in Yinchuan Plain. *Trans. Chin. Soc. Agric. Eng. (Trans. CSAE)* **2012**, *28*, 225–231.

34. Li, Y.F.; Liu, G.H. Characterizing Spatiotemporal Pattern of Land Use Change and Its Driving Force Based on GIS and Landscape Analysis Techniques in Tianjin during 2000–2015. *Sustainability* **2017**, *9*, 894. [CrossRef]

35. Zhang, W.W.; Li, H. Characterizing and Assessing the Agricultural Land Use Intensity of the Beijing Mountainous Region. *Sustainability* **2016**, *8*, 1180. [CrossRef]

36. Gong, J.Z.; Chen, W.L.; Liu, Y.S.; Wang, J.Y. The intensity change of urban development land: Implications for the city master plan of Guangzhou, China. *Land Use Policy* **2014**, *40*, 91–100. [CrossRef]

37. Chang, G.Y.; Zhang, W.X. Ecological civilization-based rethinking of large-scale immigration and land development along Shule River. *J. Lanzhou Univ. (Nat. Sci.)* **2014**, *50*, 405–409. (In Chinese)

38. Qi, J.H.; Niu, S.W.; Zhao, Y.F.; Liang, M.; Ma, L.B.; Ding, D. Responses of vegetation growth to climatic factors in Shule River Basin in northwest china: A panel analysis. *Sustainability* **2017**, *9*, 368. [CrossRef]

39. Zhang, H.; Yi, S.Z.; Wu, Y.G. Decision Support System and Monitoring of Eco-Agriculture Based on WebGIS in Shule Basin. *Energy Procedia* **2012**, *14*, 382–386. [CrossRef]

40. Huang, S.; Feng, Q.; Lu, Z.X.; Wen, X.H.; Deo, R.C. Trend Analysis of Water Poverty Index for Assessment of Water Stress and Water Management Polices: A case study in the Hexi Corridor, China. *Sustainability* **2017**, *9*, 756. [CrossRef]

41. Chen, Y.Q.; Li, X.B. Structural change of agricultural land use intensity and its regional disparity in China. *Acta. Geogr. Sin.* **2009**, *64*, 469–478. (In Chinese) [CrossRef]

42. Odum, H.T. Environmental Accounting-Emergy and Environmental Decision Making. *Child Dev.* **1996**, *42*, 1187–1201.

43. Lan, S.F.; Qin, P.; Lu, H.F. *Energy Analysis of Ecological-Economic System*; Chemical Industry Press: Beijing, China, 2002. (In Chinese)

44. Qi, J.H.; Niu, S.W.; Ma, L.B.; Wang, W.D. The characteristics and driving forces of LUCC in the middle and lower reaches of Shule River Basin. *Chin. J. Ecol.* **2014**, *33*, 2207–2220. (In Chinese)

45. Ding, H.W.; Zhao, C.; Huang, X.H. Ecological environment and desertification in Shule River Basin. *Arid. Zone. Res.* **2001**, *18*, 5–10. (In Chinese)

46. Verburg, P.H.; Crossman, N.; Ellis, E.C.; Heinimann, A.; Hostert, P.; Mertz, O.; Nagendra, H.; Sikor, T.; Erb, K.H.; Golubiewski, N.; et al. Land system science and sustainable development of the earth system: A global land project perspective. *Anthropocene* **2015**, *12*, 29–41. [CrossRef]

47. Ellis, E.C.; Ramankutty, N. Putting people in the map: Anthropogenic Biomes of the World. *Front. Ecol. Environ.* **2008**, *6*, 439–447. [CrossRef]

48. Tilman, D.; Balzer, C.; Hill, J.; Befort, B.L. Global food demand and the sustainable intensification of agriculture. *Proc. Natl. Acad. Sci. USA* **2011**, *108*, 20260–20264. [CrossRef] [PubMed]

49. Long, H.L.; Heilig, G.K.; Li, X.B.; Zhang, M. Socio-economic development and land-use change: Analysis of rural housing land transition in the transect of the Yangtse river, China. *Land Use Policy* **2007**, *24*, 141–153. [CrossRef]

50. Pan, J.H.; Su, Y.C.; Huang, Y.S.; Liu, X. Land use & landscape pattern change and its driving forces in Yumen City. *Geogr. Res.* **2012**, *13*, 1631–1639. (In Chinese)

51. Zhang, B. Empirical study on the relationship between economic development and urban construction land in China: Based on the panel data of provinces. *J. Anhui A* **2014**, *42*, 3720–3723.

52. Asselen, S.V.; Verburg, P.H. A land system representation for global assessments and land-use modeling. *Glob. Chang. Biol.* **2012**, *18*, 3125–3148. [CrossRef] [PubMed]

53. Zhang, B.L.; Yang, Q.Y.; Lu, C.Y.; Sun, P.L.; Zong, H.M. Effect on economic development of regional land use change in different development phase: Forty counties in Chongqing as the research object. *Econ. Geogr.* **2011**, *31*, 1539–1544. (In Chinese)

54. Zhou, Q.; Huang, X.J.; Pu, L.J.; Li, X.W.; Zhou, F. Intensity and mechanism in change of regional agricultural land use-a case study of former Xishan City of Wuxi City. *Resour. Environ. Yangtze Basin* **2003**, *12*, 535–540. (In Chinese)

55. Zhu, H.Y.; Li, X.B.; Xin, L.J. Intensity change in cultivated land use in China and its policy implications. *J. Nat. Resour.* **2007**, *22*, 907–915.

56. Chen, H.B.; Shao, L.Q.; Zhao, M.J.; Zhang, X.; Zhang, D.J. Grassland conservation programs, vegetation rehabilitation and spatial dependency in Inner Mongolia, China. *Land Use Policy* **2017**, *64*, 429–439. [CrossRef]

57. Ji, Y.P. Study on the development and protection of water resources in Shule River Basin. *Environ. Stud. Monit.* **2016**, *3*, 53–61. (In Chinese)

58. Sun, T.; Pan, S.B.; Li, J.R.; Den, H.Y. Analysis on the exploitation of water and land resources and its environmental effects in the Shule River Watershed. *Arid. Zone. Res.* **2004**, *21*, 313–317.

59. Huang, S.; Zhou, L.H.; Chen, Y.; Lu, H.L. Impacts of polices on eco-environment of Minqin County during the past 60 years. *J. Arid Land Resour. Environ.* **2014**, *28*, 73–78. (In Chinese)

60. Lu, D.D.; Liu, W.D. Analysis of Geo-factors behind regional development and regional policy in China. *Sci. Geogr. Sin.* **2000**, *20*, 487–493. (In Chinese)

61. Zhao, M.M.; Zhou, L.H.; Chen, Y.; Zhang, J.S.; Guo, X.L.; Wang, R. The influence of ecological policies on changes of land use and ecosystem service value in Hangjinqi, Inner Mongolia, China. *J. Desert. Res.* **2016**, *36*, 842–850. (In Chinese)

62. Gan, C.H.; Ma, L.; Nan, Q.J. Spatio-temporal characteristics and driving force of land-use change in Beijing's ecological environs—A case study of Weichang County, Hebei Province. *Chin. J. Eco-Agric.* **2007**, *15*, 165–170. (In Chinese)

63. Spalding, A.K. Exploring the evolution of land tenure and land use change in Panama: Linking land policy with development outcomes. *Land Use Policy* **2017**, *61*, 543–552. [CrossRef]

64. Wei, H. Analyze the influence of increases of population and progress of agro-technology on farmland utilization & environment. *Ecol. Econ.* **2011**, *5*, 108–112. (In Chinese)

65. Chen, Y.Q.; Li, X.B.; Tian, Y.J.; Tan, M.H. Structural change of agricultural land use intensity and its regional disparity in China. *J. Geogr. Sci.* **2009**, *19*, 545–556. [CrossRef]
66. Long, D.P.; Li, T.S.; Miao, Y.Y.; Yu, Z.S. The spatial distribution and types of the development level of Chinese agricultural modernization. *Acta. Geogr. Sin.* **2014**, *69*, 213–226. (In Chinese)

sustainability

MDPI

Article

Profiling Human-Induced Vegetation Change in the Horqin Sandy Land of China Using Time Series Datasets

Lili Xu [1,2], Zhenfa Tu [1,2,*], Yuke Zhou [3] and Guangming Yu [1,2]

[1]   Key Laboratory for Geographical Process Analysis & Simulation of Hubei Province, Wuhan 430079, China; xulls@mail.ccnu.edu.cn (L.X.); yu-guangming@mail.ccnu.edu.cn (G.Y.)

[2]   College of Urban and Environmental Sciences, Central China Normal University, Wuhan 430079, China

[3]   Key Laboratory of Ecosystem Network Observation and Modeling, Institute of Geographic Sciences and Natural Resources Research, Chinese Academy of Sciences, Beijing 100101, China; zhouyk@igsnrr.ac.cn

*   Correspondence: tuzhenfa@hotmail.com; Tel.: +86-180-626-05735

Received: 8 February 2018; Accepted: 2 April 2018; Published: 4 April 2018

Abstract: Discriminating the significant human-induced vegetation changes over the past 15 years could help local governments review the effects of eco-programs and develop sustainable land use policies in arid/semi-arid ecosystems. We used the residual trends method (RESTREND) to estimate the human-induced and climate-induced vegetation changes. Two typical regions in the Horqin Sandy Land of China were selected as study areas. We first detected vegetation dynamics between 2000–2014 using Sen's slope estimation and the Mann–Kendall test detection method (SMK) based on the Moderate Resolution Imaging Spectroradiometer (MODIS) normalized difference vegetation index (NDVI) time series, then used RESTREND to profile human modifications in areas of significant vegetation change. RESTREND was optimized using statistical and trajectory analysis to automatically identify flexible spatially homogeneous neighborhoods, which were essential for determining the reference areas. The results indicated the following. (1) Obvious vegetation increases happened in both regions, but Naiman (64.1%) increased more than Ar Horqin (16.8%). (2) Climate and human drivers both contributed to significant changes. The two factors contributed equally to vegetation change in Ar Horqin, while human drivers contributed more in Naiman. (3) Human factors had a stronger influence on ecosystems, and were more responsible for vegetation decreases in both regions. Further evidences showed that the primary human drivers varied in regions. Grassland eco-management was the key driver in Ar Horqin, while farming was the key factor for vegetation change in Naiman.

Keywords: Horqin Sandy Land; human-induced; vegetation change; RESTREND

1. Introduction

Discriminating between the driving forces of vegetation change is necessary in order to understand the interacting mechanisms between ecosystems and external drivers, and help environmental managers make effective decisions that maintain the sustainable development of ecosystems [1,2]. Complex external driving forces influence the processes of vegetation change, of which climate change and human activities are two main factors within decadal time scales [3–5]. Climate variations, such as increases or decrease in precipitation and temperature would result in a corresponding trend of greening and browning in vegetation cover [3,5,6]. Various human activities may also influence the vegetation growth by controlling or modifying ecosystem composition and distribution [3]. Worldwide, urbanization usually causes local vegetation to decrease [7–9], while eco-managements such as encouraging afforestation, grazing prohibition, and rest-rotation grazing usually cause vegetation cover to increase [5,10]. Even the price fluctuation of grain and beef in the market would cause vegetation change by influencing farmers' choices of cultivation regimes [4]. On a scale of decades, human activities may have more deep and dramatic influences on vegetation growth than climate variations [11]. The complicated interaction between human and vegetation is critical for almost all of the research studies that have been done about earth system models [12]. Discriminating human-induced vegetation change has always been a challenging issue [1,13,14], even considering the long-term perspective of palaeoecology [15,16].

Arid/semi-arid ecosystems are fragile and extremely sensitive to external interference [17,18]. Inadvertent human mismanagement, such as overgrazing, over farming, and deforestation may result in irreversible catastrophic changes for arid ecosystems. Discriminating human-induced vegetation change in a quantitative way is essential for managing worldwide arid and semi-arid ecosystems [1,19]. The Horqin Sandy Land of China is a typical semi-arid ecosystem, with a transitional zone between semi-pastoral areas and pastoral areas. Following serious degradation over recent decades, the Chinese government has carried out several grassland conservation programs (eco-programs) in this region. Profiling human-induced vegetation change in the Horqin Sandy Land is important for understanding human influences on vegetation in arid/semi-arid China, and evaluating the effects of the eco-programs.

Remote sensing technology is commonly used to monitor vegetation change on a large scale [13]. The normalized difference vegetation index (NDVI) that is obtained from remote sensing products is very sensitive to arid/semi-arid vegetation variations [20], and is usually used to monitor arid/semi-arid vegetation change [21,22]. Furthermore, the coupling analysis of time series NDVI and climate datasets could help identify the impact of human activities [1,13,23–29].

A previous study reviewed five quantitative methods to identify human-induced vegetation changes [23], i.e., model analysis, mathematical statistics, framework analysis, index assessment, and difference comparison, within which the difference comparison method (DC) stood out and commonly was used [23]. Rain-use efficiency (RUE) [24–26] and the residual trends method (RESTREND) [3,13,27,28] are two of the most effective DC methods that have been used successfully in arid and semi-arid ecosystems, and RESTREND has proved more robust than RUE [23,28,29].

Based on the assumption that climate, i.e., rainfall, is the only driver of vegetation change, the standard RESTREND [1,29] first used the NDVI as an indicator to construct regression models with rainfall, then analyzed the trend of residuals on a per pixel basis to discriminate the human-induced vegetation changes. However, Evans and Geerken (2004) then proposed that standard RESTREND would measure degeneration inaccurately, because the potential rainfall–vegetation relationships and trend analyses were all based on the same time series, which may itself contain uncertain levels of degeneration [1,30,31]. It was necessary to improve the standard RESTREND by establishing a "no change reference" to simulate the potential relationship between rainfall and vegetation.

Establishing an accurate potential vegetation–climate relationship is essential for improving RESTREND. A pixel-level vegetation–climate regression model based on a historical no change reference period, or a no change reference area in the homogeneous spatial neighborhood of the specific pixel, are two alternative strategies. However, the finite temporal accumulation of satellite datasets limits the time that can be used to define a historical no change/less change reference period. The potential vegetation–climate regression model is commonly built based on a no change area in the homogeneous spatial neighborhood [27,32]. The no change area, i.e., area without influence from human activity, refers to the maximum $\sum$ NDVI or net primary productivity (NPP) in the homogeneous spatial neighborhood. The homogeneous spatial neighborhood, within which vegetation has similar growth characteristics, could be determined in the following two ways [31–33].

A predetermined window, such as a 5 km × 5 km [33] or a 7 km × 7 km neighborhood [31] is usually suggested to define the homogeneous spatial neighborhood. However, defining a suitable size of window is a subjective task, and it needs accurate and comprehensive field investigations in order to make a suitable decision, which is a time-consuming, laborious, and regional-dependent task. Furthermore, heterogeneous vegetation patterns usually exist within a regular and limited distance, which weakens the applicability of this strategy. An irregular homogeneous neighborhood is a better choice. Prince [32] used a land capability class map, which was defined by a k-prototypes clustering technique using maps of rainfall, soils, and land use, to obtain a homogeneous spatial neighborhood. However, this method still faces two obvious problems. (1) It is hard to control the number of classes in the k-prototypes clustering algorithm, which results in uncertainty in land capability classes. (2) The type of land use within a time series might change from year to year, so it is not reasonable to use a stationary land use map to define a homogeneous area. However, it is difficult to obtain a land use map each year to update the land capability class map.

To overcome the problems of these two methods, this paper optimized the strategy of determining homogeneous spatial neighborhoods in RESTREND by statistical analysis and trajectory analysis to automatically define flexible spatially homogeneous neighborhoods. Based on this, we first used a trend analysis method to find significant vegetation change from 2000 to 2014 in the study areas, and then used the optimized RESTREND in areas with significant changes to identify significant human-induced changes during the past 15 years. By choosing two typical regions in Horqin Sandy Land as study areas, this paper aimed to discover different human–climate–vegetation change mechanisms in arid and semi-arid ecosystems with different levels of human influence.

2. Materials and Methods

2.1. Study Area

The Horqin Sandy Land spans from 41°41′ N to 46°12′ N latitude and 117°49′ E to 123°42′ E longitude, and is located in the southeast of the Inner Mongolian Autonomous Region of China (Figure 1). It is a typical transitional zone between the Inner Mongolian Plateau and the Northeast China Plain, with a temperate sub-humid and semi-arid monsoon climate [11]. The annual precipitation varies between 340–450 mm. The annual average temperature is about 5.8–6.4 °C, and the annual average wind velocity is up to 3.5–4.5 m/s. The landscape of this area is mainly consisted of sand dunes alternating with gently undulating meadows. The soil of this area is sandy, light, loose, and especially susceptible to wind erosion. The Horqin Sandy Land is a typical transitional zone between semi-pastoral areas located in the south of the study area, and pastoral areas located in the north of the study area.

We used two regions, i.e., Ar Horqin and Naiman, with different levels of human influence in the Horqin Sandy Land as our study areas to discover different human–climate–vegetation change mechanisms. The Ar Horqin region is a typical pastoral area that is in the northwest of the Horqin Sandy Land, with 71.0% of the land covered by grassland. The Naiman region is in the southeast of the Horqin Sandy Land. It is a typical agricultural area with the majority (46.5%) covered by farmland (Figure 1). Traditionally, people in the pastoral area prefer to feed sheep and cows for a living, while people in the semi-pastoral area live on farming. According to the Inner Mongolia Statistics Bureau in 2013, the population density in Naiman is 53 people/km², which is much more than in Ar Horqin (21 people/km²). It is believed that human activities in Naiman are much more intense than in Ar Horqin. Over the past two decades, the ecosystems in these two regions have changed markedly because of frequent human activities [34–37]. The Chinese government has implemented several region-wide programs in the two locations; e.g., the West Development Strategies (introduced in 1999) and the Grain for Green Program (introduced in 1999). The former strategy promotes economic development in western China, and the latter encourages farmers to tree plant in order to restore local ecological environment. Additionally, three grassland conservation programs have also been implemented in the study areas. The Beijing–Tianjin Sandstorm Source Controlling Program (introduced in 2001) and the Ecological Subsidy and Award System (introduced in 2011) were also implemented in the Ar Horqin region. The Grazing Withdrawal Program (introduced in 2003) and the Ecological Subsidy and Award System (introduced in 2011) were implemented in Naiman [38].

Discriminating the significant human-induced vegetation changes over the last 15 years could help local governments review the effects of the eco-programs and develop sustainable land use policies in the future.

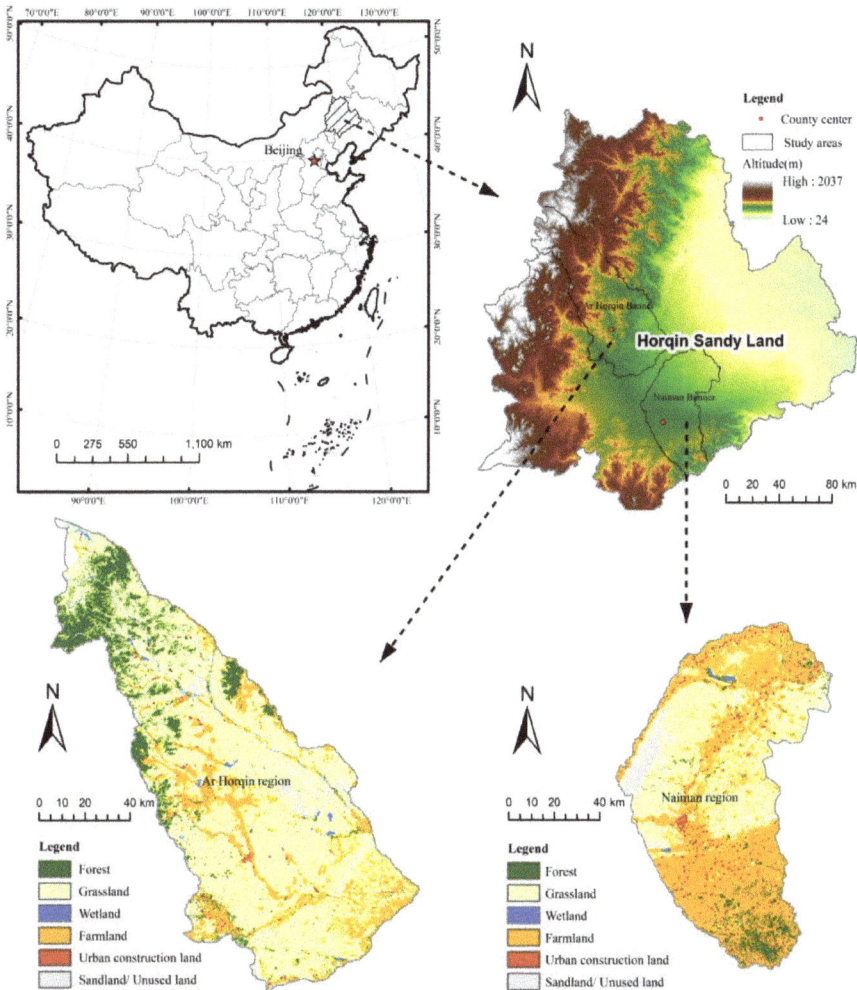

Figure 1. Location of the Horqin Sandy Land, China and the two study areas.

2.2. Data and Data Preprocessing

This study selected NDVI as the original data, because the index is very sensitive to different vegetation covers in arid/semi-arid areas. It has a strong capability for monitoring vegetation changes, land cover classifications, and the change analysis of land cover in arid and semi-arid areas [21,22,39]. The NDVI, which is a proxy for NPP, is the most commonly used indicator in RESTREND [1,40–42]. Time series Moderate Resolution Imaging Spectroradiometer (MODIS) 13Q1 NDVI datasets were downloaded from the MODIS website (http://modis.gsfc.nasa.gov/). The datasets covers 342 images from 2000 to 2014 with the same temporal resolution (16 days) and spatial resolution (250 m).

The original dataset had a hierarchical data format, within which the layer 1 and the layer 12 reflected the NDVI value and the reliability and quality of the data for each pixel, respectively.

We first removed noise from clipped and re-projected NDVI datasets using a Savitzky–Golay filter [43]. Then, we replaced high ranking pixels (i.e., ranking 0 or 1) with the original pixel value [44] and masked non-vegetated areas (e.g., deserts, water bodies, and other areas with mean annual NDVI <0.1) [45]. Here, we changed the NDVI values of high-ranking pixels, because these two types of pixels were confirmed as high reliability pixels. In layer 12 of dataset, a high ranking of '0' for pixels indicates high quality. The ranking of '1' indicates that their value was useful and could be used with reference to other QA information. Finally, the time series $\sum NDVI$ trajectory, i.e., the annual accumulated NDVIs (AA-NDVIs), was aggregated from Julian day 145 to Julian day 273 in the growing season of each year. The time series $\sum NDVI$ dataset was then produced so that the impact of frequent and strong interannual variations could be reduced.

Although the standard RESTREND only used rainfall as the climate factor, many studies suggest that other climate factors such as temperature might be key controls on vegetation growth in some regions [19]. Therefore, it was necessary to determine the principal climate factors and the suitable climate variables to build the climate–vegetation model [33,46,47]. In this paper, precipitation, temperature, and wind speed were selected as three representative climate factors to pre-analyze the best climate factors to use in the study area. Datasets for monthly average precipitation (P), monthly average temperature (T), and monthly maximum wind speed (W) were produced based on time series data (2000–2014) from 46 stations acquired from the website of the China Meteorological Data Service Center (http://data.cma.cn/). Due to the lag effect of vegetation growth, 45 indices (Table 1) with different temporal combinations were calculated for each meteorological dataset (P/T/W); then, the indices in each station were interpolated by the inverse distance weighting method to obtain 135 climate raster grids with the same resolution as the NDVI dataset.

Table 1. Forty-five indices with different temporal combinations in each meteorological dataset (precipitation, temperature, and wind speed).

Accumulated	Code of Indices [1]	Total
One month	1, 2, 3, 4, 5, 6, 7, 8, 9	9
Two months	12, 23, 34, 45, 56, 67, 78, 89	8
Three months	123, 234, 345, 456, 567, 678, 789	7
Four months	1234, 2345, 3456, 4567, 5678, 6789	6
Five months	12345, 23456, 34567, 45678, 56789	5
Six months	123456, 234567, 345678, 456789	4
Seven months	1234567, 2345678, 3456789	3
Eight months	12345678, 23456789	2
Nine months	123456789	1
Total	—	45

[1] The codes 1, 2, 3, 4, 5, 6, 7, 8, 9 represent January, February, March, April, May, June, July, August, September, respectively.

The 1:1,000,000 ecological map for Inner Mongolia, produced by the Chinese Academy of Sciences (2001), was used to identify ecological zones. An ecological zone is determined by the climate, soil, landform, and vegetation. The combination of land cover types in an ecological zone is unique, and the same land cover type in an ecological zone indicates a similar combination of plants and their growing processes.

A Level 1 Terrain-corrected (L1T) Landsat Thematic Mapper and Enhanced Thematic Mapper Plus (TM/ETM+) time series image dataset covering the study areas in summer of 2000, 2005, 2009, and 2014 was downloaded from the United States Geological Survey website (http://earthexplorer.usgs.gov/). The datasets had been processed for systematic radiometric calibration and geometric correction by incorporating ground control points while employing a digital elevation model for topographic correction. Therefore, it could be used to visually validate the results of human-induced vegetation change. The 2010 land cover data for the study area was obtained from the China Environmental Protection Agency and the Chinese Academy of Sciences. It was produced by object-based classification using charge coupled device camera data from the Chinese Huanjing-1/A/B/C satellite. The classification of land cover in the data was based on first-level categories, and the overall accuracy was up to 85% [48]. It could indicate farmland, forest, grassland, wetland, sand land/unused land, and urban construction areas, so that can be used to further target specific driven factors. In addition, the amount of chemical fertilizer and number of livestock at the end of each year in the study areas were obtained from local economic statistical datasets, i.e., the Inner Mongolia Statistics Bureau, over the study period (2001–2015). The price of corn was obtained from China Animal Industry Yearbooks (2001–2015).

2.3. Methodology

To discover different human–climate–vegetation change mechanisms in the study areas, two general steps were taken to profile human-induced vegetation change (Figure 2). (1) We detected vegetation dynamics from 2000 to 2014 to find out areas of significant change during the past 15 years, and (2) we used optimized RESTREND to discriminate human-induced changes in areas with significant change. Areas with no significant vegetation changes determined in the first step were not analyzed in the second step.

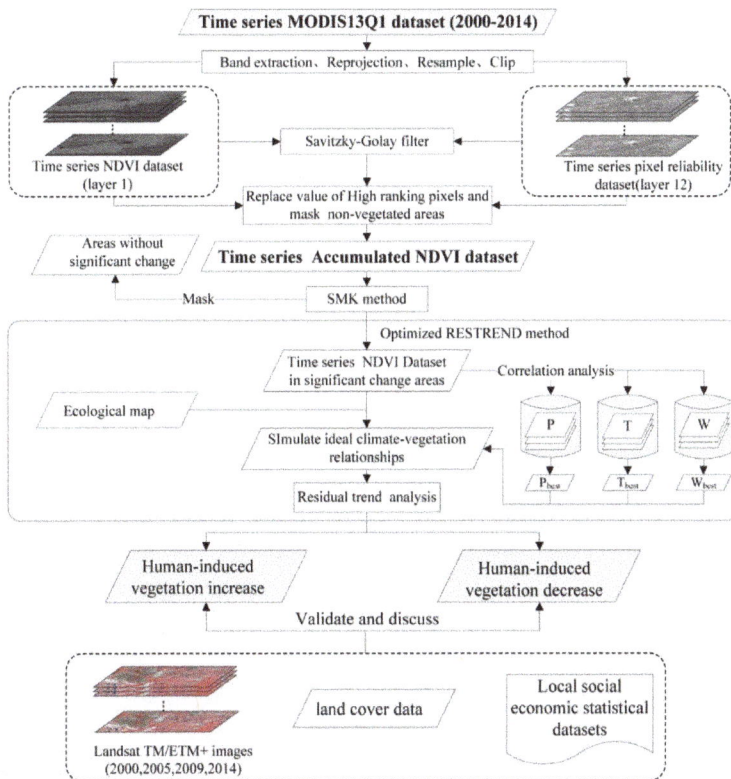

Figure 2. Framework of the study.

2.3.1. Detecting Vegetation Dynamics Using the SMK Method

Statistical methods are used frequently to detect trend dynamics of vegetation [49,50], including parametric methods such as the least squares linear regression method [1,51] and non-parametric methods such as Sen's slope estimation coordinated with the Mann–Kendall test detection (SMK) method [52,53]. In this study, the SMK method was used to determine whether a time series trajectory of $\sum NDVI$ indicated a statistically significant trend ($p < 0.05$). It is a non-parametric method that is robust against non-normality in a dataset and has good performance in error resistance [54,55], which is suitable for use in arid and semi-arid ecosystems with strong and frequent interannual vegetation variability. The trend slope can be defined by Sen's slope estimation, and the significance of the slope can be determined using the Mann–Kendall test [52,53].

Besides, the $\sum NDVI$ change rate over the study period was determined using Equations (1) and (2), and can directly express the magnitude of trend change:

$$S_{\Delta NDVI} = \left| \frac{(\beta \times t_n + a) - (\beta \times t_1 + a)}{\beta \times t_1 + a} \right| \times 100 \tag{1}$$

$$a = \overline{X} - \beta \times \overline{t} \tag{2}$$

where $S_{\Delta NDVI}$ is the $\sum NDVI$ change rate (percentage of $\sum NDVI$ increase over the study period), $\overline{X}$ is the mean of the time series of $\sum NDVIs$, $\overline{t}$ is the mean value of the dates, and t_1 and t_n are the start and end years of the time series, respectively. The significant change areas can be acquired using Equation (3):

$$|S_{\Delta NDVI}| > p \tag{3}$$

where p is the change rate threshold, which was set as 10% in this study.

2.3.2. Discriminating Human-Induced Changes Using Optimized RESTREND

In the significant change areas, we used optimized RESTREND to profile human-induced changes. The first step for optimized RESTREND was pre-analyzing the relationship between climate factors and AA-NDVIs to determine suitable climate factors and their best related period in the study areas. Second, a multiple regression model was built at the pixel level to simulate ideal climate–vegetation relationships during the study period. Finally, the residual analysis was conducted. If the changes in NDVI response were only due to climatic effects, the residuals of the predicted NDVIs and actual NDVIs in the regression model should be close to zero, and only fluctuate randomly without significant trend signals. Any significant trend in the residuals will indicate changes that are not due to climatic effects [1]. A decreasing trend in residuals represented human-induced vegetation degradation, while an increasing trend in residuals indicated that vegetation improvements were happening in the study area because of human activity. If there was no trend in the residual series, all of the vegetation changes that happened during study period were driven by climate change [28].

(1) Find the most suitable climate variables and their best-related period.

$$r = \frac{\sum(x_i - \overline{x})(y_i - \overline{y})}{\sqrt{\sum(x_i - \overline{x})^2 \cdot \sum(y_i - \overline{y})^2}}, \tag{4}$$

where r is the correlation coefficient; x_i is the value of AA-NDVIs in the ith year; $\overline{x}$ is the average value of x_i from 2000 to 2014; y_i is the value of the specific climate index (Table 1) in the *i*th year; and $\overline{y}$ is the average value of y_i from 2000 to 2014. In each pixel, there were a total of 135 (3 × 45) r results for climate factors (P/T/W). The best-related period of T, P, and W (hereafter T_{best}, P_{best}, and W_{best}, respectively) were the most suitable climate variables to build the climate–vegetation model.

(2) Simulate the ideal climate–vegetation relationship.

Treating T_{best}, P_{best}, and W_{best} as the representatives of the climate factors, we first normalized each variable using the min–max normalization method as in Equation (5), and then used multiple regression analysis to build the pixel-level climate–vegetation model as in Equation (6).

$$Norv_i = \frac{v_i - min(v)}{max(v) - min(v)} \tag{5}$$

where v is a specific time series variable (i.e., T_{best}, P_{best}, and W_{best}); v_i is the value of variable v at the ith time; $Norv_i$ is the normalized value of variable v at the ith time; and the $max(v)$ and $min(v)$ is the maximum and minimum values of time series variable v, respectively.

$$y = \beta_0 + \beta_1 P + \beta_2 T + \beta_3 W, \tag{6}$$

where y is the time series $\sum NDVIs$ of the 'reference pixel', which refers to the maximum value of $\sum NDVIs$ in the homogeneous spatial neighborhood of the target pixel; P, T, and W are the normalized values of the time series T_{best}, P_{best}, and W_{best} respectively; and β_0, β_1, β_2, and β_3, are the parameters of the model.

In this paper, the location of a corresponding reference pixel for a specific pixel x was determined by three steps (Figure 3). Step 1 found all of the neighborhood pixels that were located within the same ecological zone of x, and marked them as candidates. In step 2, candidates that could not simultaneously satisfy the following two conditions lost their qualifications: (1) there must be a significant correlation of $\sum NDVI$ trajectories between x and the candidates; (2) the differential of mean value of the $\sum NDVI$ trajectory between x and candidates must be below 3δ. The 3δ method has been used to detect land cover change in some studies [14,56], and helped determine homogeneous spatial neighborhoods in this study. All of the candidates selected by this step constituted the homogeneous spatial neighborhood of x. In step 3, the candidate with the maximum value of $\sum NDVIs$ was defined as the reference pixel.

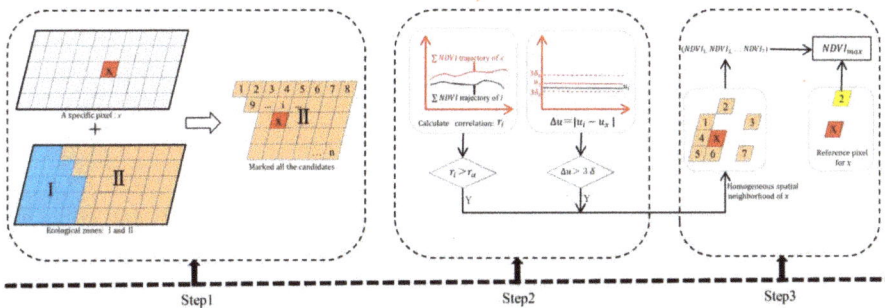

Figure 3. Three steps used to find the reference pixel to simulate ideal climate–vegetation relationships.

(3) Residual analysis to discriminate human-induced changes

Residual analysis was conducted using the SMK method though Equations (1)–(6). Residuals refer to the differential value between the potential $\sum NDVIs$ and the actual $\sum NDVIs$. A significant decreasing or increasing trend in residuals indicated that vegetation change is driven by human activities.

3. Results

3.1. Vegetation Dynamics during the Past 15 Years

Over the past 15 years, obvious vegetation increases happened in both regions, and Naiman increased more dramatically than Ar Horqin. In the Ar Horqin region, about 18.6% of the total area (2371.1 km^2) had a significant trend of change ($p < 0.05$), with 16.8% of the area increasing by more than 10%, and only 1.3% of the area decreasing more than -10%. About 66.3% of the Naiman region (5301.9 km^2) showed a significant trend change during this period, with much of the change area (64.1%) increasing by more than 10%, and only 1.1% of the area decreasing by more than -10%. About half of the change area increased between 20–40% during the study period (Table 2).

Table 2. Significant trends in the normalized difference vegetation index (*NDVI*) change rate ($S_{\Delta NDVI}$) in the two study areas during 2000–2014.

Significant Trend Change ($S_{\Delta NDVI}$)	Ar Horqin Region		Naiman Region	
	Rate (%)	Area (km^2)	Rate (%)	Area (km^2)
<−40%	0.1	17.6	0.0	3.2
−40–−20%	0.7	88.1	0.5	40.0
−20–−10%	0.5	61.5	0.6	46.8
−10–10%	0.5	61.0	1.1	89.0
10–20%	2.2	281.2	14.2	1133.0
20–40%	7.7	983.8	34.0	2717.4
>40%	6.9	877.9	15.9	1272.6
Total	18.6	2371.1	66.3	5301.9

In both regions, most positive changes happened in the southern study area showing a zonal or block distribution. The negative change areas showed a sporadic distribution in the northern study area (Figures 4a and 5a). Specifically, positive changes always happened along rivers and roads and negative changes mainly happened in the south of the Greater Khingan Range Mountain in the Ar Horqin region. In the Naiman region, negative changes were mainly located around the county center.

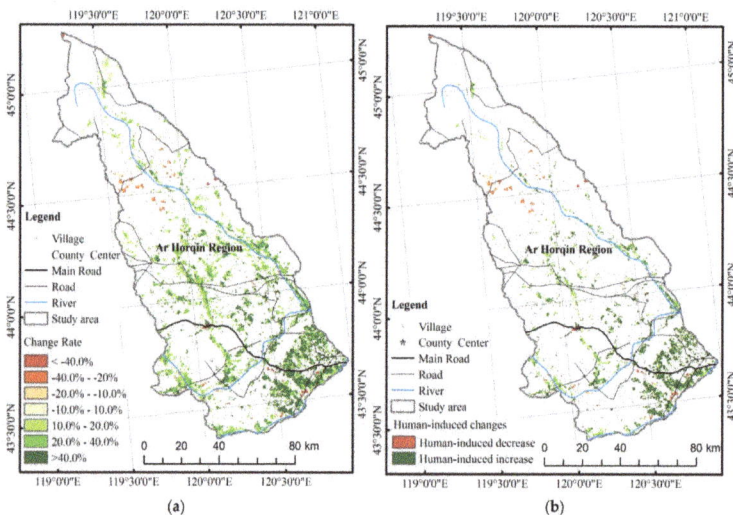

Figure 4. Vegetation change (**a**) and driving forces (**b**) in the Ar Horqin region.

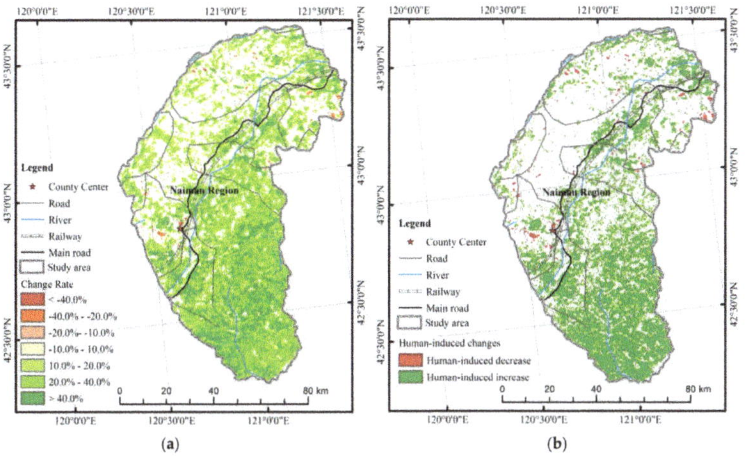

Figure 5. Vegetation change (**a**) and driving forces (**b**) in the Naiman region.

3.2. Human-Induced Vegetation Changes

3.2.1. Significant Human-Induced Change in the Ar Horqin Region

In the Ar Horqin region, both climate change and human activities contributed equally to the vegetation changes over the past 15 years, according to the trend of residuals shown in Figure 4b. About 50.1% of the significant changes resulted from human activities, while 49.9% of the significant changes were caused by climate change.

However, human activities caused more dramatic changes than climate in Ar Horqin. The results showed that vegetation in most of the human-induced change area (about 57.2%) increased by more than 40%, while climate change resulted in less rapid changes, with most of the climate-induced change area (about 56.7%) increasing between 20–40%.

The results also showed that human activities were more responsible for a vegetation decrease than climate change in the Ar Horqin region. About 85.3% of the decreases were caused by human activities. Vegetation increases over the past 15 years may be equally induced by human activities (47.3%) and climate changes (52.7%).

3.2.2. Significant Human-Induced Change in the Naiman Region

In the Naiman region, results demonstrated that human activities were the main drivers of vegetation change between 2000–2014, contributing up to 60.3% of the total significant changes (Figure 5b). Furthermore, human activities were more responsible for both negative and positive vegetation change than climate. About 84.7% of the decreases were caused by human activities, and about 59.8% of the vegetation increases were caused by human activities. Climate change resulted in more vegetation increases (40.2%) than decreases (14.3%). The results indicated that human activities had deeper and broader influences in the Naiman region compared with the Ar Horqin region.

3.2.3. Validation of Human-Induced Vegetation Change

This study aimed to discriminate significant human-induced changes in two different regions. However, the validation of vegetation change at regional scales is notoriously hard because there is seldom ground-truthed data of historical land cover changes, as these data are difficult to collect over long periods [57]. Here, we used time series Landsat images with a high spatial resolution (30 m) to validate vegetation changes and try to find further evidence of human-induced changes through visual interpretation of the images [58]. Visual interpretation of false color composite images showed that our methods effectively discriminated significant human-induced vegetation change in both regions. Figures 6 and 7 show evidence that indicates land cover changes and driving forces during the study period in a typical change area.

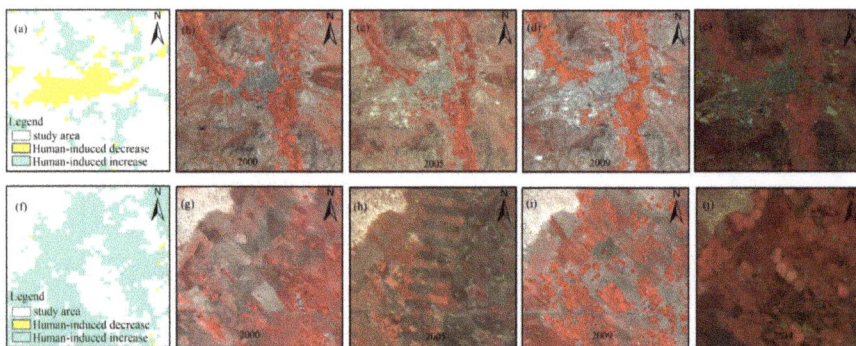

Figure 6. Typical change areas in the Ar Horqin region: (**a**) a typical human-induced vegetation decrease area from the optimized residual trends method (RESTREND) method; (**b–e**) Landsat false color image maps for (**a**) on 09/24/2000, 09/06/2005, 09/17/2009, and 09/04/2014, respectively; (**f**) a typical human-induced vegetation increase area detected by the optimized RESTREND method; (**g–j**) Landsat false color image maps for (**f**) on 09/24/2000, 09/06/2005, 09/17/2009, and 09/04/2014, respectively.

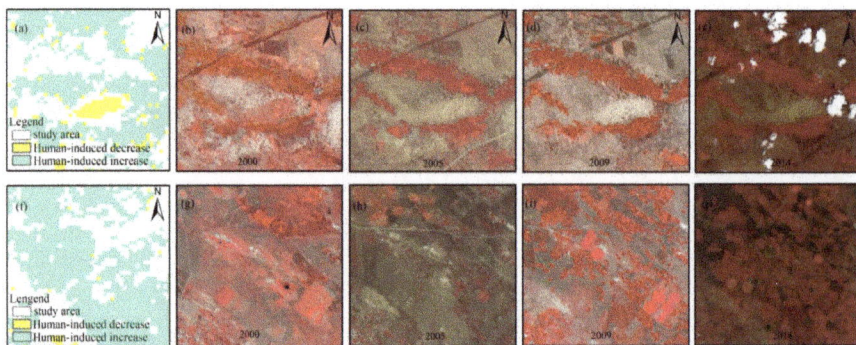

Figure 7. Typical change areas in the Naiman region (**a**) a typical human-induced vegetation decrease area from the optimized RESTREND method; (**b–e**) Landsat false color image maps for (**a**) on 09/24/2000, 09/06/2005, 09/17/2009, and 09/04/2014, respectively; (**f**) a typical human-induced vegetation increase area detected by the optimized RESTREND method; (**g–j**) Landsat false color image maps for (**f**) on 09/24/2000, 09/06/2005, 09/17/2009, and 09/04/2014, respectively.

Figure 6a–e show a decrease in vegetation that is obviously related to urban expansion. Land cover change, i.e., farmland conversion to urban area, which is solely caused by human activity, can result in decreases in vegetation. Figure 7a–e show local grassland degradation, which is also caused by human activity. The degradation area shown in (a) was surrounded by farmland. It had land cover conversion from grassland (pink in Landsat false color image) to sandy land (bright white in Landsat false color image) during 2000–2005.

An in situ investigation of vegetation increases in grassland showed that in Figures 6f–j and 7f–j, all of the increases were caused by human activity, e.g., grassland management. The conversion of natural grassland into artificial pastures was verified by the appearance of several circular grasslands, as found in Figures 6j and 7i,j. Two photographs of the circular patches are shown in Figure 8, demonstrating clear signs of the human management that resulted in a vegetation increase in these areas.

Figure 8. Photographs of in situ investigation of the circular grassland in Figure 6j, collected in August 2015.

4. Discussion

4.1. Different Driving Factors in the Study Areas

Although human factors contributed more to these changes than climate change in both regions, the specific human activities that caused the changes in the two regions were different.

In the Ar Horqin region, about 71.0% of the total area was grassland. Human activities caused 47.3% of the vegetation increase and 85.3% of the vegetation decrease. Human eco-managements (i.e., grass land restoration in Figure 8) and overgrazing were the two main human activities with positive and negative effects, respectively.

Figure 9a further supported this point of view. As we mosaicked the land cover map in 2010 by human-induced vegetation change area, it was clear that 57.9% of the human-induced changes were happening on grassland. Figure 4a also showed that most of the human-induced grassland changes were positive changes, which can be attributed to human management such as the Beijing–Tianjin Sandstorm Source Controlling Program (introduced in 2001) and the Ecological Subsidy and Award System (introduced in 2011) that were carried out in this region. Figures 5f–j and 4b indicated the effects of human management on the grassland located in the southeast corner of the Ar Horqin region. However, most of the human-induced grassland decrease that happened in the northern area shown in Figure 4b was also caused by human activities according to Figure 9a, and these decreases can be attributed to overgrazing. We believe the effect of positive human management of grassland is greater than the negative outcomes of overgrazing, although the grazing activities increased in the Ar Horqin region during the study period (Figure 10b).

Farming is not a key human factor of change in this area, because only 33.9% of human-induced changes happened on farmland, as shown in Figure 9a. The majority of the significantly changed areas of farmland are located around the county center and spread in a north–south direction along the valley (Figure 4a). However, Figures 4b and 9a showed that only part of the significant farmland changes in this area was attributed to human activities. The amount of chemical fertilizer used only had a small increase (Figure 10a), although 13.6% of the total area is farmland.

In the Naiman region, farming was the key human driver of significant vegetation change. About 46.5% of total area is farmland, and 66.3% of the human-induced significant changes happened in farmland, as shown in Figure 9b. The majority of farmland was in the southern area, and some was located along the main road in a northeast–southwest direction. Human-induced farmland changes were almost all positive, as shown in Figures 5b and 9b. The Naiman region is a typical producing area of corn, and 90% of the farmland in this region is corn. The vegetation increase that occurred in farmlands in Naiman was mainly because of market incentives and the sharp increase in using fertilizer. According to the China Animal Industry Yearbook (2001–2015), the corn price in Inner Mongolia kept rising from 0.81 yuan/kg to 2.32 yuan/kg, with an increasing rate of up to 186.4% from 2000 to 2014. We believe that it greatly increased local people's willingness to plant corn and improve corn output. In addition, statistical data also showed that the amount of chemical fertilizer used in 2013 was up to 6.7 times greater than in 2000 (Figure 10a), while the area of farmland in 2011 was up to 1.5 times more than in 2000. This could be further evidence of the human activities that positively influenced farmland changes.

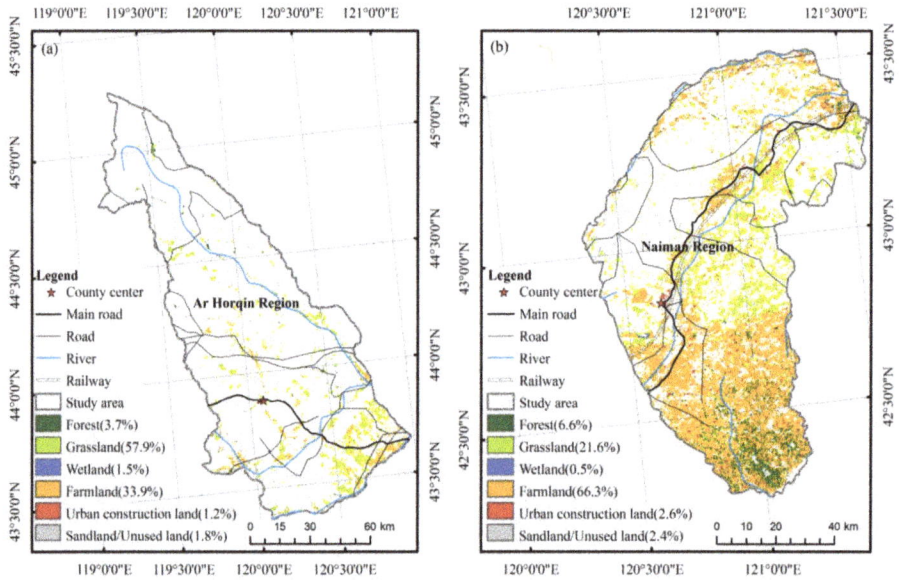

Figure 9. Human-induced changes on different land covers in Ar Horqin region (**a**) and Naiman region (**b**).

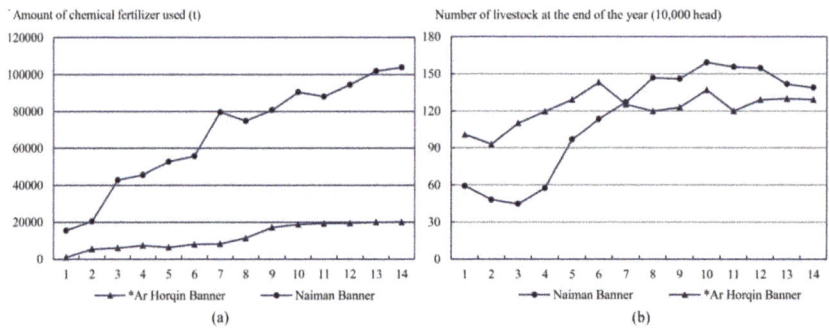

Figure 10. (**a**) Amount of chemical fertilizer and (**b**) number of livestock at the end of each year in the two regions.

About 21.6% of the significant human-induced changes happened in grassland, as shown in Figure 9b. Comparing Figures 9b and 5b, most of the human-induced grassland change was positive, which corresponded to sharp increases over time in the index of the number of livestock at the end of each year (Figure 10b). Meanwhile, the area of decreased grassland caused by overgrazing was still small. Therefore, we believe that the human management of grassland was the other important factor of vegetation change in the Naiman region, which means the Grazing Withdrawal Program (introduced in 2003) and the Ecological Subsidy and Award System (introduced in 2011) implemented in this region had positive eco-effects.

The increased number of livestock in the study area is also noteworthy (Figure 10b). Previous study had proved that the increased number of livestock at the end of each year was mainly because of market incentives [59]. It is reported that the market prices of cattle, sheep, and pork had grown rapidly in Inner Mongolia. The price of sheep raised from 14.77 yuan/kg to 65.41 yuan/kg. The price of pigs raised from 10.1 yuan/kg to 22.48 yuan/kg. Especially, the price of cattle raised from 12.88 yuan/kg to 63.29 yuan/kg, with an increasing rate up to 26.09% per year. These price increases were a great incentive for people in the Naiman region, because the Naiman region is a typical producing area of cattle. According to our field investigation, the increasing supply of forage, which was caused by the increasing number of livestock, was no longer from overgrazing. People used the subsidy given by Grazing Withdrawal Program from the government to buy forage or planted silage corn and alfalfa to give fodder to livestock.

4.2. Limitations and Further Research

This study profiled significant human-induced vegetation change in the study area. We first used the SMK method to find areas of significant change, and then used RESTREND to discriminate human-induced from climate-induced change. We used SMK to mask slightly changed and no change areas, i.e., change rates between −10% and 10%, because the variability of vegetation growth in arid/semi-arid areas is commonly high, and this extent of change was regarded as a relatively stable growth rate in the study area. However, this kind of area can also be influenced by human activities, and even further by the complicated interaction and offset mechanisms between climate change and human activities, such as increased irrigation in less rainy years, which need to be analyzed in further studies.

The strategy in optimized RESTREND had two advantages compared with previous studies [31–33]. First, the algorithm flow was automatic, and did not require human intervention, which could strengthen its applicability and ease-of-use in large areas. Second, we used a time series vegetation index to finish defining the flexible homogeneous areas, which avoided uncertainties caused by adding external time-varying data such as land use maps. Although we have detailed analysis of the human-induced vegetation change in the two regions, it is hard to validate the result directly and compare it with other methods. Here, we took a time series Landsat image dataset as the main source of validation data, and the attribution analysis mainly relied on visual interpretation. Thus, the results may contain some uncertainty. Accumulating a site investigation dataset over a long period and using a questionnaire survey about the effects of specific eco-programs are both very important in future studies.

The uncertainty of the results could be also caused by the coarse temporal resolution of meteorology datasets (i.e., one month), which was different form the time series NDVI datasets (i.e., 16 days). Using the highest correlation coefficient to choose the best-related period between NDVI and each climate factor might not be the best strategy. The partial correlation coefficient could be used in the future study to improve the process. The suddenly appearances of wildfire, diseases, and insects invasion [4,60,61] in ecosystems are neither direct human-related nor climate-related factors. Meanwhile, the drought and flood that happened during the study period may influence the linear regression relationship between climate and vegetation, which is the basic of RESTREND [5]. Results would be unreliable when these interferences happened. Breakpoints detection methods [56–58] can be used in combination with RESTREND to accurately define specific human-induced changes in the future study.

In addition, this study did not consider the possibility of changes in dominant driving factors over time as we used linear analysis methods. However, vegetation increases during past 15 years might reflect climate-induced negative changes in the first five years, and then human-induced positive changes in the following 10 years, and vice versa. A finer-scale temporal analysis would help define human-induced changes. Piecewise analysis methods [11] or other time series detection methods [62,63] can be used in further research to resolve this problem.

5. Conclusions

This paper first detected vegetation dynamics using SMK in the Ar Horqin and Naiman regions of China from 2000 to 2014, and then applied optimized RESTREND to profile significant human-induced changes in vegetation. We optimized strategies to define homogeneous spatial neighborhoods in RESTREND. An automatic algorithm was proposed to define flexible homogeneous areas based on the statistical analysis and trajectory analysis of a time series NDVI dataset.

Results showed that during past 15 years, significant vegetation increases happened in both regions, and climate change and human activities both contributed to the changes. However, more significant vegetation increases happened in the Naiman region (64.1%) than in the Ar Horqin region (16.8%). In the Ar Horqin region, climate and human activities equally contributed to vegetation change. Meanwhile, in the Naimen region, it was proved that human activities contributed more than climate factors to vegetation change. Human factors had a stronger influence on ecosystems and were responsible for the vegetation decrease in both regions. In different study areas, dominant human driving forces varied. In the Ar Horqin region, human management of grassland was the key factor, and the secondary human factor was overgrazing. Farming was not a key human factor for change in this area. In the Naiman region, farming was the key human factor for vegetation change. Positive human management in grassland was the other key driver of change.

In future studies, the improvement of RESTREND and finer-scale temporal analysis would help to accurately define specific human-induced changes. The study of a very long series of palaeoenvironmental records can also improve knowledge on the drivers of changes over long time periods in these regions [64]. The complex interactions and offset mechanisms between climate change and human activities need to be clarified so that the underestimates or overestimates of human effects can be avoided. In addition, we did not consider the trend-reversal phenomenon, which is very likely to happen over time, when analyzing the dominant driving factors. Some abrupt change detection methods can be used in further research.

Acknowledgments: This study was supported by State Key Laboratory of Resources and Environmental Information System, the National Natural Science Foundation of China (grant no. 41701474 and 41701467), the National Key Research and Development Plan of China (grant no. 2016YFC0500205), the National Basic Research Program of China (grant no. 2015CB954103) and the Key Laboratory for National Geograophy State Monitoring (National Administration of Surveying, Mapping and Geoinformation, grant no. 2017NGCM09). The authors are grateful to the anonymous reviewers for their constructive criticism and comments. We thank Leonie Seabrook, PhD, from Liwen Bianji, Edanz Group China (www.liwenbianji.cn/ac), for editing the English text of a draft of this manuscript.

Author Contributions: L.X. designed the experiments and wrote the draft version of manuscript; G.Y., Z.T. and Y.Z. contributed to the editing and reversion of the manuscript.

Conflicts of Interest: The authors declare no conflict of interest. The funding sponsors had no role in the design of the study; in the collection, analyses, or interpretation of data; in the writing of the manuscript, and in the decision to publish the results.

References

1. Evans, J.; Geerken, R. Discrimination between climate and human-induced dryland degradation. *J. Arid Environ.* **2004**, *57*, 535–554. [CrossRef]
2. Omuto, C.T.; Vargas, R.R.; Alim, M.S.; Paron, P. Mixed-effects modelling of time series NDVI-rainfall relationship for detecting human-induced loss of vegetation cover in drylands. *J. Arid Environ.* **2010**, *74*, 1552–1563. [CrossRef]
3. Marchant, R.; Richer, S.; Capitani, C.; Courtney-Mustaphi, C.; Prendergast, M.; Stump, D.; Boles, O.; Lane, L.; Wynne-Jones, S.; Vázquez, C.F.; et al. Drivers and trajectories of land cover change in East Africa: Human and environmental interactions from 6000 years ago to present. *Earth Sci. Rev.* **2018**, *178*, 322–378. [CrossRef]
4. Zhou, X.; Yamaguchi, Y.; Arjasakusuma, S. Distinguishing the vegetation dynamics induced by anthropogenic factors using vegetation optical depth and AVHRR NDVI: A cross-border study on the Mongolian Plateau. *Sci. Total Environ.* **2018**, *616–617*, 730–743. [CrossRef] [PubMed]
5. Wang, H.; Liu, G.H.; Li, Z.S.; Ye, X.; Fu, B.J.; Lv, Y.H. Impacts of Drought and Human Activity on Vegetation Growth in the Grain for Green Program Region, China. *Chin. Geogr. Sci.* **2018**. [CrossRef]
6. Pearson, R.G.; Phillips, S.J.; Loranty, M.M.; Beck, P.S.A.; Damoulas, T.; Knight, S.J.; Goetz, S.J. Shifts in Arctic vegetation and associated feedbacks under-climate change. *Nat. Clim. Chang.* **2013**, *3*, 673–677. [CrossRef]
7. Dewan, A.M.; Yamaguchi, Y. Land use and land cover change in Greater Dhaka, Bangladesh: Using remote sensing to promote sustainable urbanization. *Appl. Geogr.* **2009**, *29*, 390–401. [CrossRef]
8. Byomkesh, T.; Nakagoshi, N.; Dewan, A.M. Urbanization and green space dynamics in Greater Dhaka, Bangladesh. *Landsc. Ecol. Eng.* **2012**, *8*, 45–58. [CrossRef]
9. Rêgo, J.C.L.; Soares-Gomes, A.; Silva, F.S. Loss of vegetation cover in a tropical island of the Amazon coastal zone (Maranhão Island, Brazil). *Land Use Policy* **2017**. [CrossRef]
10. Tong, X.W.; Brandt, M.; Yue, Y.M.; Horion, S.; Wang, K.L.; Keersmaecker, W.D.; Tian, F.; Schurgers, G.; Xiao, X.M.; Luo, Y.Q.; et al. Increased vegetation growth and carbon stock in China karst via ecological engineering. *Nat. Sustain.* **2018**, *1*, 44–50. [CrossRef]
11. Xu, L.L.; Li, B.L.; Yuan, Y.C.; Gao, X.Z.; Zhang, T.; Sun, Q.L. Detecting Different Types of Directional Land Cover Changes Using MODIS NDVI Time Series Dataset. *Remote Sens.* **2016**, *8*, 495. [CrossRef]
12. Fisher, R.A.; Koven, C.D.; Anderegg, W.; Christoffersen, B.O.; Dietze, M.C.; Farrior, C.; Holm, J.A.; Hurtt, G.; Knox, R.G.; Lawrence, P.J.; et al. Vegetation Demographics in Earth System Models: A review of progress and priorities. *Glob. Chang. Biol.* **2017**, *24*, 35–54. [CrossRef] [PubMed]
13. Wessels, K.J.; Pretorius, D.J.; Prince, S.D. Reality of rangeland degradation mapping with remote sensing: The South African experience. In Proceedings of the 14th Australasian Remote Sensing and Photogrammetry Conference, Darwin, Australia, 29 September–3 October 2008.
14. Wang, Y.W.; Wang, Z.Y.; Li, R.R.; Meng, X.L.; Ju, X.J.; Zhao, Y.G.; Sha, Z.Y. Comparison of Modeling Grassland Degradation with and without Considering Localized Spatial Associations in Vegetation Changing Patterns. *Sustainability* **2018**, *10*, 316. [CrossRef]
15. Marignani, M.; Chiarucci, A.; Sadori, L.; Mercuri, A.M. Natural and human impact in Mediterranean landscapes: An intriguing puzzle or only a question of time? *Plant Biosyst.* **2017**, *151*, 900–905. [CrossRef]
16. Mercuri, A.M. Genesis and evolution of the cultural landscape in central Mediterranean: The 'where, when and how' through the palynological approach. *Landsc. Ecol.* **2014**, *29*, 1799–1810. [CrossRef]
17. Turner, B.; Benjamin, P. *Fragile Lands: Identification and Use for Agriculture: Agriculture, Environment, and Health: Sustainable Development in the 21st Century*; University of Minnesota Press: Minneapolis, MN, USA; London, UK, 1994.
18. Wessels, K.J.; Prince, S.D.; Frost, P.E.; Zyl, D.V. Assessing the effects of human-induced land degradation in the former homelands of northern South Africa with a 1 km AVHRR NDVI time-series. *Remote Sens. Environ.* **2004**, *91*, 47–67. [CrossRef]
19. Guo, Z.L.; Huang, N.; Dong, Z.B.; Pelt, R.S.V.; Zobeck, T.M. Wind Erosion Induced Soil Degradation in Northern China: Status, Measures and Perspective. *Sustainability* **2014**, *6*, 8951–8966. [CrossRef]
20. Petit, C.; Scudder, T.; Lambin, E. Quantifying processes of land-cover change by remote sensing: Resettlement and rapid land-cover changes in south-eastern Zambia. *Int. J. Remote Sens.* **2001**, *22*, 3435–3456. [CrossRef]

21. Tian, F.; Fensholt, R.; Verbesselt, J.; Grogan, K.; Horion, S.; Wang, Y.J. Evaluating temporal consistency of long-term global NDVI datasets for trend analysis. *Remote Sens. Environ.* **2015**, *163*, 326–340. [CrossRef]

22. Lyon, J.G.; Yuan, D.; Lunetta, R.S.; Elvidge, C.D. A change detection experiment using vegetation indices. *Photogramm. Eng. Remote Sens.* **1998**, *64*, 143–150.

23. Shi, X.; Wang, W.; Shi, W. Progress on quantitative assessment of the impacts of climate change and human activities on cropland change. *J. Geogr Sci.* **2016**, *26*, 339–354. [CrossRef]

24. Houerou, H.N.L. Rain-Use Efficiency: A Unifying Concept in Arid-Land Ecology. *J. Arid Environ.* **1984**, *7*, 213–247.

25. Prince, S.D.; Colstoun, D.; Brown, E.; Kravitz, L.L. Evidence from rain-use efficiencies does not indicate extensive sahelian desertification. *Glob. Chang. Biol.* **1998**, *4*, 359–374. [CrossRef]

26. Dardel, C.; Kergoat, L.; Hiernaux, P.; Grippa, M.; Mougin, E.; Ciais, P.; Nguyen, C.C. Rain-Use-Efficiency: What It Tells Us about the Conflicting Sahel Greening and Sahelian Paradox. *Remote Sens.* **2014**, *6*, 3446–3474. [CrossRef]

27. Li, A.; Wu, J.; Huang, J. Distinguishing between human-induced and climate-driven vegetation changes: A critical application of RESTREND in inner Mongolia. *Landsc. Ecol.* **2012**, *27*, 969–982. [CrossRef]

28. Wessels, K.J.; Van Den Bergh, F.; Scholes, R.J. Limits to detectability of land degradation by trend analysis of vegetation index data. *Remote Sens. Environ.* **2012**, *125*, 10–22. [CrossRef]

29. Wessels, K.J.; Prince, S.D.; Malherbe, J.; Small, J.; Frost, P.E.; Vanzyl, D. Can human-induced land degradation be distinguished from the effects of rainfall variability? A case study in South Africa. *J. Arid Environ.* **2007**, *68*, 271–297. [CrossRef]

30. Wessels, K.J.; Prince, S.D.; Zambatis, N.; MacFadyen, S.; Frost, P.E.; Vanzyl, D. Relationship between herbaceous biomass and 1-km^2 Advanced Very High Resolution Radiometer (AVHRR) NDVI in Kruger National Park, South Africa. *Int. J. Remote Sens.* **2006**, *27*, 951–973. [CrossRef]

31. Zhuo, L.; Cao, X.; Chen, J.; Chen, Z.X.; Shi, P.J. Assessment of grassland ecological restoration project in Xilin Gol grassland. *Acta Geogr. Sin.* **2007**, *62*, 471–480. (In Chinese)

32. Prince, S.D.; Becker-Reshef, I.; Rishmawi, K. Detection and mapping of long-term land degradation using local net production scaling: Application to Zimbabwe. *Remote Sens. Environ.* **2009**, *113*, 1046–1057. [CrossRef]

33. He, C.Y.; Tian, J.; Gao, B.; Zhao, Y.Y. Differentiating climate- and human-induced drivers of grassland degradation in the Liao River Basin, China. *Environ. Monit. Assess.* **2015**, *187*, 1–14. [CrossRef] [PubMed]

34. Han, Z.; Wang, T.; Yan, C.; Yan, C.Z.; Liu, Y.B.; Liu, L.C.; Li, A.M.; Du, H.Q. Change trends for desertified lands in the Horqin Sandy Land at the beginning of the twenty-first century. *Environ. Earth Sci.* **2010**, *59*, 1749–1757. [CrossRef]

35. Yan, Q.L.; Zhu, J.J.; Hu, Z.B.; Sun, O.J. Environmental impacts of the shelter forests in Horqin Sandy land, northeast China. *J. Environ. Qual.* **2011**, *40*, 815. [CrossRef] [PubMed]

36. Zhang, G.; Dong, J.W.; Xiao, X.M.; Hu, Z.M.; Sheldon, S. Effectiveness of ecological restoration projects in Horqin Sandy Land, China Based on SPOT-VGT NDVI data. *Ecol. Eng.* **2012**, *38*, 20–29. [CrossRef]

37. Yan, Q.L.; Zhu, J.J.; Zheng, X.; Jin, C.J. Causal effects of shelter forests and water factors on desertification control during 2000–2010 at the Horqin Sandy Land region, China. *J. For. Res.* **2015**, *26*, 33–45. [CrossRef]

38. Chen, H.B.; Shao, L.Q.; Zhao, M.J.; Zhang, X.; Zhang, D.J. Grassland conservation programs, vegetation rehabilitation and spatial dependency in Inner Mongolia, China. *Land Use Policy* **2017**, *64*, 429–439. [CrossRef]

39. Xu, L.L.; Li, B.L.; Yuan, Y.C.; Gao, X.Z.; Zhang, T. A. Temporal-Spatial Iteration Method to Reconstruct NDVI Time Series Datasets. *Remote Sens.* **2015**, *7*, 8906–8924. [CrossRef]

40. Higginbottom, T.P.; Symeonakis, E. Assessing Land Degradation and Desertification Using Vegetation Index Data: Current Frameworks and Future Directions. *Remote Sens.* **2014**, *6*, 481–493. [CrossRef]

41. Landmann, T.; Dubovyk, O. Spatial analysis of human-induced vegetation productivity decline over eastern Africa using a decade (2001–2011) of medium resolution MODIS time-series data. *Int. J. Appl. Earth Obs.* **2014**, *33*, 76–82. [CrossRef]

42. Ibrahim, Y.Z.; Balzter, H.; Kaduk, J.; Tucker, C.J. Land Degradation Assessment Using Residual Trend Analysis of GIMMS NDVI3g, Soil Moisture and Rainfall in Sub-Saharan West Africa from 1982 to 2012. *Remote Sens.* **2015**, *7*, 5471–5494. [CrossRef]

43. Savitzky, A.; Golay, M.J.E. Smoothing and Differentiation of Data by Simplified Least Squares Procedures. *Anal. Chem.* **1964**, *36*, 1627–1639. [CrossRef]

44. Chen, J.; Jönsson, P.; Tamura, M.; Gu, Z.H.; Matsushita, B.; Eklundh, L. A simple method for reconstructing a high-quality NDVI time-series data set based on the Savitzky-Golay filter. *Remote Sens. Environ.* **2004**, *91*, 332–344. [CrossRef]

45. Fang, J.; Piao, S.; He, J.; Ma, W.H. Increasing terrestrial vegetation activity in China, 1982–1999. *Sci. China Life Sci.* **2004**, *47*, 229–240. [CrossRef]

46. Song, Y.; Ma, M.G. A statistical analysis of the relationship between climatic factors and the Normalized Difference Vegetation Index in China. *Int. J. Remote Sens.* **2011**, *45*, 374–382. [CrossRef]

47. Du, J.Q.; Shu, J.M.; Yin, J.Q.; Yuan, X.J.; Jiaerheng, A.; Xiong, S.S.; He, P.; Liu, W.L. Analysis on spatio-temporal trends and drivers in vegetation growth during recent decades in Xinjiang, China. *Int. J. Appl. Earth Obs.* **2015**, *38*, 216–228. [CrossRef]

48. Xu, L.L.; Li, B.L.; Yuan, Y.C.; Gao, X.Z.; Liu, H.J.; Dong, G.H. Changes in China's cultivated land and the evaluation of land requisition-compensation balance policy from 2000 to 2010. *Resour. Sci.* **2015**, *37*, 1543–1551. (In Chinese)

49. Fensholt, R.; Rasmussen, K.; Nielsen, T.T.; Mbow, C. Evaluation of earth observation based long term vegetation trends-Intercomparing NDVI time series trend analysis consistency of Sahel from AVHRR GIMMS, Terra MODIS and SPOT VGT data. *Remote Sens. Environ.* **2009**, *113*, 1886–1898. [CrossRef]

50. Wang, H.Y.; Li, Z.Y.; Gao, Z.H.; Wu, J.J.; Sun, B.; Li, C.L. Assessment of land degradation using time series trend analysis of vegetation indictors in Otindag Sandy land. In Proceedings of the 35th IOP Conference Series: Earth and Environmental Science, Beijing, China, 22–26 April 2013; p. 17.

51. Olsson, L.; Eklundh, L.; Ardo, J. A recent greening of the Sahel-trends, patterns and potential causes. *J. Arid Environ.* **2005**, *63*, 556–566. [CrossRef]

52. Ahmedou, O.C.A.; Nagasawa, R.; Osman, A.E.; Hattori, K. Rainfall variability and vegetation dynamics in the Mauritanian Sahel. *Clim. Res.* **2009**, *38*, 75–81. [CrossRef]

53. Li, Z.; Huffman, T.; Mcconkey, B.; Townley-Smith, L. Monitoring and modeling spatial and temporal patterns of grassland dynamics using time-series MODIS NDVI with climate and stocking data. *Remote Sens. Environ.* **2013**, *138*, 232–244. [CrossRef]

54. Beurs, K.M.; de Henebry, G.M. A statistical framework for the analysis of long image time series. *Int. J. Remote Sens.* **2005**, *26*, 1551–1573. [CrossRef]

55. Cai, B.F.; Yu, R. Advance and evaluation in the long time series vegetation trends research based on remote sensing. *J. Remote Sens.* **2009**, *13*, 1170–1186. (In Chinese)

56. Zhu, Z.; Woodcock, C.E. Continuous change detection and classification of land cover using all available Landsat data. *Remote Sens. Environ.* **2014**, *144*, 152–171. [CrossRef]

57. Jamali, S.; Jönsson, P.; Eklundh, L.; Ardo, J.; Seaquist, J. Detecting changes in vegetation trends using time series segmentation. *Remote Sens. Environ.* **2015**, *156*, 182–195. [CrossRef]

58. De Jong, R.; Verbesselt, J.; Zeileis, A.; Schaepman, M.E. Shifts in global vegetation activity trends. *Remote Sens.* **2013**, *5*, 1117–1133. [CrossRef]

59. Wei, Y.; Lin, H.L.; Wang, X.Y. Analysis of the impact of price factors on livestock output in China. *J. Gansu Radio TV Univ.* **2016**, *26*, 83–86. (In Chinese)

60. Li, F.; Lawrence, D.M.; Bond-Lamberty, B. Human impacts on 20th century fire dynamics and implications for global carbon and water trajectories. *Glob. Planet. Chang.* **2018**, *162*, 18–27. [CrossRef]

61. Veenendaal, E.M.; Torelloraventos, M.; Miranda, H.S.; Margaretesato, N.; Oliveras, I.; Langevelde, F.V.; Asner, G.P.; Lloyd, J. On the relationship between fire regime and vegetation structure in the tropics. *New Phytol.* **2018**, *218*, 153–166. [CrossRef] [PubMed]

62. Li, T.; Wang, J.D.; Zhou, H.M.; Wang, J. Automatic detection of forest fire disturbance based on dynamic modeling from MODIS time-series observations. *Int. J. Remote Sens.* **2018**, *39*, 3801–3815. [CrossRef]

63. Song, Y.; Jin, L.; Wang, H. Vegetation Changes along the Qinghai-Tibet Plateau Engineering Corridor since 2000 Induced by Climate Change and Human Activities. *Remote Sens.* **2018**, *10*, 95. [CrossRef]

64. Yang, L.H.; Zhou, J.; Zhong, P.L.; Long, H.; Zhang, J.R. Lateglacial and Holocene dune evolution in the Horqin dunefield of northeastern China based on luminescence dating. *Palaeogeogr. Palaeoclimatol. Palaeoecol.* **2010**, *296*, 44–51. [CrossRef]

sustainability

MDPI

Article

Modelling Soil Carbon Content in South Patagonia and Evaluating Changes According to Climate, Vegetation, Desertification and Grazing

Pablo Luis Peri [1,2,*], Yamina Micaela Rosas [3], Brenton Ladd [4,5], Santiago Toledo [2], Romina Gisele Lasagno [1] and Guillermo Martínez Pastur [3]

[1] Instituto Nacional de Tecnología Agropecuaria (INTA); 9400 Río Gallegos, Argentina; lasagno.romina@inta.gob.ar

[2] Universidad Nacional de la Patagonia Austral (UNPA)-CONICET, 9400 Río Gallegos, Argentina; toledo.santiago@inta.gob.ar

[3] Laboratorio de Recursos Agroforestales, Centro Austral de Investigaciones Científicas (CADIC CONICET); 9410 Ushuaia, Argentina; yamicarosas@gmail.com (Y.M.R.); cadicforestal@gmail.com (G.M.P.)

[4] School of Biological, Earth and Environmental Sciences, University of New South Wales, Sydney 2052, Australia; brenton.ladd@gmail.com

[5] Escuela de Agroforestería, Universidad Científica del Sur; Lima 33, Perú

* Correspondence: peri.pablo@inta.gob.ar

Received: 29 November 2017; Accepted: 1 February 2018; Published: 8 February 2018

Abstract: In Southern Patagonia, a long-term monitoring network has been established to assess bio-indicators as an early warning of environmental changes due to climate change and human activities. Soil organic carbon (SOC) content in rangelands provides a range of important ecosystem services and supports the capacity of the land to sustain plant and animal productivity. The objectives in this study were to model SOC (30 cm) stocks at a regional scale using climatic, topographic and vegetation variables, and to establish a baseline that can be used as an indicator of rangeland condition. For modelling, we used a stepwise multiple regression to identify variables that explain SOC variation at the landscape scale. With the SOC model, we obtained a SOC map for the entire Santa Cruz province, where the variables derived from the multiple linear regression models were integrated into a geographic information system (GIS). SOC stock to 30 cm ranged from 1.38 to 32.63 kg C m^{-2}. The fitted model explained 76.4% of SOC variation using as independent variables isothermality, precipitation seasonality and vegetation cover expressed as a normalized difference vegetation index. The SOC map discriminated in three categories (low, medium, high) determined patterns among environmental and land use variables. For example, SOC decreased with desertification due to erosion processes. The understanding and mapping of SOC in Patagonia contributes as a bridge across main issues such as climate change, desertification and biodiversity conservation.

Keywords: soil carbon; grasslands; livestock; climate; native forest; land use

1. Introduction

Scientists and land managers of natural ecosystems acknowledge the importance of long-term monitoring systems for evaluating responses to disturbances (climate change or human activities) and providing baselines to evaluate potential changes [1,2]. In this context, since 2002, a long-term monitoring system (defined as a repeated field-based empirical measurements collected continuously and analyzed for at least 10 years) was established to monitor natural ecosystems and to produce scientific research focused on ecosystem function and ecosystem services, as well as on trends in biodiversity and the interactions between natural environments and land-use activities throughout southern Patagonia, Argentina [3].

Sustainable management of rangeland (rangeland can be defined as extensive areas of land that are occupied by native herbaceous or shrubby vegetation which are grazed by domestic or wild herbivores) for livestock production is an important economic activity in Southern Patagonia. Herbivores are known to be key drivers of soil processes in rangelands [4]. Changes in herbivore pressure (e.g., stocking rate) can have important consequences for ecosystem functioning [5]. Grazing can alter soil carbon stocks by changing the quality (e.g., dung, urine and litter inputs) and/or quantity (e.g., by causing compensatory regrowth in vegetation and/or by changing patterns of biomass allocation in the standing vegetation) of carbon that enters the soil. Grazing may also affect soil carbon stocks by altering rates of organic matter decomposition [4,6–9]. Soil organic carbon (SOC) content is important for ecosystem service provision, for example by supporting biodiversity [10], increasing soil aggregation, limiting soil erosion, and increasing water holding capacities [11]. Soil carbon also supports the capacity of the land to sustain plant and animal productivity and this potential depends on how rangelands are managed for livestock production [12]. Soil carbon is therefore a useful indicator for assessing the sustainability of livestock production on rangelands.

Rangelands are important economically and also culturally in Patagonia as rangelands provide the people of the region with a sense of place [13]. Given that maintenance of soil carbon is so important for the long term economic viability of rangelands it is surprising that little scientific research has focused on soil properties and on how these soil properties relate to grazing and land management more generally in these ecosystems. In Patagonia over the last 70 years, we have witnessed extensive degradation of once productive steppe ecosystems (desertification) [14]. As a result, stakeholders in Southern Patagonia developed a certification scheme for sustainable land management in the region [15]. However, all of the indicators are qualitative and quantitative indicators of rangeland condition are needed. SOC is one such possible indicator [16]. Data on carbon storage in forests, grasslands and shrublands at plant and stand level in Patagonia have been reported [6,7,17–19]. However, similar data for Patagonian rangelands are notable for their absence.

Long-term grazing intensity in arid and semiarid regions, such as southern Patagonia, may affect soil C stocks. The effects of grazing on soil C stocks likely also interact with other environmental variables that drive soil C [20]. Thus, variations in soil carbon pool may be influenced by climatic and topographic conditions. Understanding how these variables interact with grazing to alter soil carbon stocks at a regional scale is critical for understanding the impacts of land use decisions on both the sustainability of rangeland management and on atmospheric carbon concentrations and the climate.

The objectives in this study were (i) to model SOC stocks to 30 cm at a regional scale using climatic, topographic and vegetation variables; and (ii) to establish a regional baseline for SOC so that SOC can be used as an indicator of rangeland condition and therefore of the sustainability of land management. We hypothesize that (1) SOC would be lower where environmental conditions are harsh (low soil moisture conditions and high altitudes) and (2) that adverse environmental conditions would have a larger effect on soil carbon stocks than land use (stocking rate) at the regional scale in Patagonia.

2. Material and Methods

The study was conducted in of the PEBANPA (Parcelas de Ecología y Biodiversidad de Ambientes Naturales en Patagonia Austral-Biodiversity and Ecological long-term plots in Southern Patagonia) network of permanent plots [3]. There are 145 sites in the PEBANPA network of permanent plots (Figure 1), all of which were used in this analysis. Further detail of environmental conditions across Santa Cruz Province can be found in Peri et al. [3].

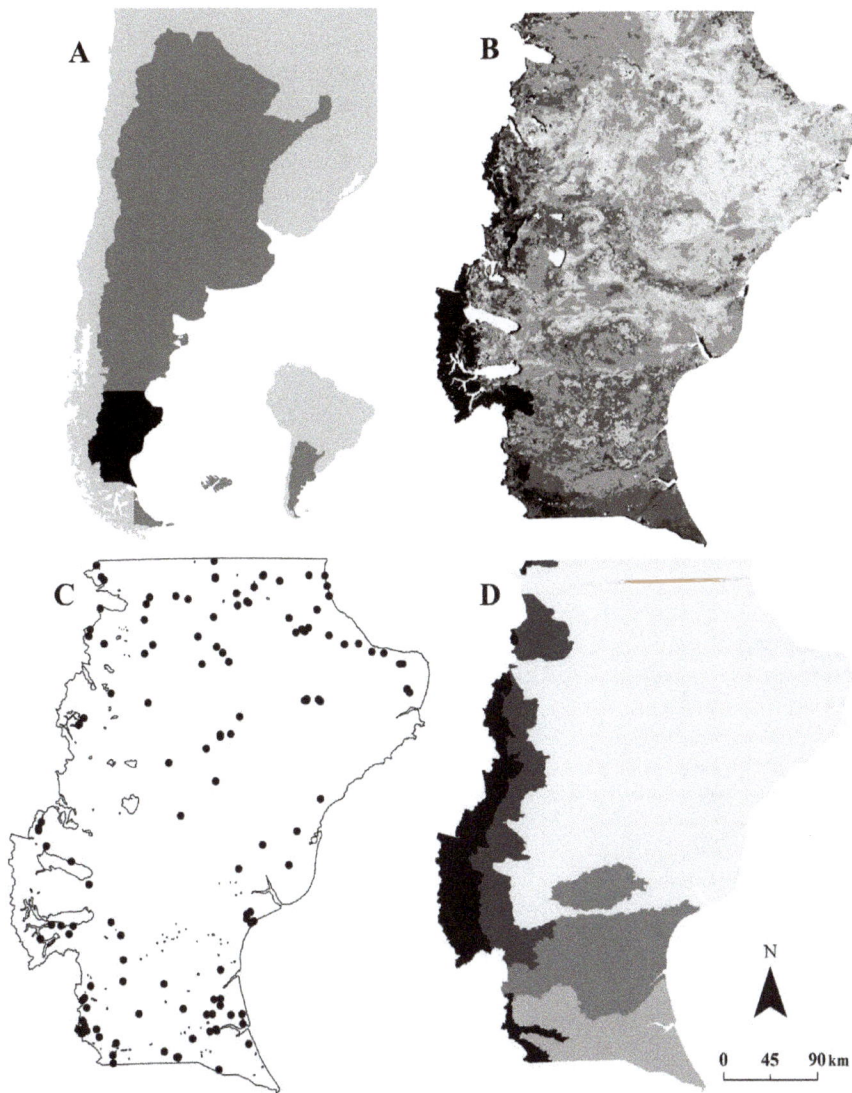

Figure 1. Characterization of the study area: (**A**) location of Argentina (dark grey) and Santa Cruz province (black); (**B**) Desertification (black = none, very dark grey = slight degraded, dark grey = moderate desertification, grey = moderate to severe desertification, light grey = severe desertification, very light grey = very severe desertification [21]; (**C**) sample sites (black dots) and main water bodies in the zone of the Parcelas de Ecología y Biodiversidad de Ambientes Naturales en Patagonia Austral (PEBANPA) plots; (**D**) main ecological areas (light grey = dry steppe, grey = humid steppe, medium grey = shrub-lands, dark grey = sub-Andean grasslands, black = forests and alpine vegetation) [22].

2.1. Soil Organic Carbon

For all 145 sites, we extracted data of SOC concentration (% C) and soil bulk density (BD) from the PEBANPA database (see Peri et al. [3] for details of the methodology). At each site, soil samples were collected from nine randomly selected points within a 20 m × 40 m quadrat using a hand auger (30 cm depth). Coarse root debris > 2 mm from soil samples had been removed by sieving. To reduce the number of chemical analyses we pooled individual soil samples into combined samples. From the nine samples collected within each quadrat, we created three composite samples so that each composite sample contained an equal proportion of soil from three auger holes ($n = 3$ for each site). The samples were finely ground to below 2 μm using a tungsten-carbide mill. Measurements of SOC concentration were derived from the dry combustion (induction furnace) method. Soil BD was estimated using the cylindrical core method ($n = 3$) by collecting a known volume of soil using a metal tube pressed into the soil (intact core), and determining the weight after drying. Knowing soil BD and depth of soil layers (0 to 30 cm) (Z), we applied the following equation to calculate the soil carbon stock:

$$\text{SoilCstock} = \%C\frac{g}{g} \times \text{BD}\left(\frac{g}{cm^3}\right) \text{Xdepthtobedrock(cm)}. \tag{1}$$

2.2. GIS-Derived Independent Variables

Climatic, topographic, landscape and land-use variables were obtained for each sampling point (Table 1). The methods used to generate the GIS-derived independent variables were also described in Peri et al. [3].

Table 1. Explanatory variables used in soil carbon stock analysis.

Category	Description	Code	Unit	Data Source
	mean annual temperature	AMT	°C	WorldClim [1]
	mean diurnal range	MDR	°C	WorldClim [1]
	isothermality	ISO	%	WorldClim [1]
	temperature seasonality	TS	°C	WorldClim [1]
	max temperature of warmest month	MAXWM	°C	WorldClim [1]
	min temperature of coldest month	MINCM	°C	WorldClim [1]
	temperature annual range	TAR	°C	WorldClim [1]
	mean temperature of wettest quarter	MTWEQ	°C	WorldClim [1]
	mean temperature of driest quarter	MTDQ	°C	WorldClim [1]
Climate	mean temperature of warmest quarter	MTWAQ	°C	WorldClim [1]
	mean temperature of coldest quarter	MTCQ	°C	WorldClim [1]
	mean annual precipitation	AP	mm years^{-1}	WorldClim [1]
	precipitation of wettest month	PWEM	mm years^{-1}	WorldClim [1]
	precipitation of driest month	PDM	mm years^{-1}	WorldClim [1]
	precipitation seasonality	PS	%	WorldClim [1]
	precipitation of wettest quarter	PWEQ	mm years^{-1}	WorldClim [1]
	precipitation of driest quarter	PDQ	mm years^{-1}	WorldClim [1]
	precipitation of warmest quarter	PWAQ	mm years^{-1}	WorldClim [1]
	precipitation of coldest quarter	PCQ	mm years^{-1}	WorldClim [1]
	global potential evapo-transpiration	EVTP	mm years^{-1}	CSI [2]
	elevation	ELE	m.a.s.l.	DEM [3]
	slope	SLO	%	DEM [3]
Topography	aspect	ASPC	cosine	DEM [3]
	aspect	ASPS	sine	DEM [3]
	distance to water bodies	DWB	km	SIT Santa Cruz [4]
	distance to rivers	DR	km	SIT Santa Cruz [4]
	normalized difference vegetation index	NDVI	dimensionless	MODIS [5]
	net primary productivity	NPP	gr C m^{-2} year^{-1}	MODIS [6]
Landscape and	desertification	DES	degree	CENPAT [7]
land-use	ecological area	EA	dimensionless	SIT Santa Cruz [4]
	stocking rate	SR	(ewe/ha/year)	SIT Santa Cruz [4]
	carrying capacity	RF	(ewe/ha/year)	SIT Santa Cruz [4]

[1] Hijmans et al. [23]; [2] Consortium for Spatial Information (CSI) [24]; [3] Farr et al. [25]; [4] SIT-Santa Cruz (http://www.sitsantacruz.gob.ar); [5] ORNL DAAC [26]; [6] Zhao et al. [27]; [7] Del Valle et al. [21].

2.3. Modelling and Data Analyses

A pre-selection of variables was performed based on Pearson's correlation indices obtained from paired analyses and considering the strength of the linear relationship (−1 to +1) and a *p*-value less than 0.05, with a confidence level of 95%.

For modelling, we used a stepwise multiple regression to identify which variables among these uncorrelated variables helped to explain SOC variation at landscape level. We employed a *p* value of <0.05 for the significance of each variable to be included into the model, analyzing the utility of the inclusion of the constant in the model, and used 100 steps for the final model selection. The model was evaluated through the standard error (SE) of estimation (the r^2-adj), defined as the average of the difference between predicted versus observed values, and the mean absolute error (AE) defined as the average of the difference between predicted versus the observed absolute values (Statgraphics Centurion software, Statpoint Technologies, The Plains, VA, USA).

To test the model across different gradients we performed a calibration procedure, using the same database employed for the modelling (observed vs. modelled). The first test was carried out by analyzing the mean and absolute errors (differences between observed and modelled values of SOC expressed as kg m^2). Secondarily, we tested the model performance by comparing SOC across different gradients of natural and human related variables: (i) vegetation types; (ii) stocking rates; (iii) soil covers (bare soil, shrubs, dwarf-shrubs, grasses, herbs, trees) (see further description of data and calculations in Peri et al. [27]).

With the SOC model, we obtained a SOC map for the entire Santa Cruz province (Argentina), where the variables derived from the multiple linear regression models were integrated into a geographical information system (GIS) using ArcMap 10.0 software [28]. For the SOC map, SOC values were assigned to three categories: low (0.01–4.47), medium (4.48–5.46), and high soil carbon stock (5.47–16.24 kg C m^{-2}). The limits of each SOC class were defined to contain an equal quantity of pixels for the whole province. For each SOC class (low, medium and high) we calculated the mean values and standard deviation of 25 continuous variables (including climate, topographic and landscape variables, see Table 1) using data from the entire province. In addition, the mean values and standard deviation of SOC were also calculated for discrete variables of the animal stocking rate, forage receptivity, ecological area, and desertification.

3. Results

Across Santa Cruz province, SOC stock to 30 cm depth ranged from 1.38 to 32.63 kg C m^{-2}. Climate variables presented correlation indexes between 0.07 and 0.99. Some climate variables were greatly influenced by the landscape variables (e.g., precipitation of driest quarter was strongly negatively correlated with the NDVI index, 0.78). Landscape variables presented correlation indexes between 0.06 and 0.88. Potential evapo-transpiration was strongly and negatively correlated with the desertification index, 0.66. Finally, topography variables were the group with the lowest correlation indices (0.06 to 0.39). Most of the variables were highly correlated to SOC stock using the Pearson's correlation index (Table 2), where NDVI (0.600, *p* < 0.001) was the most correlated. The variables ASPC, DWB, ASPS, MTWEQ, ELE and DR were not significant correlated with SOC stock using the Pearson's correlation index (Table 2).

The stepwise multiple regression selected three variables for the modelling: isothermality (ratio of average day variation in temperature divided by annual variability in temperature) (ISO, %), precipitation seasonality (PS, %) and normalized vegetation index (NVDI, dimensionless). These variables presented the best statistics, a high correlation with SOC stock, and low correlation among them (ISO × PS = −0.562, *p* < 0.001; ISO × NDVI = 0.384, *p* < 0.001; PS × NDVI = −0.471, *p* < 0.001). The inclusion of the constant in the model decreased the goodness of fit of the model;

for this reason we decided to not include a constant in the modelling. The fitted model (r^2-adj = 0.764; F = 156.1; SE = 4.08; AE = 2.71) explained 76.4% of variation in SOC values, and was expressed as:

$$\text{SOC (kg m}^2) = 0.11 \times \text{ISO} - 0.10 \times \text{PS} + 12.03 \times \text{NDVI} \qquad (2)$$

Table 2. Pearson's correlation index used in soil carbon stock (SOC) analysis. (see Table 1 for variables definition).

Category	Variables	SOC Correlation	SOC p-Value
Climate	AMT	−0.40	<0.001
	MDR	−0.32	<0.001
	ISO	0.35	<0.001
	TS	−0.51	<0.001
	MAXWM	−0.44	<0.001
	MINCM	−0.27	=0.001
	TAR	−0.44	<0.001
	MTWEQ	−0.10	=0.252
	MTDQ	−0.42	<0.001
	MTWAQ	−0.44	<0.001
	MTCQ	−0.30	<0.001
	AP	0.48	<0.001
	PWEM	0.41	<0.001
	PDM	0.54	<0.001
	PS	−0.42	<0.001
	PWEQ	0.42	<0.001
	PDQ	0.52	<0.001
	PWAQ	0.56	<0.001
	PCQ	0.39	<0.001
	EVTP	−0.50	<0.001
Topography	ELE	−0.06	=0.502
	SLO	0.24	=0.005
	ASPC	−0.16	=0.052
	ASPS	0.13	=0.118
	DWB	−0.15	=0.073
	DR	0.01	=0.868
Landscape and land-use	NDVI	0.60	<0.001
	NPP	0.53	<0.001
	DES	−0.46	<0.001

The map of the adjusted SOC model showed a continuous decline from the northeast and central areas of Santa Cruz province where most forests and shrublands are growing to the south and southwest where rangelands dominate (Figure 2).

The characteristics of the climatic and topographic variables according to the SOC map developed for the entire study area and the different SOC map quantities (low, medium, high) determined patterns of change among environmental and land use variables (Table 3). Mean Annual Temperature (MAT) influenced SOC. SOC values are higher at lower temperatures compared to the average for the entire province (7.8 °C). Other related temperature variables followed the same pattern (TS, MAXWM, MINCM, MTDQ, MTWAQ, MTCQ). However, MTWEQ did not present a clear pattern of variation. Seasonal and daily variations of temperature (MDR and TAR) and isothermality (ISO) did not greatly influence SOC (Table 3). Rainfall (MAP) also influenced SOC. SOC values increased with precipitation. The correlation between SOC and other rainfall variables (PWEM, PDM, PWEQ, PDQ, PWAQ, PCQ) followed a similar pattern. The other studied climatic indices (EVTP and GAI) followed the combined patterns of the temperature and rainfall variables, where SOC values decreased with the evapotranspiration and aridity. SOC values were generally low in the mountain environments with SOC values generally increasing below 460 m above sea level (m.a.s.l.), and the topographic variable slope did not correlate with changes in SOC quantity. As normalized difference vegetation index (NDVI) and net primary productivity (NPP) increased so did SOC stock to 30 cm (Table 3).

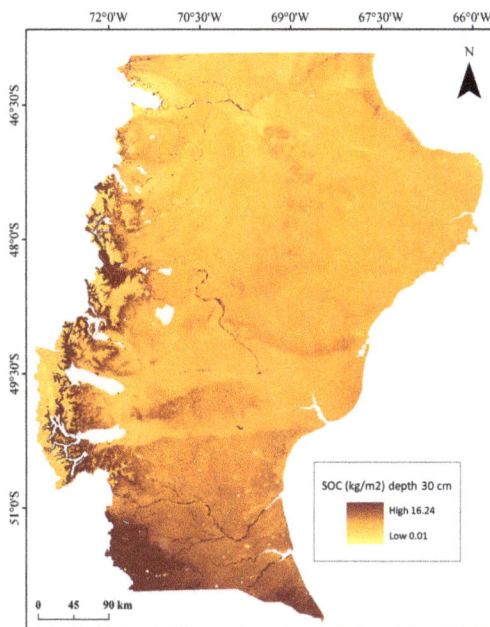

Figure 2. Soil organic carbon stock (30 cm depth) in Santa Cruz province, South Patagonia, Argentina.

Table 3. Mean (standard deviation) values of climatic, topographic and vegetation variables classified according to the soil carbon classes: low (0.01–4.47 kg C m^{-2}), medium (4.48–5.46 kg C m^{-2}) and high (5.47–16.24 kg C m^{-2}). (see Table 1 for variables definition).

Variable	Total	Low	Middle	High
AMT	7.77 (2.40)	8.57 (2.66)	8.48 (2.06)	6.25 (1.59)
MDR	10.33 (0.65)	10.45 (0.67)	10.49 (0.54)	10.07 (0.67)
ISO	46.1 (1.5)	45.3 (1.1)	45.8 (1.1)	47.2 (1.5)
TS	4.47 (0.44)	4.67 (0.43)	4.59 (0.32)	4.16 (0.38)
MAXWM	19.56 (3.16)	20.77 (3.47)	20.50 (2.53)	17.42 (2.10)
MINCM	(-2.65) (2.20)	(−2.09) (2.43)	(−2.15) (2.12)	(−3.71) (1.54)
TAR	22.21 (1.76)	22.86 (1.73)	22.65 (1.38)	21.13 (1.60)
MTWEQ	5.67 (2.95)	6.03 (3.17)	4.58 (1.93)	6.42 (3.21)
MTDQ	9.81 (3.72)	12.46 (3.45)	10.06 (2.83)	6.90 (2.42)
MTWAQ	13.21 (2.84)	14.34 (3.10)	14.04 (2.28)	11.25 (1.84)
MTCQ	1.85 (2.11)	2.45 (2.34)	2.42 (1.93)	0.68 (1.43)
AP	245.92 (181.38)	251.45 (219.82)	222.28 (140.29)	262.68 (169.31)
PWEM	30.15 (18.90)	32.42 (22.31)	27.47 (14.82)	30.40 (18.15)
PDM	13.61 (12.52)	13.21 (15.48)	12.48 (9.88)	15.13 (11.30)
PS	24.41 (6.57)	29.26 (6.55)	23.58 (4.54)	20.35 (4.97)
PWEQ	79.81 (53.24)	84.29 (62.82)	72.67 (41.44)	82.36 (52.35)
PDQ	46.38 (41.07)	46.00 (51.11)	42.46 (32.03)	50.41 (36.41)
PWAQ	53.62 (42.92)	48.65 (54.03)	49.29 (33.29)	62.92 (36.76)
PCQ	67.33 (46.03)	71.97 (53.70)	62.10 (35.32)	67.56 (45.92)
EVTP	807.88 (101.56)	848.78 (107.69)	839.90 (78.57)	735.15 (71.02)
GAI	0.33 (0.36)	0.33 (0.45)	0.28 (0.30)	0.37 (0.28)
ELE	468.83 (383.85)	556.10 (411.60)	388.62 (366.35)	460.13 (348.21)
SLO	5.00 (5.76)	5.07 (5.99)	4.49 (5.00)	5.43 (6.14)
NDVI	0.21 (0.12)	0.13 (0.06)	0.17 (0.04)	0.32 (0.14)
NPP	1275.68 (684.62)	971.12 (306.50)	1124.78 (422.34)	1708.11 (899.45)

Vegetation types also correlated with differences in SOC (3.8–5.5 kg C m^{-2} shrubs, 5.9–6.8 kg C m^{-2} grasslands, 12.1–12.3 kg C m^{-2} forests), while animal stocking rate decreased SOC values along the studied gradient (10.9 kg C m^{-2} in enclosures versus 4.6–6.7 kg C m^{-2} with medium and high stocking rate) (Table 4). Finally, SOC decreased with desertification gradient (10.6 kg C m^{-2} without presence, 8.3 kg C m^{-2} low, 5.2 kg C m^{-2} at moderate levels of desertification, and 4.4 kg C m^{-2} at sites where desertification was pronounced) due to erosion processes.

Table 4. Mean values (standard deviation, SD) and areas of soil carbon content (kg C m^{-2}) at 30 cm depth sorted by discrete variables in Santa Cruz province, Patagonia, Argentina.

Variable	Category	Mean (SD)	Area (km^2)
Desertification	No desertification	6.82 (3.38)	15,061
	Slight degraded	7.52 (2.47)	12,085
	Moderate desertification	6.52 (1.80)	34,135
	Moderate to severe desertification	5.11 (1.30)	84,011
	Severe desertification	4.72 (0.93)	63,502
	Very severe desertification	4.26 (0.61)	29,828
simple ecosystem classiffication	Humid magellanic grass steppe	9.64 (1.14)	6056
	Andean vegetation	7.51 (3.70)	15,815
	Dry magellanic grass steppe	7.29 (0.77)	11,796
	Mata negra thicket	6.35 (0.77)	28,374
	Sub-andean grassland	5.88 (1.74)	19,540
	Central plateau	4.62 (0.73)	131,911
	Shrub steppe San Jorge Gulf	4.59 (0.71)	11,990
	Mountains and plateaus	3.76 (0.87)	13,125
potential stocking rate (ewes/ha/yr)	<0.1	4.56 (0.46)	17,119
	0.1 a 0.20	4.59 (0.71)	84,363
	0.13 a 0.25	4.11 (0.68)	18,654
	0.16 a 0.30	3.76 (0.86)	13,166
	0.20 a 0.30	5.83 (1.22)	71,911
	0.3 a 0.4	7.40 (2.84)	27,493
	>0.4	9.71 (1.17)	5917
Actual Stocking rate (ewes/ha/yr)	0.07	4.62 (0.73)	131,758
	0.13	5.64 (1.55)	60,872
	0.18	4.60 (0.71)	11,886
	0.22	7.29 (0.76)	11,707
	0.3	7.52 (3.69)	15,738
	0.5	9.66 (1.14)	5886

The calibration of the model showed an average error of −0.01 kg m^2, and an absolute error of 2.72 kg m^2. When the performance of the model was tested across different natural and management related gradients, it can be observed that the error dispersion isn't homogeneous (Table 5). In general, errors increased with SOC quantity. When vegetation type was considered, lower error values were observed in arid grasslands and shrublands, while greater errors were found in humid grasslands and forests.

When bare soil cover increased, the error in SOC predictions decreased. When herb and tree covers decreased, the error in SOC predictions also decreased. Prediction error didn't vary systematically with stocking rate, or with shrub and/or grass cover (Table 5).

Table 5. Model performance analysis using a calibration of the soil carbon content (kg C m^{-2}) sorted by discrete variables: (i) vegetation types, (ii) stocking rate, (iii) soil covers (bare soil, shrubs, dwarf-shrubs, grasses, herb, trees).

Vegetation	N	Carbon	Modelled	Mean Error	Absolute Error
Shrub steppe	23	4.54	5.52	−0.98	2.27
Dwarf shrub steppe	28	3.98	4.77	−0.79	1.81
Shrub steppe	5	3.06	3.84	−0.78	1.82
Grass steppe	55	5.35	5.92	−0.57	2.15
Nothofagus antarctica forest	10	12.46	12.08	0.38	4.33
Nothofagus pumilio forest	12	13.73	12.29	1.44	4.12
Grassland	11	11.24	6.67	4.57	5.75
Nothofagus betuloides forest	1	20.25	12.34	7.91	7.91
Stocking rate (ewes/ha/yr)					
<0.1	38	8.85	8.15	0.70	3.08
0.1–0.17	41	4.43	4.60	−0.17	1.71
0.18–0.25	30	5.79	6.76	−0.96	2.97
>0.25	36	7.41	7.18	0.23	3.27
% bare soil					
<16.7	35	11.06	9.76	1.31	4.68
16.7–30.0	36	5.51	5.82	−0.31	2.10
30.1–48.0	39	5.11	5.57	−0.47	1.98
>48.0	35	4.97	5.45	−0.49	2.22
% shrubs					
<1.0	35	6.91	7.11	−0.20	2.48
1–6.0	39	6.17	6.30	−0.13	2.43
6.1–21.3	35	4.32	5.25	−0.93	2.01
>21.3	36	9.03	7.81	1.22	3.96
% dwarf-shrubs					
<0.3	35	7.85	8.01	−0.16	2.64
0.4–3.0	38	7.28	6.74	0.54	2.89
3.1–20.0	38	6.62	6.65	−0.03	3.24
>20.0	34	4.58	5.01	−0.43	2.02
% grasses					
<10.9	37	8.18	8.84	−0.66	3.83
11.0–22.5	36	5.19	5.14	0.05	1.70
22.6–35.9	36	5.28	6.19	−0.91	2.16
>36.0	36	7.75	6.23	1.51	3.16
% Herbs					
<1.8	35	4.83	4.77	0.06	2.01
1.9–4.3	38	4.86	5.53	−0.68	2.01
4.4–14.0	37	6.99	7.09	−0.11	2.60
>14.0	35	9.90	9.13	0.77	4.32
% trees					
none	119	5.27	5.43	−0.16	2.33
0.1–50.0	16	11.50	11.84	−0.33	4.63
>50.0	10	14.72	12.38	2.34	4.29
Total general	145	6.61	6.62	−0.01	2.72

4. Discussion

Our model for SOC prediction was able to account for 76% of the variation of this soil property across the study area, with values ranging from 1.38 to 32.63 kg C m^{-2}. In the present study, SOC stock to 30 cm was mainly a function of climate and vegetation. As we have already shown [29,30], the prediction and mapping of soil carbon at the macro scale was possible using freely available geospatial data. The correlation between SOC and climate variables (isothermality and seasonality precipitation) may reflect the influence of climate variables on semi-arid ecosystem productivity, which are mainly related to water limitation. This highlights the importance of long term monitoring to know the processes that determine the magnitude of SOC variation, and to forecast how it may operate as climate and land use changes in the future. The effect of stocking rate on SOC was minimal in this analysis. However, grazing may indirectly affect SOC by modifying the type of vegetation cover [7]. In this study, vegetation cover, as represented by a Normalized Difference Vegetation Index (NDVI), also was a strong predictor of SOC in the fitted model. This was consistent with Kunkel et al. [31] who reported that NDVI-predicted soil carbon distribution in semi-arid montane ecosystem.

The characteristics of the climatic and topographic variables according to the SOC map developed for the entire study area and the different SOC map quantities (low, medium, high) determined patterns among the environmental and land use variables. Temperature (AMT) influenced SOC by increasing the quantity at lower temperatures compared to the average SOC stock for the entire province. Some models and observations suggest that high latitude forests and grasslands may behave as a C source in response to increased decomposition of soil organic matter resulting from temperature increases [32,33]. Thus, the temperature sensitivity of decomposing organic matter in soil partly determines how much carbon will be transferred to the atmosphere because of global warming. This is consistent with Peri et al. [29] who reported that soil respiration rates were correlated strongly to air and soil temperatures by evaluating seasonal dynamics in contrasting grasslands across gradients of climate (rainfall), long term grazing intensity (moderate and high stocking rates over the last 80 years) and land uses (silvopastoral system, primary forest and grassland) in Southern Patagonia. The interaction of climate change with C pools in high latitude ecosystems may be particularly important because climate change is expected to be greatest at high latitudes. For instance, in Southern Patagonia (Santa Cruz and Tierra del Fuego provinces), mean maximum annual temperature is predicted to increase by 2–3 °C by 2080 between 46° and 52°30′ SL [34], and this will have significant effects on Patagonian ecosystems.

Rainfall and other related rainfall variables (seasonality) also influenced SOC by increasing the quantity with precipitation. It has been demonstrated that increased variability in rainfall and soil water content significantly affected SOC in this grassland [35]. In the semiarid temperate steppe in northern China soil water availability was more important than temperature in regulating soil and microbial respiratory processes, microbial biomass and their responses to climate change [36]. The strong and direct relationship between rainfall and SOC may be related to the ANPP and mean soil water content. In the present work net primary productivity (NPP) influenced positively SOC quantity compared to the average for the province. There is evidence that ecosystem C inputs from ANPP can be directly affected by altered rainfall variability, independent of precipitation quantity [35]. Thus, precipitation constrains plant production and decomposition in arid ecosystems, with a greater response of plant production relative to decomposition [37]. However, it is probable that SOC is controlled by the complex interaction of environmental and biotic factors. At the regional scale, in this study, patterns of SOC were positively associated with mean annual precipitation and negatively correlated with mean annual temperature across a diverse range of soils and vegetation types. This has been also documented in grasslands in North America [38].

We found that SOC values were low in high altitude mountain environments (SOC values generally increasing below 460 m.a.s.l.). This may be due to changes in climatic variables, precipitation, temperature and vegetation types along altitudinal gradients that influence in consequence the quality, quantity and turnover of soil organic matter. For example, Mulugeta and Itanna [39] determined that soil carbon stocks and turnovers in various vegetation types was directly proportional to the mean annual precipitation and inversely proportional to the mean annual temperature prevailing along the elevation gradient.

Vegetation types also showed differences in SOC, being higher in forest than in shrublands. The C efflux from soils in these rangeland ecosystems was higher than in subtropical savanna grasslands of southern Texas, USA [40] and C effluxes measured in Patagonian forest and woodland ecosystems [29]. This highlights the importance of environmental conditions (mainly soil water availability and temperature), input of organic residues, soil microbial biomass, and soil properties on the magnitude on soil respiration among different ecosystems.

In this regional study, increased animal stocking rates decreased SOC values, consistent with Peri [7]. We believe that low SOC at high stocking rates may be due to both low vegetation cover (or high bare soil cover) and low ANPP. Also, Peri et al. [29] reported that litter cover, litter depth, and soil carbon concentration (C %) in the uppermost soil layers, in both the dry and humid Magellanic steppe areas, were lower under heavy long term stocking rates than sites under moderate grazing intensity. Bahn et al. [41] indicated that the degree to which soil CO_2 efflux is coupled to soil C content may be largely determined by the reductions of supply by removal of aboveground biomass through grazing. Grassland ecosystems with high soil organic matter may promote organic matter decomposition (microbial activity) by continuous addition of litter and root turnover, thereby increasing soil respiration rates.

Finally, the development of sustainable land management practices for Patagonian Rangeland could benefit from a paleo-ecological perspective. The results and underlying data presented here are heavily focused on the importance of domestic animal grazing as a key driver of SOC over the last 100 years, when sheep-farming by European settlers began. However, an applied palaeoecology perspective could address specific ecological and environmental questions highly relevant to nature conservation, allowing us to assess the naturalness, rarity and fragility of Patagonia's current ecosystems. This in turn could provide a sound basis prioritizing conservation and ecosystem restoration investments [42]. For example, palaeoecological studies indicated that few vegetation types in western Europe are natural [43]. In northern Patagonia, Schäbitz [44] reported changes in vegetation in late Holocene due to a drying climate that favored semi-arid vegetation. In Santa Cruz province (Southern Patagonia), Horta et al. [45] showed also vegetation change from grass steppe to shrub steppe during the Middle Holocene, most likely related to climate changes and the subsequent consequences for human occupational dynamics. Thus while it is clear that current environmental conditions, including patterns of human land use provide a useful baseline against which land management can be assessed, it is also clear that additional research focused on the paleoecology of the region could enrich this baseline and facilitate the development of effective strategies for sustainable land management.

5. Conclusions

Soil carbon storage across sites was influenced by a large number of interacting variables the most important of which were climatic conditions and plant productivity. Best practice ecosystem management (grazing) can increase net carbon storage in grasslands, shrublands, wetlands, and forests, but economic incentives to maintain or increase soil C stocks are needed. Understanding the causes of variation and mapping of SOC in Patagonia is a first step which allows for assessment of the sustainability of land management at the local scale, which can help to not only increase resilience of rangeland regionally in Patagonia, but also help address issues at the global scale, such as climate change, desertification, and biodiversity conservation.

Better understanding the results of existing long-term studies such as soil carbon stocks, and also realizing the existence of permanent plots in Patagonia for future research (e.g., PEBANPA network) will hopefully contribute to solve regional ecological and socio-economic challenges in the sustainable use of our native ecosystems. However, research and administrative institutions, and farmers must cooperate and have a sustained commitment to finance and maintaining these large unique long-term plots and research platforms.

Acknowledgments: The present research was supported by the INTA and UNPA.

Author Contributions: Pablo Luis Peri and Guillermo Martínez Pastur conceived and designed the experiments, and wrote the paper; Romina G. Lasagno and Santiago Toledo performed the experiments; Yamina M. Rosas and Brenton Ladd mainly analyzed the data and contributed analysis tools.

Conflicts of Interest: The authors declare no conflict of interest.

References

1. Lindenmayer, D.B.; Likens, G.E. Adaptive monitoring: A new paradigm for long-term research and monitoring. *Trends Ecol. Evol.* **2009**, *24*, 482–486. [CrossRef] [PubMed]
2. Lindenmayer, D.B.; Likens, G.E. The science and application of ecological monitoring. *Biol. Conserv.* **2010**, *143*, 1317–1328. [CrossRef]
3. Peri, P.L.; Lencinas, M.V.; Bousson, J.; Lasagno, R.; Soler, R.; Bahamonde, H.; Martínez Pastur, G. Biodiversity and ecological long-term plots in Southern Patagonia to support sustainable land management: The case of PEBANPA network. *J. Nat. Conserv.* **2016**, *34*, 51–64. [CrossRef]
4. Bardgett, R.D.; Wardle, D.A. Herbivore mediated linkages between aboveground and belowground communities. *Ecology* **2003**, *84*, 2258–2268. [CrossRef]
5. Schlesinger, W.H.; Reynolds, J.F.; Cunningham, G.L.; Huenneke, L.F.; Jarrell, W.M.; Virginia, R.A.; Whitford, W.G. Biological feedbacks in global desertification. *Science* **1990**, *247*, 1043–1048. [CrossRef] [PubMed]
6. Peri, P.L.; Lasagno, R.G. Biomass, carbon and nutrient storage for dominant grasses of cold temperate steppe grasslands in southern Patagonia, Argentina. *J. Arid Environ.* **2010**, *74*, 23–34. [CrossRef]
7. Peri, P.L. Carbon Storage in Cold Temperate Ecosystems in Southern Patagonia, Argentina. In *Biomass and Remote Sensing of Biomass*; Atazadeh, I., Ed.; InTech Publisher: Rijeka, Croacia, 2011; pp. 213–226, ISBN 978-953-307-490-0.
8. Ziter, C.; MacDougall, A.S. Nutrients and defoliation increase soil carbon inputs in grassland. *Ecology* **2013**, *94*, 106–116. [CrossRef] [PubMed]
9. Bahamonde, H.; Gargaglione, V.; Peri, P.L. Sheep faeces decomposition and nutrient release across an environmental gradient in Southern Patagonia. *Ecol. Austral* **2017**, *27*, 18–28.
10. Sheil, D.; Ladd, B.; Silva, L.C.R.; Laffan, S.W.; Heist, M.V. How are soil carbon and tropical biodiversity related? *Environ. Conserv.* **2016**, *43*, 231–241. [CrossRef]
11. Miller, R.W.; Donahue, R.L. *Soils: An Introduction to Soils and Plant Growth*, 6th ed.; Prentice Hall: Englewood Cliffs, NJ, USA, 1990.
12. Doran, J.W.; Parkin, T.B. Quantitative indicators of soil quality: A minimum data set. In *Methods for Assessing Soil Quality*; Doran, J.W., Jones, A.J., Eds.; Soil Science Society of America, Special Publication: Madison, WI, USA, 1996; pp. 25–37.
13. Martínez Pastur, G.; Peri, P.L.; Lencinas, M.V.; Garcia-Llorente, M.; Nartin-Lopez, B. Spatial patterns of cultural ecosystem services provision in Southern Patagonia. *Landsc. Ecol.* **2016**, *31*, 383–399. [CrossRef]
14. Golluscio, R.A.; Deregibus, V.; Paruelo, J.M. Sustainability and range management in the Patagonian steppe. *Ecol. Austral* **1998**, *8*, 265–284.
15. Borrelli, P.; Boggio, F.; Sturzenbaum, P.; Paramidani, M.; Heinken, R.; Pague, C.; Stevens, M.; Nogués, A. *GRASS: Grassland Regeneration and Sustainability Standard*; Food and Agriculture Organization: Rome, Italy, 2013. Available online: http://www.fao.org/nr/sustainability/grassland/best-practices/projects-detail/en/c/237687/ (accessed on 10 October 2017).
16. Lal, R. Soil quality and food security: The global perspective. In *Soil Quality and Soil Erosion*; Lal, R., Ed.; CRC Press: Boca Raton, FL, USA, 1999; pp. 3–16.
17. Peri, P.L.; Gargaglione, V.; Martínez Pastur, G. Dynamics of above- and below-ground biomass and nutrient accumulation in an age sequence of *Nothofagus antarctica* forest of Southern Patagonia. *For. Ecol. Manag.* **2006**, *233*, 85–99. [CrossRef]
18. Peri, P.L.; Gargaglione, V.; Martínez Pastur, G. Above- and belowground nutrients storage and biomass accumulation in marginal *Nothofagus antarctica* forests in Southern Patagonia. *For. Ecol. Manag.* **2008**, *255*, 2502–2511. [CrossRef]

19. Peri, P.L.; Gargaglione, V.; Martínez Pastur, G.; Lencinas, M.V. Carbon accumulation along a stand development sequence of *Nothofagus antarctica* forests across a gradient in site quality in Southern Patagonia. *For. Ecol. Manag.* **2010**, *260*, 229–237. [CrossRef]

20. Frank, A.B.; Tanaka, D.L.; Hofmann, L.; Follett, R.F. Soil carbon and nitrogen of Northern great plains grasslands as influenced by long-term grazing. *J. Range Manag.* **1995**, *48*, 470–474. [CrossRef]

21. Del Valle, H.F.; Elissalde, N.O.; Gagliardini, D.A.; Milovich, J. Status of desertification in the Patagonian region: Assessment and mapping from satellite imagery. *Arid Land Res. Manag.* **1998**, *12*, 95–121. [CrossRef]

22. Gonzalez, L.; Rial, P. *Guía Geográfica Interactiva de Santa Cruz*; INTA: Buenos Aires, Argentina, 2004.

23. Hijmans, R.J.; Cameron, S.E.; Parra, J.L.; Jones, P.G.; Jarvis, A. Very high resolution interpolated climate surfaces for global land areas. *Int. J. Climatol.* **2005**, *25*, 1965–1978. [CrossRef]

24. Zomer, R.J.; Trabucco, A.; Bossio, D.A.; VanStraaten, O.; Verchot, L.V. Climate change mitigation: A spatial analysis of global land suitability for clean development mechanism afforestation and reforestation. *Agric. Ecosyst. Environ.* **2008**, *126*, 67–80. [CrossRef]

25. Farr, T.G.; Rosen, P.A.; Caro, E.; Crippen, R.; Duren, R.; Hensley, S.; Kobrick, M.; Paller, M.; Rodriguez, E.; Roth, L.; et al. The shuttle radar topography mission. *Rev. Geophys.* **2007**, *45*, RG2004. [CrossRef]

26. ORNL DAAC. *MODIS Collection 5 Land Products Global Subsetting and Visualization Tool*; ORNL DAAC: Oak Ridge, TN, USA, 2008.

27. Zhao, M.; Running, S.W. Drought-induced reduction in global terrestrial net primary production from 2000 through 2009. *Science* **2010**, *329*, 940–943. [CrossRef] [PubMed]

28. Environmental Systems Research Institute (ESRI). *ArcGIS Desktop: Release 10*; Environmental Systems Research Institute Inc.: Redlands, CA, USA, 2011.

29. Peri, P.L.; Ladd, B.; Lasagno, R.G.; Martínez Pastur, G. The effects of land management (grazing intensity) vs. the effects of topography, soil properties, vegetation type, and climate on soil carbon concentration in Southern Patagonia. *J. Arid Environ.* **2016**, *134*, 73–78. [CrossRef]

30. Ladd, B.; Laffan, S.W.; Amelung, W.; Peri, P.L.; Silva, L.C.R.; Gervassi, P.; Bonser, S.P.; Navall, M.; Sheil, D. Estimates of soil carbon concentration in tropical and temperate forest and woodland from available GIS data on three continents. *Glob. Ecol. Biogeogr.* **2013**, *22*, 461–469. [CrossRef]

31. Kunkel, M.L.; Flores, A.N.; Smith, T.J.; McNamara, J.P.; Benner, S.G. A simplified approach for estimating soil carbon and nitrogen stocks in semi-arid complex terrain. *Geoderma* **2011**, *165*, 1–11. [CrossRef]

32. Oechel, W.C.; Hastings, S.J.; Vourlitis, G.; Jenkins, M.; Riechers, G. Recent change of Arctic tundra ecosystems from a net carbon dioxide sink to a source. *Nature* **1993**, *361*, 520–523. [CrossRef]

33. Davidson, E.A.; Trumbore, S.E.; Amundson, R. Biogeochemistry: Soil warming and organic carbon content. *Nature* **2000**, *408*, 789–790. [CrossRef] [PubMed]

34. Kreps, G.; Martínez Pastur, G.; Peri, P.L. *Cambio Climático en Patagonia Sur: Escenarios Futuros en el Manejo de los Recursos Naturales*; Ediciones INTA: Buenos Aires, Argentina, 2012; ISBN 978-987-679-137-3.

35. Knapp, A.K.; Fay, P.A.; Blair, J.M.; Collins, S.L.; Smith, M.D.; Carlisle, J.D.; Harper, C.W.; Danner, B.T.; Lett, M.S.; McCarron, J.K. Rainfall variability, carbon cycling, and plant species diversity in a Mesic grassland. *Science* **2002**, *298*, 2202–2205. [CrossRef] [PubMed]

36. Liu, W.; Zhang, Z.; Wan, S. Predominant role of water in regulating soil and microbial respiration and their responses to climate change in a semiarid grassland. *Glob. Chang. Biol.* **2009**, *15*, 184–195. [CrossRef]

37. Sala, O.E.; Parton, W.J.; Joyce, L.A.; Lauenroth, W.K. Primary production of the central grassland region of the United States. *Ecology* **1988**, *69*, 40–45. [CrossRef]

38. Burke, I.C.; Yonker, C.M.; Parton, W.J.; Cole, C.V.; Flach, K.; Schimel, D.S. Texture, climate, and cultivation effects on soil organic matter content in U.S. grassland soils. *Soil Sci. Soc. Am. J.* **1989**, *53*, 800–805. [CrossRef]

39. Mulugeta, L.; Itanna, F. Soil carbon stocks and turnovers in various vegetation types and arable lands along an elevation gradient in southern Ethiopia. *Geoderma* **2004**, *123*, 177–188.

40. McCulley, R.L.; Boutton, T.W.; Archer, S.R. Soil respiration in a subtropical savanna parkland: Response to water additions. *Soil Sci. Soc. Am. J.* **2007**, *71*, 820–828. [CrossRef]

41. Bahn, M.; Rodeghiero, M.; Anderson-Dunn, M.; Dore, S.; Gimeno, C.; Drösler, M.; Williams, M.; Ammann, C.; Berninger, F.; Flechard, C.; et al. Soil respiration in European grasslands in relation to climate and assimilate supply. *Ecosystems* **2008**, *11*, 1352–1367. [CrossRef] [PubMed]

42. Marignani, M.; Chiarucci, A.; Sadori, L.; Mercuri, A.M. Natural and human impact in Mediterranean landscapes: An intriguing puzzle or only a question of time? *Plant Biosyst.* **2017**, *151*, 900–905. [CrossRef]

43. Birks, H.J.B. Contributions of Quaternary palaeoecology to nature conservation. *J. Veg. Sci.* **1996**, *7*, 89–98. [CrossRef]

44. Schäbitz, F. Holocene climatic variations in northern Patagonia, Argentina. *Palaeogeogr. Palaeoclimatol. Palaeoecol.* **1994**, *109*, 287–294. [CrossRef]

45. Horta, L.R.; Marcos, M.A.; Bozzuto, D.L.; Mancini, M.V.; Sacchi, M. Paleogeographic and paleoenvironmental variations in the area of the Pueyrredón, Posadas and Salitroso lakes, Santa Cruz Province, Argentina, during the Holocene and its relationship with occupational dynamics. *Palaeogeogr. Palaeoclimatol. Palaeoecol.* **2016**, *449*, 541–552. [CrossRef]

sustainability

MDPI

Article

Agricultural Oasis Expansion and Its Impact on Oasis Landscape Patterns in the Southern Margin of Tarim Basin, Northwest China

Yi Liu [1,2,3,4], Jie Xue [4,5,*], Dongwei Gui [4,5,*], Jiaqiang Lei [4,5], Huaiwei Sun [6], Guanghui Lv [1,2] and Zhiwei Zhang [7]

[1] Institute of Arid Ecology and Environment, Xinjiang University, Urumqi 830046, China; yiliu319@126.com (Y.L.); ler@xju.edu.cn (G.L.)
[2] Key Laboratory of Oasis Ecology, Xinjiang University, Urumqi 830046, China
[3] University of Chinese Academy of Sciences, Beijing 100049, China
[4] State Key Laboratory of Desert and Oasis Ecology, Xinjiang Institute of Ecology and Geography, Chinese Academy of Sciences, Urumqi 830011, China; desert@ms.xjb.ac.cn
[5] Cele National Station of Observation and Research for Desert-Grassland Ecosystems, Cele 848300, China
[6] School of Hydropower and Information Engineering, Huazhong University of Science and Technology, Wuhan 430074, China; huaiweisun@whu.edu.cn
[7] Xizang Agriculture and Animal Husbandey College, Tibet University, Lhasa 850000, China; aiwoweige@163.com
* Correspondence: xuejie11@mails.ucas.ac.cn (J.X.); guidwei@ms.xjb.ac.cn (D.G.); Tel.: +86-0991-7885-317 (J.X. & D.G.)

Received: 20 April 2018; Accepted: 5 June 2018; Published: 11 June 2018

Abstract: Oasis landscape change and its pattern dynamics are considered one of the vital research areas on global land use and landscape change in arid regions. An agricultural oasis is the main site of food security and ecosystem services in arid areas. Recently, the dramatic exploitation of agricultural oases has affected oasis stability, inducing some ecological and environmental issues such as water shortage and land degradation. In this study, the Qira oasis on the southern margin of Tarim Basin, Northwest China, was selected as a study area to examine the spatiotemporal changes in an agricultural oasis and the influence on oasis landscape pattern. Based on the integration of Thematic Mapper, Enhanced Thematic Mapper Plus, and GF-1 images, the agricultural Qira oasis has rapidly increased, with annual change rates of −0.3%, 1.6%, 3.7%, and 1.5% during 1970–1990, 1990–2000, 2000–2013, and 2013–2016, respectively. With the agricultural oasis expansion, the agricultural land has increased from 91.10 km² in 1970 to 105.04 km² in 2016. The percentage of farmland area has increased by 15.3% in 2016 compared with that in 1970. The natural vegetation is decreasing owing to the reclamation of desert–oasis ecotone. The oasis landscape change and pattern are mainly affected by agricultural expansion under water-saving technological utilization, land use policy, and regional economic development demand. The expansion of agricultural oasis is alarming due to human overexploitation. Thus, the government should adjust the layout of agricultural development and pay considerable attention to the oasis environment sustainability. This study can provide a valuable reference on the impact of climate change and human activities on a landscape.

Keywords: oasis; landscape change; agricultural oasis expansion; Tarim Basin

---

1. Introduction

The landscape is shaped by a combination of natural and anthropogenic factors [1,2]. Landscape change and its dynamics are one of the major research areas in global environmental variation and landscape ecology [3–7]. Recently, landscape change has received worldwide concern from different perspectives including the context, direction, rate, and spatiotemporal variation [6,8]. In arid or semiarid regions, the oasis as a unique landscape coexists with deserts like the Gobi, allowing vegetation to flourish and humans to thrive due to the presence of runoff, spring water, and groundwater [9–11]. The oasis is the basis of human settlement and economic development; it supports more than 95% of the population and more than 90% of social wealth, with only 5–6% of the land surface in arid and semiarid regions of China [6,12]. Therefore, research on oasis change or dynamics has become a key topic for sustainability [10,13].

According to its formation mechanism, oases are classified into natural or artificial types [14]. Natural oases include shrubby grassland, water bodies, and desert areas. Artificial oases contain agricultural oases, shelter forests, and residential areas [12,14,15]. In recent decades, there has been a significant expansion of oasis land, particularly agricultural oasis, in the arid regions of Northwest China [16]. An agricultural oasis refers to cultivated land irrigated by human activities, and provides food security [10]. With the expansion of agricultural oasis to increase grain yields, the scale of agricultural oasis is limited by the availability of irrigation water in arid regions [17]. If agricultural oasis expansion exceeds the water carrying capacity, the natural landscape will be threatened, resulting in a series of environmental issues [17,18]. Therefore, the dynamics of agricultural oasis landscape patterns need to be analyzed to determine the sustainable scale.

Previous studies on oasis change mainly focused on land use and vegetation change in desert oasis areas [2,15,19,20], the ecological effects and the suitable scale of agricultural oasis [17,21,22], the relationship between the oasis dynamics and water resources [14,23], and the patterns and reasons of agricultural oasis expansion [10]. These studies analyzed the changes and dynamics of oasis landscapes in arid and semiarid regions. However, less attention has been given to identifying the trends and driving forces of agricultural oasis and their impact on oasis landscape patterns through qualitative and quantitative analyses.

To systematically understand the expansion dynamics of agricultural oasis and its impact on oasis landscape patterns, this study selected the Qira oasis, a typical natural and agricultural oasis on the southern margin of Tarim Basin, Northwest China, as a study area to examine agricultural oasis change and its impact on landscape patterns using multisource satellite images. The main objectives of this study are to (1) detect the spatiotemporal changes of agricultural oasis in the Qira oasis from 1970 to 2016; (2) characterize the trends of agricultural oasis and analyze its influence on oasis landscape patterns; (3) identify the direction and driving factors of oasis landscape change; and (4) discuss feasible recommendations for agricultural oasis sustainability.

2. Study Area

The Qira oasis is located within the central part of the southern Taklamakan Desert and the northern foot plain of the Kunlun Mountains (80°43′ E–80°53′ E, 36°57′ N–37°05′ N), covering an area of approximately 157 km^2 (Figure 1). It is a typical alluvial fan, in which the agricultural and natural oases are the main landscape units [24,25]. The water supply in the Qira oasis area depends mainly on Qira river runoff, which comes from a high-altitude valley of the Kunlun Mountains, flowing through the Qira oasis, and finally discharging into the extremely arid Taklimakan desert [24,25]. Its annual runoff was 1.27×10^8 m^3 during 1985–2010. The climate is hyperarid with an average annual temperature of 11.9 °C, annual precipitation of 35 mm, and annual potential evaporation of 2600 mm [26]. The soil is mainly classified as aeolian sandy soil. The oasis has a soil moisture content of 0.15–0.35% on the surface at a depth of 0–20 cm [27]. The farmland is the main land use type, with wheat, maize, and cotton being the main crops. The desert–oasis ecotone with 20–40% vegetation coverage is distributed in the northwestern edge of the Qira oasis [28]. The key vegetation types contain *Tamarix chinensis*, *Calligonum*, and *Alhagi sparsifolia*. In recent decades, with the expansion of the agricultural oasis, the landscape patterns have changed significantly, creating a large number of ecological and environmental issues. Therefore, research on the dynamics of agricultural oasis and landscape patterns in the Qira oasis is crucial for better-informed landscape management decisions.

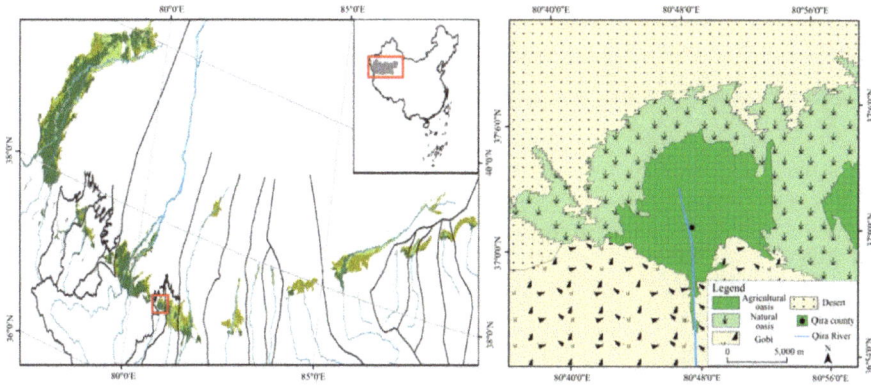

Figure 1. Location of Qira oasis in the southern Tarim Basin (adapted from [24]).

3. Materials and Methods

3.1. Data Sources

In this study, the data used to examine agricultural oasis expansion and landscape dynamics in the Qira oasis were based on five time periods of images. Land use images of the Xinjiang Uyghur Autonomous Region in 1970 with a scale of 1:50,000 were provided by the Surveying and Mapping Bureau of Xinjiang Uyghur Autonomous Region. Landsat Thematic Mapper (TM) and Enhanced Thematic Mapper Plus (ETM+) provided images in 1990 and 2000 with 30 × 30 m nominal resolution. The images in 2013 and 2016 were acquired GF-1 satellite. The scale of the images was 1:100,000. The images in summer or autumn were selected to determine vegetation phenology [7]. The TM, ETM+, and GF-1 images were interpreted by a human–machine interactive approach [10]. The land use/cover in the oasis landscape was divided into six types, including farmland, water area, residential area, shrubby grassland, forestry area, and desert area with low vegetation cover (<5%) (Table 1). The mean interpretative accuracy of the land use/cover images was greater than the kappa coefficient of 95% [29,30].

Table 1. Explanation for the landscape units in the Qira oasis.

Landscape Type	Explanation
Farmland	Agricultural area including annual crops, crop residues, vegetables, and bare soils
Water area	Entity water including rivers, lakes, reservoirs, and wetland
Residential area	Built-up area including buildings and facilities
Shrubby grassland	High-coverage shrubby grassland with vegetation cover >60%, medium-coverage shrubby grassland with vegetation cover of 20–60%, and low-coverage shrubby grassland with vegetation cover of 5–20%
Forestry area	Shelter forest for protecting farmland from aeolian sand disaster
Desert area	Sandy land with vegetation cover <5%, barren rock, and Gobi

To analyze the impacts of the climate change and socioeconomic factors on agricultural oasis expansion and landscape dynamics, the meteorological (i.e., temperature and precipitation) and hydrological data in 1958–2010 were collected from the Qira meteorological station and Qira hydrological station, respectively. The socioeconomic data including population, gross domestic product, income, irrigation, and crop yields were acquired from the Xinjiang Statistical Yearbook

(2000–2013) on the Qira county level. These data were used as ancillary data to interpret the main reason, driving forces, and trend of oasis landscape change in the Qira study area.

3.2. Detection Methods of the Oasis Landscape Changes

The satellite images interpreted are generated by the five resulting maps in 1970, 1990, 2000, 2013, and 2016 using GIS. These images can identify the change occurrence and temporal trajectories of land cover types. Variation in land use types is one of the main indicators for showing the landscape change and dynamics. Oasis landscape change can be described by the annual change rate (R) for each oasis land cover types [31], which was calculated as follows:

$$R = \left(\sqrt[t_2-t_1]{A_2/A_1} - 1 \right) \times 100\%, \tag{1}$$

where A_1 and A_2 refer to the land-use area for each land cover types of Qira oasis based on the satellite images interpreted at the initial and last stages of a time period, respectively, and $t_2 - t_1$ represents the time length of the initial and last stage period.

In addition, the analysis of oasis landscape pattern can be characterized by the average annual transition probabilities (P) between the agricultural oasis and other landscape units, which was calculated as follows [7]:

$$P = \frac{A_{ij}}{T} / \sum_j A_{ij} \times 100\%, \tag{2}$$

where A_{ij} is the area of land use type i of Qira oasis from the initial period to the last stage j in a time period and T is the number of years from the initial period to the last stage.

4. Results

4.1. Oasis Landscape Change and Agricultural Oasis Expansion

Figures 2 and 3 show the oasis landscape change from 1970 to 2016. The Qira oasis area decreases slightly at an annual rate of −0.69% in 1970–1990, but has increased rapidly with annual change rates of 0.48%, 0.95%, and 0.08% during 1990–2000, 2000–2013, and 2013–2016, respectively. For the entire period of 1970–2016, the Qira oasis first shrank (1970–1990) and then gradually sprawled (1990–2016). By contrast, the desert area increased significantly from 214.55 km^2 in 1970 to 260.33 km^2 in 1990, but the area dramatically decreased by 21.86% from 260.33 km^2 in 1990 to 149.42 km^2 in 2016 (Figure 4). The speed of oasis expansion is faster than that of land desertification.

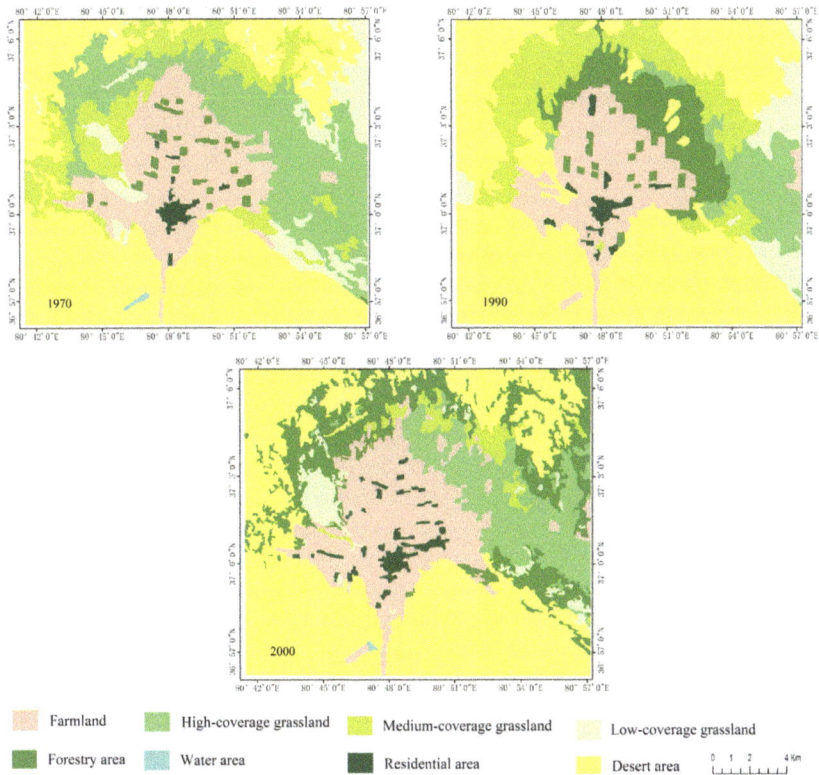

Figure 2. Landscape pattern and agricultural oasis expansion in the Qira oasis from 1970 to 2000.

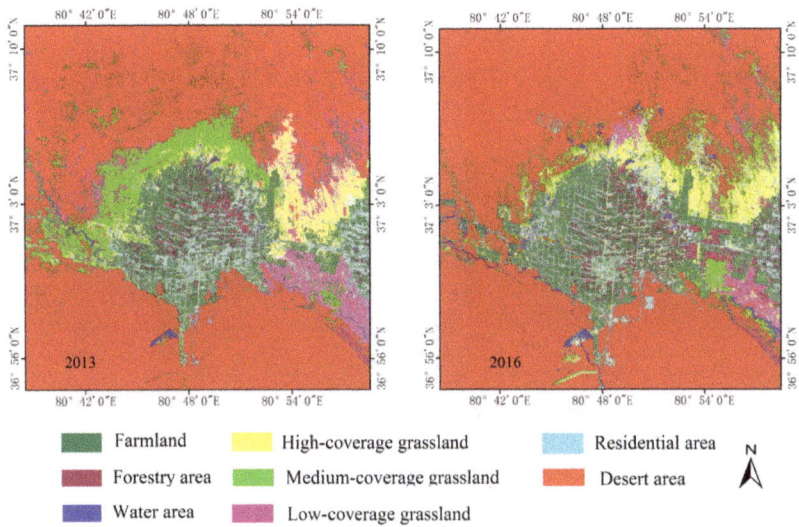

Figure 3. Landscape pattern and agricultural oasis expansion in the Qira oasis from 2013 to 2016.

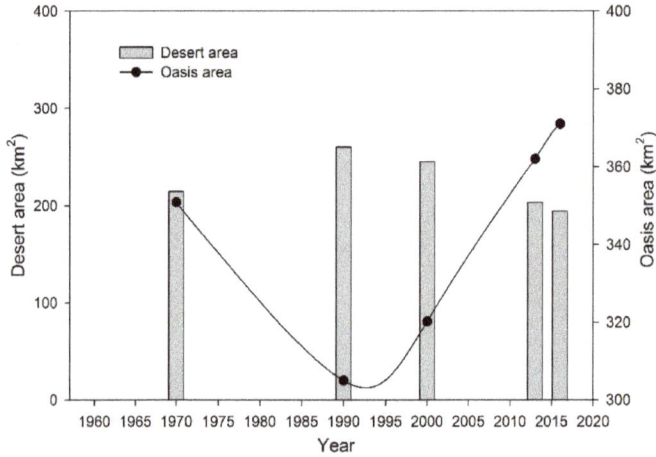

Figure 4. Changes of desert and oasis landscapes during 1970–2016.

In the oasis landscape units, the most significant change in the Qira oasis belongs to the agricultural oasis expansion. The agricultural oasis rapidly increases with annual change rates of −0.3%, 1.6%, 3.7%, and 1.5% during 1970–1990, 1990–2000, 2000–2013, and 2013–2016, respectively. With the expansion of agricultural oasis, the agricultural land has increased from 91.10 km^2 in 1970 to 105.04 km^2 in 2016. The percentage of farmland area has increased by 15.3% since 1970 (Figure 5a). In the spatial variation, the agricultural oasis expansion mainly extends toward the northwest of the oasis during 1970–2000, whereas the farmland expansion from 2000 to 2016 was in the northwest and northeast directions.

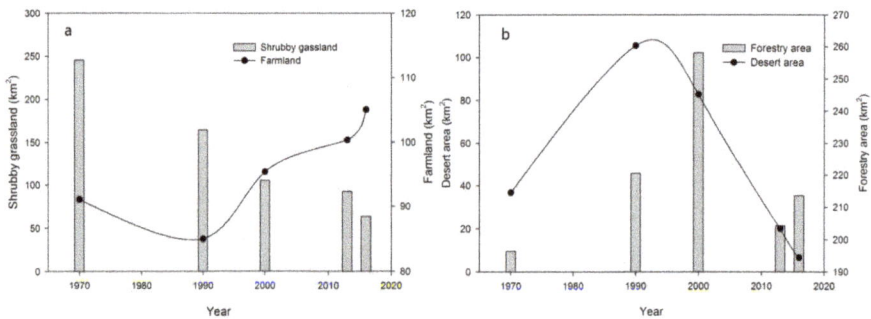

Figure 5. Land-use changes in the Qira oasis during 1970–2016: (**a**) farmland area and shrubby grassland area; (**b**) forestry area and desert area.

4.2. Response of Oasis Land Cover Pattern to the Agricultural Oasis Expansion

As the most important natural vegetation for wind prevention and sand resistance, shrubby grassland has rapidly declined with annual change rates of −1.98%, −4.37%, −2.85%, and 11.60% in 1970–1990, 1990–2000, 2000–2013, and 2013–2016, respectively. Compared with that in 1970, the shrubby grassland area in the Qira oasis has decreased by 74.01% in 2016 (Figure 5a). In addition, the forestry area has rapidly increased with annual change rates of 8.15% and 8.36% in 1970–1990 and 1990–2000, respectively. However, the forestry area significantly decreased with annual change rate of −6.08% from 2000 to 2016 (Figure 5b). In 2016, the forestry area has drastically declined owing to agricultural reclamation and overexploitation.

In terms of the oasis land cover change, there is no significant variation in water areas and built-up areas in the Qira oasis. The expansion of agricultural oasis has impacted the change of the oasis landscape pattern. The average annual transition probability from desert to agricultural oasis has increased by 26.82%, that from shrubby grassland to agricultural oasis by 38.04%, and that from forestland to agricultural oasis by 35.13%. Thus, the natural vegetation is partly lost to agricultural expansion and occupation. The protective function of natural vegetation against desertification and soil degradation may be weakened. With respect to the decrease of natural vegetation, such a trend is alarming for the sustainable development of the Qira oasis in the future.

4.3. Driving Force Analysis of Oasis Landscape Pattern under Agricultural Oasis Expansion

In the Qira oasis area, the natural condition, anthropogenic activities, and human factors, such as population, food security demand, water resource availability, and land use policy, are affecting the oasis landscape change [7,8,10].

The annual mean temperature and precipitation are affirmed with obvious positive impacts on the agricultural oasis expansion at significance level $p < 0.01$ using the panel data model [10]. Figure 6 shows the change of annual mean temperature and precipitation in the Qira oasis study area. These two meteorological parameters illustrate an increasing trend. The increase in temperature benefits cotton plantations, while the increase in precipitation increases the possibility of crop planting. Thus, these natural conditions can induce the reclamation of agricultural oasis.

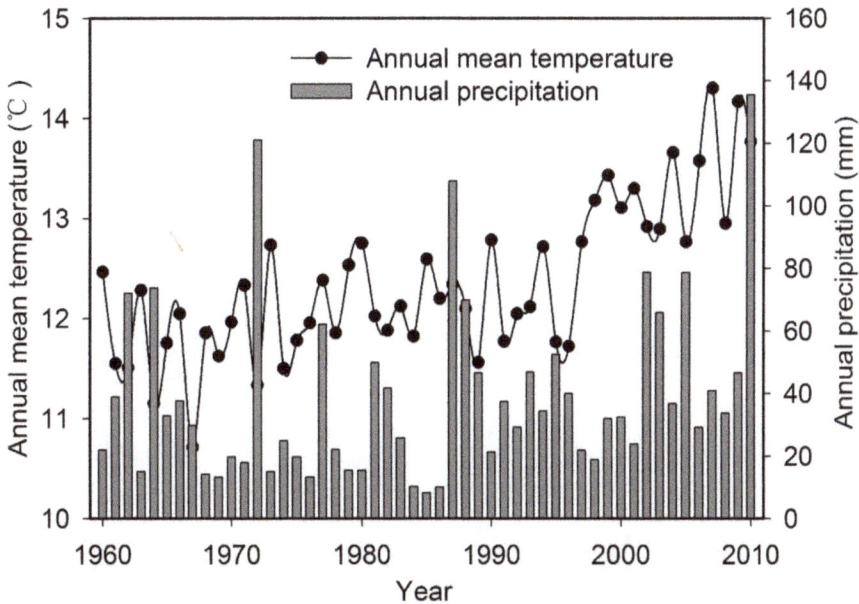

Figure 6. Annual mean temperature and precipitation in the Qira oasis.

The continuous increase in human population is considered as the main factor in agricultural oasis expansion [32,33]. The population in this area has increased from 24,100 in 1955, to 44,900 in 1990, to 55,568 in 2010 (Figure 7). Compared with 1990, the population in 2010 has increased by 23.76%. The continuous growth in human population will inevitably lead to an increase in arable land to support grain yield (Figure 7). Therefore, new agricultural oasis reclamation is needed for human survival at the expense of the natural ecosystem.

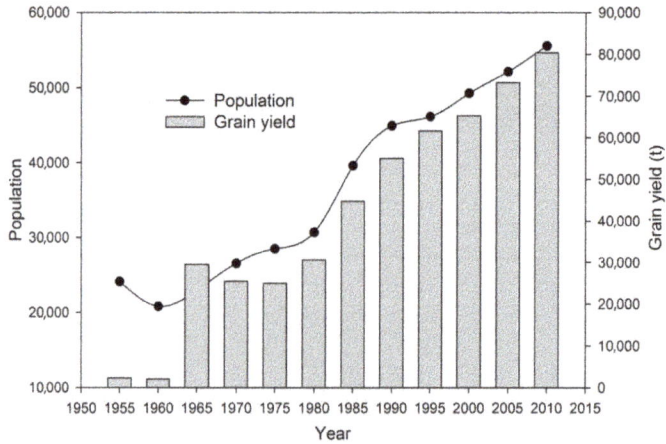

Figure 7. Population and grain yield in the Qira oasis.

The limited water resources directly determine the scale of the oasis expansion [17,19]. The water availability in the Qira oasis area mainly relies on Qira River discharge due to strong evaporation (2700 mm/year) and highly vulnerable ecosystems [2]. Approximately 97.7% of the total water consumption is used to meet irrigation needs. Moreover, 82.1% of the irrigation for agriculture comes from the Qira River, while 17.9% of the remaining water is extracted from groundwater [33]. According to the hydrological data from 1960 to 2010, the runoff in Qira River declined at a rate of 0.003×10^8 m^3/year (Figure 8). This makes it difficult to support new agricultural oasis expansion and threatens the sustainability of the Qira oasis. However, water-saving methods, such as the application of drip irrigation technology and improvement of agricultural mechanization level, can improve the expansion of Qira agriculture oasis with less irrigation needed.

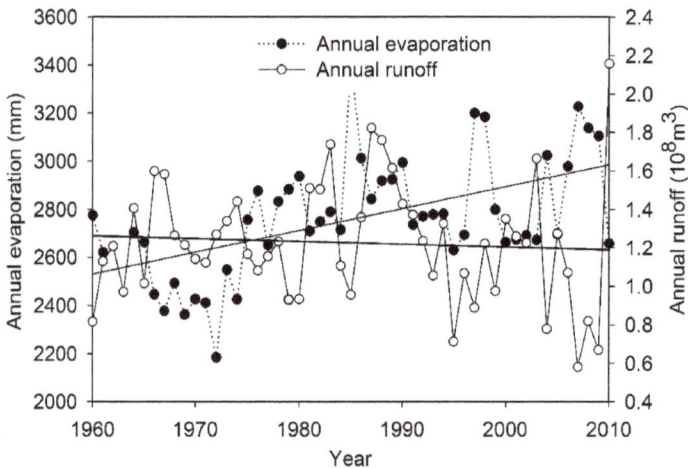

Figure 8. Annual evaporation and runoff in the Qira oasis.

The policy of opening up wasteland has played a key role in the Qira oasis sustainability [34]. Land reclamation is an alternative way to obtain abundant food and income due to poor living conditions. In addition, new reservoirs have been built in the upper reaches of the Qira River for agricultural irrigation. A decreasing groundwater table in the frontier area of Qira oasis leads to the die-off of natural vegetation (such as grassland and desert shrub) at the oasis landscape level [2]. However, the natural vegetation of the oasis, especially the desert–oasis ecotone as an important site for ecosystem services, is important in order to protect the ecotone biodiversity against excessive wasteland reclamation and abusive levels of grazing [35].

5. Discussion

This study shows that the Qira oasis landscape has changed significantly during 1970–2016. The increase in agricultural oasis profoundly impacts its landscape pattern and stability. The dynamics of agricultural oasis is the most important cause of oasis landscape change [8]. Given the human activities and climate change, the size and scale of the Qira oasis is highly dynamic under agricultural oasis expansion. This finding is similar to previous research in other oases, such as the Jinta [8] and Hotan oases [7], in which it was reported that human activities and climate changes have significantly changed the water resource distribution, leading to a rapid expansion of agricultural oasis.

The relationship between agricultural oasis change and its landscape pattern is very complex. In arid and semiarid regions, there is still no consensus to reasonably interpret the oasis landscape change and its driving forces at a regional level [8,36,37]. Furthermore, considering the influence of climate change and human activities, it is difficult to determine the influence of agricultural oasis expansion on the oasis landscape [10]. Song and Zhang [10] found that the precipitation and temperature drives the agricultural oasis expansion in the Heihe River Basin of China, while Fu [38] reported that human activities mostly accounted for the rapid expansion of agricultural oasis. However, this study shows that socioeconomic factors such as population increase, technological advancements in water resource utilization for surface water exploitation and groundwater pumping [7], economic development, and land use policy in opening up wastelands are responsible for the expansion of the Qira agriculture oasis. The results are basically consistent with the work of Zhou [39], who points out that rapid agricultural and economic development lead to expansion of agricultural oasis through occupying the oasis–desert ecotone.

While agricultural oasis expansion is necessary for ensuring food security, it significantly decreases the proportion of natural grasslands and forestlands. The implication for ecological and environmental development of the Qira oasis is alarming. The decline in grassland and forestland due to farmland reclamation will lead to desertification risks. Once destroyed, natural vegetation is difficult to re-establish [2]. Therefore, attention should be paid to the protection of natural vegetation in the arid oasis area.

The Chinese government has launched policies for ecological protection, such as Grain for Green and ecological red lines. These policies play a key role in promoting the efficient use of land resources and improving oasis landscape sustainability in arid regions in Northwest China, especially in the Qira oasis. However, as agricultural expansion leads to increased income, it is difficult to balance economic development and ecological protection. Currently, the land use policy in the Qira oasis is still directed at wasteland reclamation, and policies focused on oasis diversity, oasis ecological safety, and land and water management are lacking. Therefore, future studies should focus on agricultural oasis expansion and its ecological effect on oasis landscape stability and sustainability.

6. Conclusions

In this study, the Qira oasis on the southern margin of the Tarim Basin, Northwest China was selected as a study area to examine the spatiotemporal change of agricultural oasis and determine its influence on oasis landscape pattern based on TM, ETM+, and GF-1 images. Furthermore, the driving force of oasis landscape pattern under agricultural oasis expansion was analyzed by using hydrometeorological and socioeconomic data.

The results show that the agricultural oasis in the Qira oasis has increased rapidly, with annual change rates of −0.3%, 1.6%, 3.7%, and 1.5% in 1970–1990, 1990–2000, 2000–2013, and 2013–2016, respectively. The agricultural land has increased from 91.10 km^2 in 1970 to 105.04 km^2 in 2016. The percentage of farmland area increased by 15.3% between 1970 and 2016. The natural vegetation is decreasing owing to the reclamation of the desert–oasis ecotone. The oasis landscape change pattern is mainly affected by agricultural expansion under water-saving technological utilization, land use policy, and demand for regional economic development. The expansion of agricultural oasis is alarming due to human overexploitation.

Understanding oasis landscape change is essential for its sustainable management. The expansion of agricultural irrigation in the oasis has dramatically affected the landscape stability and ecological sustainability. Therefore, effective solutions (e.g., government policies directed at controlling the overexpansion of agricultural oasis) should be proposed. Future studies should focus on the ecological effect of and water shortage created by agricultural oasis expansion.

Author Contributions: All authors contributed to the design and writing of this manuscript. Y.L. and J.X. carried out the data analysis and prepared the first draft of the manuscript; D.G., J.L. and G.L. are the graduate advisors of Yi Liu and contributed an overall internal review; H.S. and Z.Z. provided important advice on the manuscript.

Acknowledgments: This work was financially supported by the National Natural Science Foundation of China (41601595, 41471031), the "Western Light" program of the Chinese Academy of Science (2017-XBQNXZ-B-016), the State Key Laboratory of Desert and Oasis Ecology, Xinjiang Institute of Ecology and Geography, Chinese Academy of Sciences (1100002394), the Task 2 of the Key Service Project 5 for the Characteristic Institute of CAS (TSS-2015-014-FW-5-3), and the Project of Science and Technology Service Network Initiative of CAS (KFJ-SW-STS-176).

Conflicts of Interest: The authors declare no conflict of interest.

References

1. Verburg, P.H.; Soepboer, W.; Veldkamp, A.; Limpiada, R.; Espaldon, V.; Mastura, S.A. Modeling the spatial dynamics of regional land use: The CLUE-S model. *Environ. Manag.* **2002**, *30*, 391–405. [CrossRef] [PubMed]
2. Bruelheide, H.; Jandt, U.; Gries, D.; Thomas, F.M.; Foetzki, A.; Buerkert, A.; Wang, G.; Zhang, X.M.; Runge, M. Vegetation changes in a river oasis on the southern rim of the Taklamakan Desert in China between 1956 and 2000. *Phytocoenologia* **2003**, *33*, 801–818. [CrossRef]
3. Nagendra, H.; Munroe, D.K.; Southworth, J. From pattern to process: Land-scape fragmentation and the analysis of land use/land cover change. *Agric. Ecosyst. Environ.* **2004**, *101*, 111–115. [CrossRef]
4. Turner, B.L. The sustainability principle in global agendas: Implications for understanding land-use/cover change. *Geogr. J.* **1997**, *163*, 133–140.
5. Lambin, E.F.; Geist, H.J. *Land-Use and Land-Cover Change: Local Processes and Global Impacts*; The IGBP Series; Springer: Berlin, Germany, 2006.
6. Jia, B.Q.; Zhang, Z.Q.; Ci, L.J.; Ren, Y.P.; Pan, B.R.; Zhang, Z. Oasis land-use dynamics and its influence on the oasis environment in Xinjiang, China. *J. Arid Environ.* **2004**, *56*, 11–26. [CrossRef]
7. Amuti, T.; Luo, G. Analysis of land cover change and its driving forces in a esert oasis landscape of Xinjiang, northwest China. *Solid Earth* **2014**, *5*, 1071–1085. [CrossRef]
8. Xie, Y.C.; Gong, J.; Sun, P.; Gou, X.H. Oasis dynamics change and its influence on landscape pattern on Jinta oasis in arid China from 1963a to 2010a: Integration of multi-source satellite images. *Int. J. Appl. Earth Obs.* **2014**, *33*, 181–191. [CrossRef]
9. Liu, J.Y.; Zhang, Z.X.; Xu, X.L.; Kuang, W.H.; Zhou, W.C.; Zhang, S.W.; Li, R.D.; Yan, C.Z.; Luo, G.P.; Feng, Y.X.; et al. Sustainable land-use patterns forarid lands: A case study in the northern slope areas of the Tianshan Mountains. *J. Geogr. Sci.* **2010**, *20*, 510–524.
10. Song, W.; Zhang, Y. Expansion of agricultural oasis in the Heihe River Basin of China: Patterns, reasons and policy implications. *Phys. Chem. Earth Parts A/B/C* **2015**, *89–90*, 46–55. [CrossRef]
11. Gui, D.; Xue, J.; Liu, Y.; Lei, J.; Zeng, F. Should oasification be ignored when examining desertification in Northwest China? *Solid Earth Discuss.* **2017**, under review. [CrossRef]
12. Han, D.L.; Meng, X.Y. Recent Progress of Research on Oasis in China. *Chin. Geogr. Sci.* **1999**, *9*, 199–205. [CrossRef]
13. Cheng, W.; Zhou, C.; Liu, H.; Zhang, Y.; Jiang, Y.; Zhang, Y.; Yao, Y. The oasis expansion and eco-environment change over the last 50 years in Manas River Valley, Xinjiang. *Sci. China Ser. D* **2006**, *49*, 163–175. [CrossRef]

14. Xue, J.; Gui, D.W.; Zhao, Y.; Lei, J.Q.; Zeng, F.J.; Feng, X.L.; Mao, D.L.; Shareef, M. A decision-making framework to model environmental flow in oasis areas using Bayesian networks. *J. Hydrol.* **2016**, *540*, 1209–1222. [CrossRef]
15. Guo, M.; Yu, W.B.; Ma, M.G.; Li, X. Study on the oasis landscape fragmentation in northwestern China by using remote sensing data and GIS: A case study of Jinta oasis. *Environ. Geol.* **2008**, *54*, 629–636.
16. Bai, J.; Chen, X.; Li, L.; Luo, G.; Yu, Q. Quantifying the contributions of agricultural oasis expansion, management practices and climate change to net primary production and evapotranspiration in croplands in arid northwest China. *J. Arid Environ.* **2014**, *100*, 31–41. [CrossRef]
17. Ling, H.B.; Xu, H.L.; Fu, J.Y.; Fan, Z.L.; Xu, X.W. Suitable oasis scale in a typical continental river basin in an arid region of China: A case study of the Manas River Basin. *Quat. Int.* **2013**, *286*, 116–125. [CrossRef]
18. Zhang, H.; Wu, H.W.; Zheng, Q.H.; Yu, Y.H. A preliminary study of oasis evolution in the Tarim Basin, Xinjiang, China. *J. Arid Environ.* **2003**, *55*, 545–553.
19. Zhou, D.; Wang, X.; Shi, M. Human Driving Forces of Oasis Expansion in Northwestern China during the Last Decade—A Case Study of the Heihe River Basin. *Land Degrad. Dev.* **2017**, *28*, 412–420. [CrossRef]
20. Wang, Y.; Gao, J.X.; Wang, J.S.; Qiu, J. Value assessment of ecosystem services in nature reserves in Ningxia, China: A response to ecological restoration. *PLoS ONE* **2014**, *9*, e89174. [CrossRef] [PubMed]
21. Su, Y.Z.; Zhao, W.Z.; Su, P.X.; Zhang, Z.H.; Wang, T.; Ram, R. Ecological effects of desertification control and desertified land reclamation in an oasis-desert ecotone in an and region: A case study in Hexi Corridor, northwest China. *Ecol. Eng.* **2007**, *29*, 117–124. [CrossRef]
22. Zhang, X.F.; Zhang, L.H.; He, C.S.; Li, J.L.; Jiang, Y.W.; Ma, L.B. Quantifying the impacts of land use/land cover change on groundwater depletion in Northwestern China—A case study of the Dunhuang oasis. *Agric. Water Manag.* **2014**, *146*, 270–279. [CrossRef]
23. Siebert, S.; Nagieb, M.; Buerkert, A. Climate and irrigation water use of a mountain oasis in northern Oman. *Agric. Water Manag.* **2007**, *89*, 1–14. [CrossRef]
24. Xue, J.; Gui, D.W.; Lei, J.Q.; Zeng, F.J.; Mao, D.L.; Zhang, Z.W. Model development of a participatory Bayesian network for coupling ecosystem services into integrated water resources management. *J. Hydrol.* **2017**, *554*, 50–65. [CrossRef]
25. Xue, J.; Gui, D.W.; Lei, J.Q.; Zeng, F.J.; Mao, D.L. A hybrid Bayesian network approach for trade-offs between environmental flows and agricultural water using dynamic discretization. *Adv. Water Resour.* **2017**, *110*, 445–458. [CrossRef]
26. Xu, L.; Mu, G.; Ren, X.; Wan, D.; He, J.; Lin, Y. Oasis microclimate effect on the dust deposition in Cele oasis at southern Tarim Basin, China. *Arab. J. Geosci.* **2016**, *9*, 1–7.
27. Mao, D.L.; Lei, J.Q.; Li, S.Y. Characteristics of meteorological factors over different landscape types during dust storm events in Cele, Xinjiang, China. *J. Meteorol. Res.* **2014**, *28*, 576–591. [CrossRef]
28. Lin, Y.; Mu, G.; Xu, L.; Zhao, X. The origin of bimodal grain-size distribution for aeolian deposits. *Aeolian Res.* **2016**, *20*, 80–88. [CrossRef]
29. Liu, J.Y.; Kuang, W.H.; Zhang, Z.X.; Xu, X.L.; Qin, Y.W.; Ning, J.; Zhou, W.C.; Zhang, S.W.; Li, R.D.; Yan, C.Z.; et al. Spatiotemporal characteristics, patterns, and causes of land-use changes in China since the late 1980s. *J. Geogr. Sci.* **2014**, *24*, 195–210. [CrossRef]
30. Liu, J.Y.; Liu, M.L.; Tian, H.Q.; Zhuang, D.F.; Zhang, Z.X.; Zhang, W.; Tang, X.M.; Deng, X.Z. Spatial and temporal patterns of China's cropland during 1990–2000: An analysis based on Landsat TM data. *Remote Sens. Environ.* **2005**, *98*, 442–456. [CrossRef]
31. Pontius, J.R.G.; Shusas, E.; McEachern, M. Detecting important categoricalland changes while accounting for persistence. *Agric. Ecosyst. Environ.* **2004**, *101*, 251–268. [CrossRef]
32. Zhou, D.C.; Luo, G.P.; Lu, L. Processes and trends of the land use change in Aksu watershed in the central Asia from 1960 to 2008. *J. Arid Land* **2010**, *2*, 157–166.
33. Hotan Water Resources Planning. 2013; Xinjiang Tarim River Basin Management Bureau. Available online: http://www.tahe.gov.cn (accessed on 4 June 2018).
34. Shen, Y.; Lein, H.; Xi, C. Water conflicts in Hetian District, Xinjiang, during the Republic of China period. *Water Hist.* **2016**, *8*, 77–94.
35. Mao, D.; Lei, J.; Zhao, Y.; Zhao, J.; Zeng, F.; Xue, J. Effects of variability in landscape types on the microclimate across a desert–oasis region on the southern margins of the Tarim Basin, China. *Arid Land Res. Manag.* **2016**, *30*, 89–104. [CrossRef]

36. Li, X.Y.; Xiao, D.N.; He, X.Y.; Chen, W.; Song, D.M. Dynamics of typical agricultural landscape and its relationship with water resource in inland Shiyang River watershed, Gansu Province, northwest China. *Environ. Monit. Assess.* **2006**, *123*, 199–217.
37. Yu, D.S.; Wu, S.X.; Nan, J. Spatial patterns and driving forces of land use change in China during the early 21st century. *J. Geogr. Sci.* **2010**, *20*, 483–494.
38. Fu, L.; Zhang, L.H.; He, C.S. Analysis of agricultural land use change in the middle reach of the Heihe River Basin, Northwest China. *Int. J. Environ. Res. Public Health* **2014**, *11*, 2698–2712. [CrossRef] [PubMed]
39. Zhou, S.; Huang, Y.F.; Yu, B.F.; Wang, G.Q. Effects of human activities on the eco-environment in the middle Heihe River Basin based on extended environmental Kuznets curve model. *Ecol. Eng.* **2015**, *76*, 14–26. [CrossRef]

sustainability <small>(logo)</small>

MDPI

Article

Analyzing Trends of Dike-Ponds between 1978 and 2016 Using Multi-Source Remote Sensing Images in Shunde District of South China

Fengshou Li [1], Kai Liu [1,*], Huanli Tang [2], Lin Liu [3,4,*] and Hongxing Liu [4,5]

1   Guangdong Key Laboratory for Urbanization and Geo-simulation, Guangdong Provincial Engineering
    Research Center for Public Security and Disaster, School of Geography and Planning,
    Sun Yat-Sen University, Guangzhou 510275, China; fesoon@163.com
2   Guangzhou Zengcheng District Urban and Rural Planning and Surveying and Mapping Geographic
    Information Institute, Guangzhou 511300, China; tanghl@mail2.sysu.edu.cn
3   Center of Geo-Informatics for Public Security, School of Geographic Sciences, Guangzhou University,
    Guangzhou 510006, China
4   Department of Geography and Geographic Information Science, University of Cincinnati,
    Cincinnati, OH 45221, USA; Hongxing.Liu@uc.edu
5   Department of Geography, the University of Alabama, Tuscaloosa, AL 35487, USA
*   Correspondence: liuk6@mail.sysu.edu.cn (K.L.); Lin.Liu@uc.edu (L.L.);
    Tel.: +86-020-8411-3044 (K.L.); +1-513-556-3429 (L.L.); Fax: +86-020-8411-3057 (K.L. & L.L.)

Received: 27 August 2018; Accepted: 26 September 2018; Published: 30 September 2018

Abstract: Dike-ponds have experienced significant changes in the Pearl River Delta region over the past several decades, especially since China's economic reform, which has seriously affected the construction of ecological environments. In order to monitor the evolution of dike-ponds, in this study we use multi-source remote sensing images from 1978 to 2016 to extract dike-ponds in several periods using the nearest neighbor classification method. A corresponding area weighted dike-pond invasion index (AWDII) is proposed to describe the spatial evolution of dike-ponds, both qualitatively and quantitatively. Furthermore, the evolution mechanisms of dike-ponds are determined, which can be attributed to both natural conditions and human factors. Our results show that the total area of dike-ponds in 2016 was significantly reduced and fragmentation had increased compared with the situation in 1978. The AWDII reveals that Shunde District has experienced three main phases, including steady development, rapid invasion and a reduction of invasion by other land use types. Most dike-ponds have now converted into built-up areas, followed by cultivated lands, mainly due to government policies, rural area depopulation, and river networks within Shunde. Our study indicates that the AWDII is applicable towards the evaluation of the dynamic changes of dike-ponds. The rational development, and careful protection, of dike-ponds should be implemented for better land and water resource management.

Keywords: dike-ponds; land use changes; DISP; Landsat; Shunde District

---

1. Introduction

Dike-ponds are a predominant and typical traditional agricultural production mode that was developed in the Pearl River Delta region of south China by local farmers [1]. Pond-breeding fish form in the low-lying, artificially dug fields, while clay in the dikes is used for planting mulberry-dominated trees, which thus form the basic dike-pond pattern. Dike-ponds represent the integration of terrestrial ecosystems (dike) and freshwater ecosystems (pond), and they have become one of the most important types of ecological landscapes in dense river network areas [2,3]. The land-water interaction between dikes and ponds is very significant because of the unique natural conditions of the area where dike-ponds are located, as well as their individual system characteristics. Dike-ponds not only effectively solve some natural environment problems that often occur in low-lying areas, and in turn preserve ecological balance, but they also create huge benefits in terms of driving economic growth [4]. The energy exchange, conversion, and material recycling processes within the system are intense and frequent. Dike-ponds are considered a model of traditional Chinese agriculture practices, and they are an important agricultural cultural heritage in the world [5,6].

Researchers have long sought a better understanding of the various aspects of dike-ponds. The study of dike-ponds began in 1980s whereby the basic structure and functions of dike-ponds was explained [4,7,8]. In addition, scholars also discussed the land–water interactions that occur between dikes and ponds. Such researchers measured the productivity of dike-ponds based on their natural environment, energy flow, material flow and the input and output of the major types of dike-pond (mulberry-dike-fish-pond, fruit-dike-fish-pond, sugar-dike-fish-pond and flower-dike-fish-pond) [9,10]. According to the available research on dike-ponds, reformation of the original dike-pond was conducted in low-lying lands such as Southeast Asia, North America, Australia, and other tropical and subtropical regions to introduce three-dimensional planting [11,12]. In the 1990's, the Pearl River Delta region became a pioneer in reform and opening up, but was deemed to excessively pursue economic benefits. The situation led to the phenomenon of heavy fishpond breeding and light base planting. In-depth studies of dike-ponds fell into stagnation then, where the focus switched to merely analyzing the economic benefits that existing dike-ponds bring about [13–16]. At the beginning of 21th century, the traditional dike-pond had been severely damaged due to the high level of economic development in the Pearl River Delta. A series of subsequent ecological and environmental problems have aroused the attention of the public. For example [17–19], by being mainly focused on the repair and renovation of dike-ponds by exploring the mechanisms related to their construction and regulation. Zhao [20] and Wu [21] demonstrated the impact of the socio-economic development of dike-ponds, while other scholars [22,23] evaluated the ecological environment quality of dike-ponds. As a field of research, the agricultural landscape of dike-ponds provides a reference for the continuation of traditional ecological service functions, the restoration of agricultural ecosystems and the sustainable development of regional ecological economies [24–27].

At present, the study of dike-ponds includes a combination of field survey data of local areas, using geographic information systems (GIS) and remote sensing (RS) spatial analysis methods. The assistance of RS technology has the advantage of the timeliness of data acquisition in field investigations and their wide and large geographic coverage [28–30]. Some existing studies refer to the monitoring of land use changes both in terms of their distribution and area with multi-temporal data [30,31]. Ye used data from the Landsat TM from 1990, 2000, and 2006 to analyze dike-pond changes in the Pearl River Delta [32]. He divided the region into four types of dike-ponds based on a comprehensive expansion coefficient. Wang used data from three periods in 1990, 2000, and 2008 to extract the dike-pond distribution, while determining the dynamic change processes experienced by dike-ponds in the study area [33]. However, most of these analyses involved short time spans and were performed at only a single scale, making it difficult to obtain spatial information of multi-temporal dike-pond areas at larger scales, as well as the spatial distribution of dike-ponds before the 1980's. Some studies stated that the Key Hole (KH) of Declassified Intelligence Satellite Photographs (DISP) series can scan both generally and in detail [34,35]. For the period from the 1960s to 1970s, the Pearl River Delta was a key area of interest for the United States, who archived considerable DISP data covering this region. Considering the available data, DISP images were selected as a source of RS data to determine the river network distribution and classify land use. Many literature works concluded that multi-source data can improve the image resolution and accuracy of land use classification [36–39]. Taking the Landsat Thematic Mapper(TM) images provided by the National Aeronautics and Space Agency (NASA) as a main data source, the evolution of dike-ponds from the 1980's to the present can be realized with the help of KH-9 images [40–42].

Landscape patterns and their dynamic change processes are crucial parts of landscape ecology [43]. Human activities and their environmental effect are frequently discussed in the global change research [44]. The dike-ponds, which are typical ecological landscapes, are especially studied and explored with the help of pollen and palaeoecological records, within an interdisciplinary study [45–47]. Scholars have attempted to measure the conversion of land use quantitatively with feasible indexes, such as the freedom, dispersion, and goodness of construction land expansion [48–50]. Previous studies refer to other indices like the landscape expansion index (LEI) to better reveal spatio-temporal land use dynamics [51]. Unfortunately, the land use types that these indices determine remain relatively limited. For other types of land use which are not always considered, such as wetlands, and it is hard to use such indices to measure their dynamic change. Indeed, there exist very few publications regarding indices that effectively reflect the development trends of wetlands, especially dike-ponds. Therefore, evaluation methods and/or applicable models need to be further formulated. In the present study, we propose a new index called the area weighted dike-pond invasion index (AWDII), which is a variant of the landscape invasion index (LII). The modified parameter, which is based on the LEI, improves our ability to effectively evaluate dike-ponds.

The estimation and monitoring of dike-pond changes are necessary to establish an integrated land and water resource management system [52,53]. As a typical agricultural mode in the Pearl River Delta, it is of great practical significance to implement ecological agriculture construction and sustainable development of regional social economy in developed regions. Therefore, in this study we assemble a series of RS images that span 38 years and which have sufficient spatial resolution to classify land use and extract the distribution of dike-ponds for each period of research. A corresponding evaluation indicator is constructed to analyze the spatial evolution of dike-pond distributions, both qualitatively and quantitatively. By combining the natural environment characteristics in Shunde District with social economics, the factors which led to the dike-pond changes were performed.

2. Materials and Methods

2.1. Study Area

The study area, Shunde, is located at $22°40'15''$–$23°0'40''$ N and $113°1'35''$–$113°23'30''$ E in the Southern part of Foshan City, Guangdong Province. It covers an area of about 806 km^2 (Figure 1). The area has a warm, humid, south Asian subtropical monsoon climate with sufficient sunshine, abundant precipitation, and other unique natural conditions. The territory has a flat land dominated by alluvial plains and rivers, while sporadic hillocks can be found in Southern and Western regions. The Northwest land is slightly higher than the Southwest, and the average elevation of most areas is 0.2 to 2 m above sea level [54]. Several rivers flow across Shunde District, including Xijiang River and the Shunde District Waterway. There are currently about 16 tributaries, and the total length of them reaches 210 km.

Shunde District has jurisdiction over 10 towns (Chencun, Junan, Xingtan, Longjiang, Lecong, Beijiao, Daliang, Ronggui, Lunjiao, and Leliu). These 10 towns are distributed on both sides of the main river channel in Shunde, which has become one of the main ways to communicate with the outside and has provided a link and means of interaction among the towns in the district.

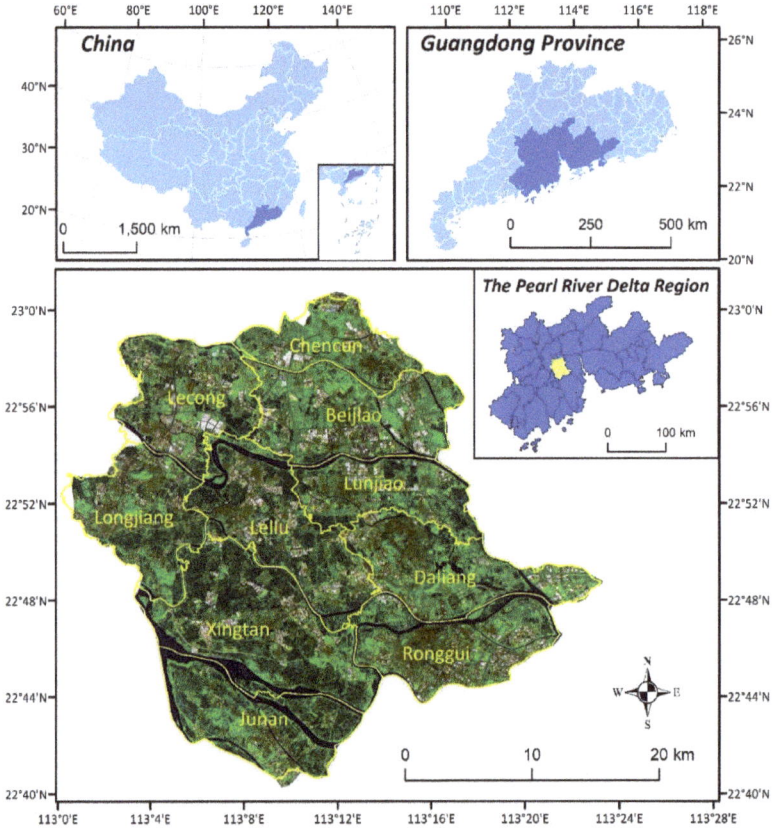

Figure 1. Location of the study area.

2.2. Data and Pre-Processing

In this study, Declassified Intelligence Satellite Photographs (DISP) and Landsat MSS/TM/OLI images were used. To obtain a large time range of recorded RS data as far as possible, the Landsat MSS data were acquired on 2 November 1978, and the archived DISP data KH-9 is close to MSS data on time. KH-9 are single-band images with high spatial resolution, between 6 and 9 m, while the Landsat MSS images have low spatial resolution of about 78 m over four bands. To obtain enhanced images with both high spatial resolution and multispectral features, image fusion was performed. Among the available KH-9 images, we used two cloud-free images of good quality obtained on 17 December 1975 and 18 January 1976 (i.e., 1 month apart), respectively, which were combined to obtain an image that covered the entire study area (i.e., Shunde District). Dominated by agricultural developments between 1975 and 1978, Shunde District can be regarded as being constant due to only small changes occurring in land use over the three-year period. Therefore, image fusion was undertaken with the KH-9 image and the MSS image that was acquired 3 years later in 2 November 1978. Through several comparison experiments, it was found that the Brovey method resulted in a fused image of higher quality, which hence served as the data source for the subsequent analysis in 1978. To avoid inconsistency in resolutions, the resolution of Landsat images in 1978 was normalized to that of KH-9 image in 1975. The required image pre-processing varied according to the range of single scene images from the different data sources, the projection information of the images, and their application purposes (Table 1).

Table 1. Summary of the data used and pre-processing steps taken in this study.

Satellite	Sensor	Acquisition Time	Geometric Correction	Image Mosaic	Image Clipping	Image Fusion
DISP	KH-4	1967-12-18	√ *	√	√	√
	KH-9	1975-12-17/1976-01-18	√ *	√	√	√
Landsat	MSS	1978-11-02	√	×	√	√
	TM	1988-12-10, 1993-11-22, 2000-09-14, 2005-07-18, 2011-06-01	√	×	√	×
Landsat	OLI	2016-12-07	√	×	√	×

Note: √ means the pre-processing step was implemented; × means the pre-processing step was not implemented; * means image rotation was implemented before geometric correction.

2.3. Classification and Accuracy Assessment

There are differences in the characteristics of the target categories between the fused and individual Landsat TM images. Interpretation of the two types of RS images in the fused MSS and KH-9 data (R:4 G:3 B:2) and Landsat TM image (R:7 G:5 B:3) band combinations are respectively displayed in Table 2 and described in Table 3.

Table 2. Comparison between the fused images (MSS and KH-9) and Landsat TM.

Categories	Fusion Image Features (R:4 G:3 B:2)	Landsat TM Image Features (R:7 G:5 B:3)
Dike-pond		
Built-up area		
Cultivated land		

Table 2. *Cont.*

Categories	Fusion Image Features (R:4 G:3 B:2)	Landsat TM Image Features (R:7 G:5 B:3)
Forest		
River		

Table 3. Description of the target categories.

Categories		Descriptions
Dike-pond	1	A combination of light blue or dark blue regular patches and surrounding yellow linear features which are meshed
	2	A combination of dark blue regular patches and surrounding green linear features which are meshed
Built-up area	1	Gray-black, uneven internal shade, irregular shape
	2	Gray-purple, uneven internal shade, irregular shape
Cultivated land	1	Light yellow or brown bar, uniform color shade, regular patches
	2	Green, uniform color shade, relatively regular patches
Forest	1	Brown, relatively uniform internal shade, darker in the middle part
	2	Green, relatively uniform internal shade, darker in the middle part
River	1	Light blue, uneven width, uniform internal shade, long curved lines
	2	Dark blue, uneven width, uniform internal shade, long curved lines

Note: 1 means fusion image (R:4 G:3 B:2); 2 means Landsat TM image (R:7 G:5 B:3).

ECognition Developer 9.0 was used for multi-scale image segmentation, the scale of which varied according to the spatial resolution of the image and the object characteristics. Through repeated experiments of setting the segmentation parameters, the segmentation scales of the seven period images were finally determined to be 70, 30, 15, 15, 15, 20, and 100, while other relevant parameters include the hue index (0.9), shape index (0.1), tightness index (0.5), and smoothness index (0.5). A certain number of classification features were selected from the mean, standard deviation, object length shape area, etc. of each object. Combination of these features was obtained to maximize the classification distance, and we used the nearest neighbor classification method to realize automatic classification of the RS images. Land use classification results were acquired after manually correcting the automatic classification result of each image. Classification validation is an integral and essential part of any image-classification process [55]. Therefore, the accuracy of the classification results was assessed by randomly collecting a certain number of sample objects for various types of ground objects to avoid contingency and improve the quality of accuracy evaluation. For the land use maps of earlier

196

years, the relevant topographic maps, and photographs were considered to choose validation samples. The overall classification accuracy and global Kappa coefficient for each period from 1978 to 2016 were carried out.

Figure 2 shows the workflow of the image pre-processing, classification, and analysis.

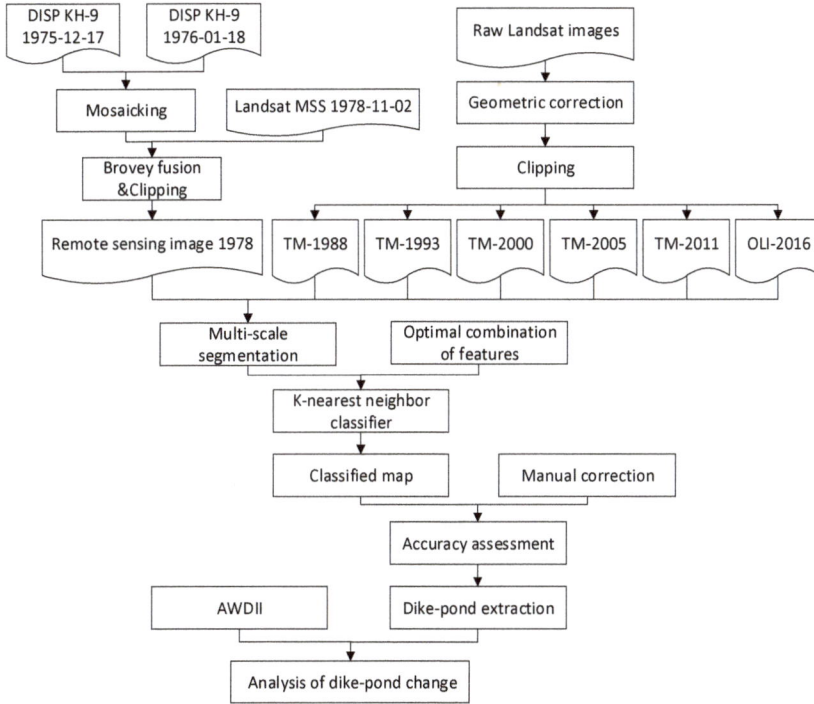

Figure 2. Workflow of the research.

2.4. AWDII

The landscape expansion index (LEI) was proposed to identify three types of urban expansion, i.e., infilling type, edge-expansion type, and outlying type to analyze the dynamic changes of two or more temporal phases of landscape patterns [56]. At present, the LEI is usually applied to spatial patterns and process descriptions of construction land expansion [51,57,58]; however, it is difficult to use them to determine dynamical changes of landscape patterns that contain wetlands because they mostly fluctuate rather than present a single trend. In addition to the proposed definition and quantification index LEI [59], we also propose a landscape invasion index (LII) that can be used for cases where a certain land type invades another land type, and which can describe the spatial pattern and extent of the invasion within a specific stage. The LII can be calculated using Equation (1):

$$LII = \frac{(A_i - A_0)}{(A_i + A_0)} \tag{1}$$

where A_0 is the original patch area, and A_i is the area of patch occupied by a certain landscape adjacent to the original patch. When $A_0 = 0$, there is no original patch adjacent to the invading patch. At that time, $LII = 1$, that is, it belongs to the outlying invasion type. When $A_i = 0$, $LII = -1$, and the original patch does not invade any other landscape. When LII is between -1 and 1, the invasion mode is the edge-invasion type.

This study considers dike-ponds as the sole wetland research object. Therefore, the area weighted dike-pond invasion index (AWDII) is proposed, which can be calculated using Equation (2):

$$AWDII = \sum_{i=1}^{n} (LII_i \times \frac{a_i}{A})$$ (2)

where LII_i is the LII of the invading patch i, n is the total number of patches in the invaded landscape, a_i is the area of the invading patch i, and A is the total invaded area of the landscape. According to Equation (1), the value range of LII is $[-1, 1]$, which implies that the range of $AWDII$ is also $[-1, 1]$. The larger $AWDII$ is, the larger the area of the invading dike-pond is relative to the original dike-pond area is, or in other terms, the more severely the dike-pond is being invaded.

3. Results

3.1. Land Use Classification Results and Accuracy Analysis

To analyze the land use distribution in Shunde, five land use types were identified including built-up areas, cultivated lands, dike-ponds, forests, and rivers. Figure 3 shows the superposition of an RS image with the results of the dike-pond extraction, which illustrates the distribution of the main research objects of this paper in each chosen period. The results of the seven periods (1978, 1988, 1993, 2000, 2005, 2011, and 2016) indicate that the total area of dike-ponds in 2016 was significantly reduced and more fragmented compared with 1978, i.e., after approximately 40 years of development. Comparing the classification results from 1978 to 2016, the trends of the dike-pond distribution changes were qualitatively obtained.

The dike-ponds extended from West to East between 1978 and 1988. In 1978, the dike-ponds were mainly in the Western, Central, and Southern regions of Shunde District, which were continuously distributed and densely packed. In 1988, the dike-ponds in these areas stayed constant and some began to appear in the Eastern parts of Shunde District, though overall they have a more scattered distribution. The distribution changed little from 1988 to 1993. Compared to1988, the dike-ponds in the Wstern, Central, and Southern regions remained lumped and densely distributed in 1993, while those in Eastern regions were scattered and distributed relatively compactly. Locally, the conversion to other land use types from dike-ponds and the contrary circumstances simultaneously existed. The fragmentation of dike-ponds in the Western and Central regions increased significantly between 1993 and 2000. The contiguousness of the dike-ponds in Shunde District was broken in 2000, where the largest changes occurred in the Western and Northern parts. It can be observed from the image from 2000 that these changes are mainly due to urban land expansion that occupied regions previously populated by dike-ponds. The land changes in the Central and Southern areas were relatively small, and they still contained the most concentrated distributions within the entire study area.

The changing trends of dike-pond distribution were basically the same both from 2000 to 2005 and from 2005 to 2011, that is, the distribution range of dike-ponds continued to shrink and the degree of fragmentation increased. Since 2000, the contiguous area of the dike-pond landscape in Shunde District was broken by other types of land use, and its area declined year by year. Dike-ponds in the Western, North-Central, and Eastern regions experienced the largest changes. By 2011, they were only sporadically distributed in these areas. In the South and South-Central parts, dike-ponds also shrank and the degree of fragmentation increased. However, compared with other regions, they were still more densely distributed, and became the main areas of dike-pond occurrence. The distribution pattern of dike-ponds has not changed substantially from 2011 to 2016, although its total area has further decreased due to invading, built-up areas. Throughout the whole study area, the fragmentation of dike-ponds became more obvious, especially in the Northwest and Eastern regions. Dike-ponds remained tightly spaced along the river networks, but were sparsely distributed in the Northern and Southern parts of Shunde. As of 2016, the total area of dike-ponds was less than one third of the entire study area.

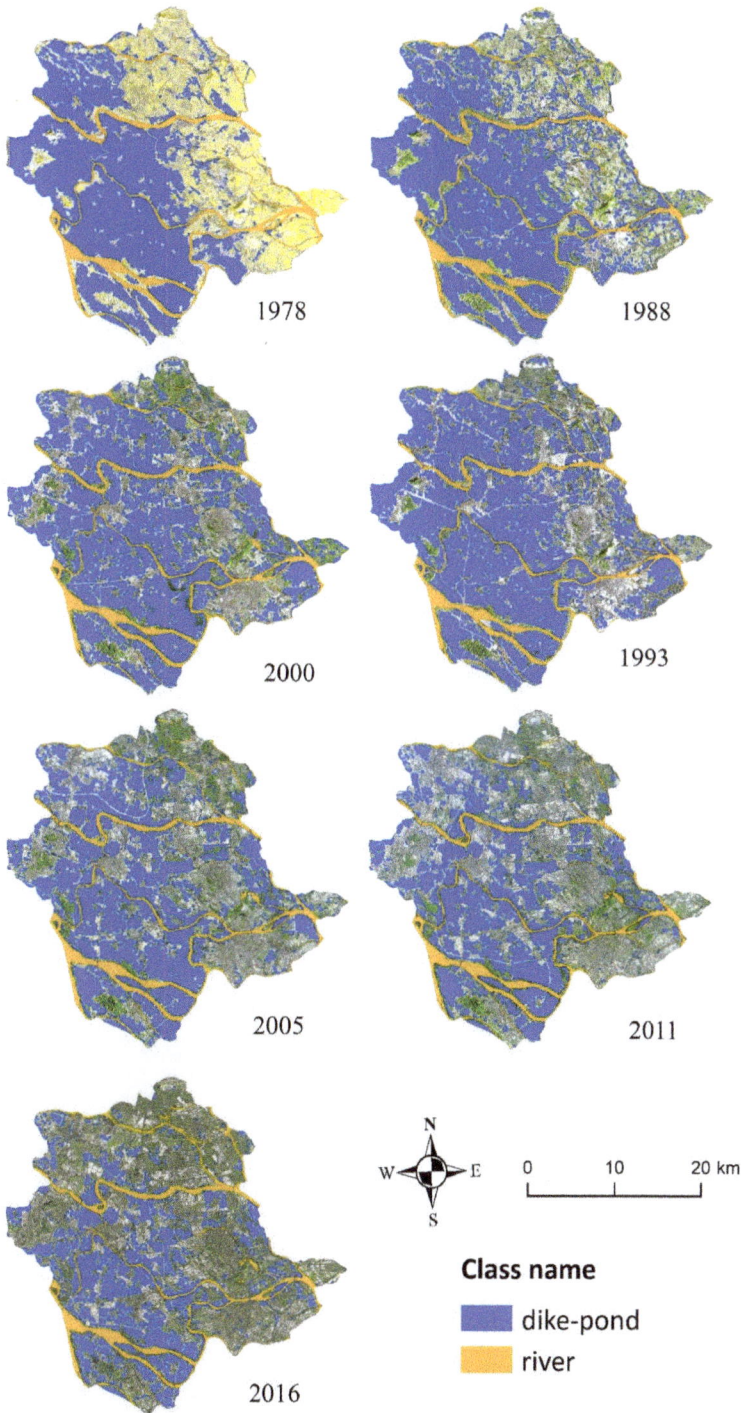

Figure 3. Distribution of dike-pond in Shunde District from 1978 to 2016.

The classification results were assessed by randomly collecting a certain number of sample objects, i.e., various types of ground objects, to determine the overall classification accuracy and global Kappa coefficient for each period from 1978 to 2016 (Table 4). Error matrix calculation information is shown in Appendix A. The results show that the overall accuracy of the RS image classifications during the seven periods from 1978 to 2016 is higher than 89%, and the Kappa coefficient is more than 0.86. The high classification accuracy demonstrates that the use of object-oriented methods and artificially assisted corrections to classify images can improve the classification efficiency compared with pure manual interpretation, and ensure the reliability of the classification results.

Table 4. Overall accuracy assessment of the classification results.

Accuracy Measures	Overall Accuracy (%)	Kappa Coefficient
1978	91.55	0.894
1988	92.92	0.911
1993	93.30	0.916
2000	90.94	0.886
2005	89.49	0.868
2011	92.90	0.910
2016	91.29	0.890

RS images from the seven periods from 1978 to 2016 were taken to calculate the area of each land use type and their relative proportion to the total area. The changes in land use over the past 38 years are given in Figure 4. Between 1978 and 2000, the area of dike-pond wetlands dominated all land use types in the study area, with an area ratio of over 45% and a maximum of 50%. The area of dike-pond wetlands increased first, reaching a peak in 1993, and then decreased, to become the second most prevalent land use type after 2000. Between 1978 and 1993, the proportion of cultivated land in the study area was in the second most dominant type, while the combined cultivated land plus dike-ponds in the area was more than 70%, which reached a peak of 86.62% in 1978. This reflects the dominant agricultural economy before the 1990's in Shunde.

Over the past 38 years, the number of built-up areas in Shunde District has continued to increase, and it became the most prevalent land use type in the whole study area in 2005, accounting for 54.43% of the total area in 2016. The steady declining trend of forests reflect their increasing level of exploitation, while the area of rivers stayed stable at around 7.5% over the last 38 years. Since the area of forests and rivers was much smaller than the other land types, the absolute value of their area change was not very considerable. It can be observed from Figure 4 that the total area of forests and rivers remained relatively steady during the four-decade period.

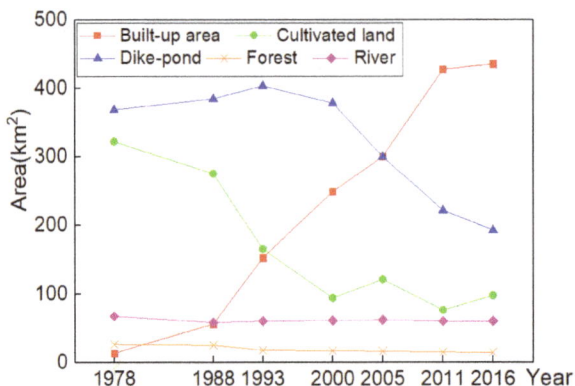

Figure 4. Changing area of land use types in Shunde District between 1978 and 2016.

3.2. Dike-Pond Trends

Next, we took the towns of Shunde District as land use units. With the land use classification results acquired for the seven periods from 1978 to 2016, information pertaining to the administrative scope of each town in Shunde District was obtained so that the distribution of dike-ponds in each town could be extracted. We calculated the area of dike-ponds in each town and their proportion of the overall dike-pond area in the district, which were also compared and analyzed for the different periods.

Figure 5 shows the relative dike-pond areas of all towns in each period. From 1978 to 2016, the dike-pond areas of Xingtan and Leliu are ranked first and second, respectively, among the 10 towns. The area of dike-ponds in Xingtan accounts for more than 20% of the total area of dike-ponds in the study area, while the proportion in Leliu were above 14%, where both decreased first and then increased, and which were accompanied by fluctuations in their relative proportions. Xingtan and the Leliu, which are located in the Southwest and Central part of Shunde, respectively, were within the main distribution area of dike-ponds in the overall scale analysis. Over the past 38 years, the area of dike-ponds in Chencun ranged from 0.86 to 8.51 km^2, which is relatively small compared with other towns. It accounts for less than 3% of the total area of dike-ponds in Shunde and as low as 0.4% in 2011. Chencun is located in the Northern part of Shunde, and it has fewer areas compared with other towns. After 2000, the area of dike-ponds in each town showed a downward trend except for Ronggui, where its dike-pond area in 2016 was higher than that in 2011.

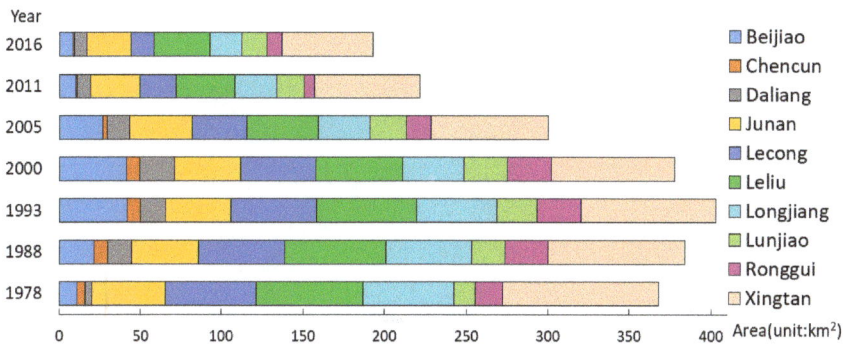

Figure 5. Stacked bar graph of dike-pond area of the 10 towns in Shunde District from 1978 to 2016.

In can be concluded that the dike-pond distribution in the different towns was asymmetric according to the spatial analysis. Xingtan and Leliu in Shunde District contain the largest dike-pond areas, while Chencun has the smallest dike-pond area between 1978 and 2016.

We have shown here the difference between the dike-pond proportions of each town. In 1978, the dike-pond proportions of Xingtan and Longjiang were larger than the other towns, and were similar at 78.94% and 78.99%, respectively. In the process of extracting the dike-ponds, the spatial resolution and spectral characteristics of the images inevitably affected the accuracy of the classification results. Therefore, a difference of 0.05% in the dike-pond area proportion of Xingtan and Longjiang in 1978 can be neglected, and are hence considered to be approximately equal.

During the period from 1978 to 1988, the dike-pond area proportions of Xingtan and Longjiang were both more than 70%, and ranked in the top two. In 1993, the dike-pond proportion of Lecong began to surpass Longjiang and Xingtan, and was ranked first in Shunde. However, it fell down slightly to 60.49% followed after Xingtan with a decline of 2%. After 2000, the dike-pond proportion in Xingtan stayed the highest, though its proportion gradually decreased, and it was seen that Lecong and Junan were ranked second in 2000 and 2005, respectively. Leliu remained the second highest proportion after 2005. From our results, we predict that Xingtan might be the town with the largest proportion of dike-ponds in the near future.

According to the above analysis, it can be summarized that the dike-pond area proportion of each town has changed in two ways over the past 38 years. The 10 towns were divided into two types with Class A, the proportion of which continuously decreased, and Class B, the proportion of which increased first and then decreased. The different trends are shown in Figure 6.

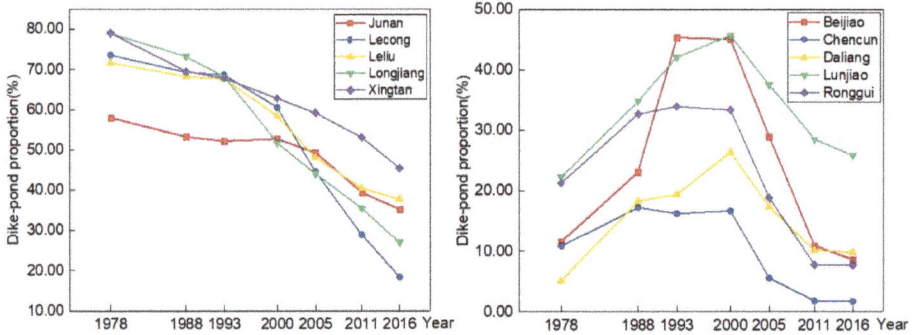

Figure 6. Two different evolution trends of dike-pond proportion in each town.

Based on the statistical data of dike-pond area in each town, the year (i.e., time) was used as the independent variable (x) and the dike-pond area was the dependent variable (y, unit: km^2). Linear and quadratic polynomial functions were constructed to fit the evolution trends of the two types for the 10 towns, as classified in Table 5.

Table 5. Two types of dike-pond evolution fitting function on town.

Types	Towns	Fitted Curves	Functions	R^2
	Junan		$y = -0.0056x + 11.689$	0.8344
	Lecong		$y = -0.0153x + 31.022$	0.8763
Class A	Leliu		$y = -0.0101x + 20.703$	0.9229
	Longjiang		$y = -0.0147x + 29.843$	0.9666
	Xingtan		$y = -0.0083x + 17.115$	0.9767

Table 5. *Cont.*

Types	Towns	Fitted Curves	Functions	R^2
	Beijiao		$y = -0.0009x^2 + 3.5686x - 3561.3$	0.7888
	Chencun		$y = -0.0003x^2 + 1.0258x - 1020.6$	0.8155
Class B	Daliang		$y = -0.0004x^2 + 1.7216x - 1719.2$	0.8328
	Lunjiao		$y = -0.0005x^2 + 2.1476x - 2144.5$	0.9023
	Ronggui		$y = -0.0005x^2 + 2.0255x - 2016.9$	0.8664

Each function fitting result had a high correlation coefficient, R^2, ($0.79 \leq R^2 \leq 0.98$), which was used to evaluate how well each function could effectively reflect the dike-pond area changes in each town. The dike-pond area proportion of the Class A towns decreased with different curvatures, in which Junan had the smallest curvature and Lecong had the largest. We can see from Figure 6 that the deceleration rate of dike-pond areas in each period was not the same. Among the five fitting curves, Xingtan has the highest degree of fitting, indicating that it decreased at a more uniform rate. As for the Class B towns, the rate of increase and decrease of the dike-pond areas in a given town was inconsistent from 1978 to 2016, and there was also a difference in the rates of decline between different towns. The curve of Beijiao was more symmetrical, that is, the rates of increase and decrease were both relatively consistent. Daliang and Lunjiao had similar increase and decrease rates. Comparatively, the dike-pond areas in Ronggui fell faster than those in Chencun.

We marked two types of towns on the map of Shunde District (Figure 7) based on the data in Figure 5. This figure shows that the two types have obvious spatial differences, where Class A towns are located in Western, Central and Southern parts, and Class B towns are located in the East and North of Shunde. A combination of the spatial distribution map and the statistical data of the dike-pond area proportion in each town quantitatively reflect the evolution processes of the spatial distribution patterns of dike-ponds in Shunde. From 1978 to 2016, the dike-ponds were mainly distributed in the Western, Central and Southern regions, and they spread to the East and North between 1978 and 2000. After 2000, the dike-pond area in the Eastern and Northern parts decreased at a faster rate, but the dike-pond area proportion in the Northeast gradually returned to the historical level of 1978.

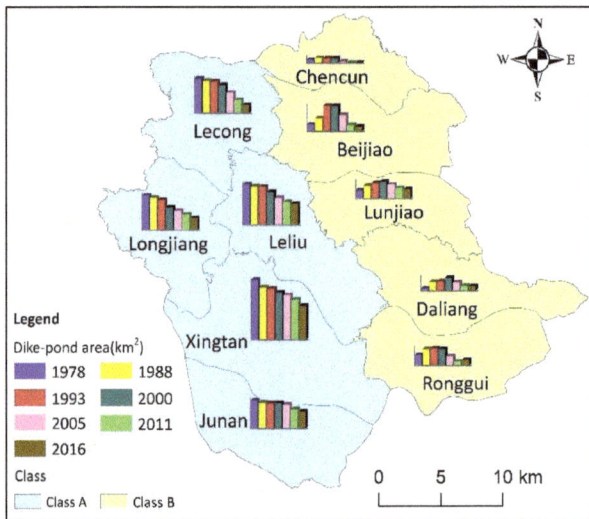

Figure 7. Spatial distribution of the Class A and Class B towns in Shunde.

3.3. Invasion Evaluation of Dike-Pond

In this study, we used the AWDII to measure the spatial pattern and extent of dike-ponds being invaded by other land use types in each period. By overlaying and analyzing the conversion of dike-ponds and non-dike-ponds between different years, the transformation of dike-pond areas in each period was obtained (Figure 8). We first calculated the LII values of each patch, and then determined the AWDII of each area, which was used to evaluate the amount of invasion experienced by dike-ponds in each region (Table 6).

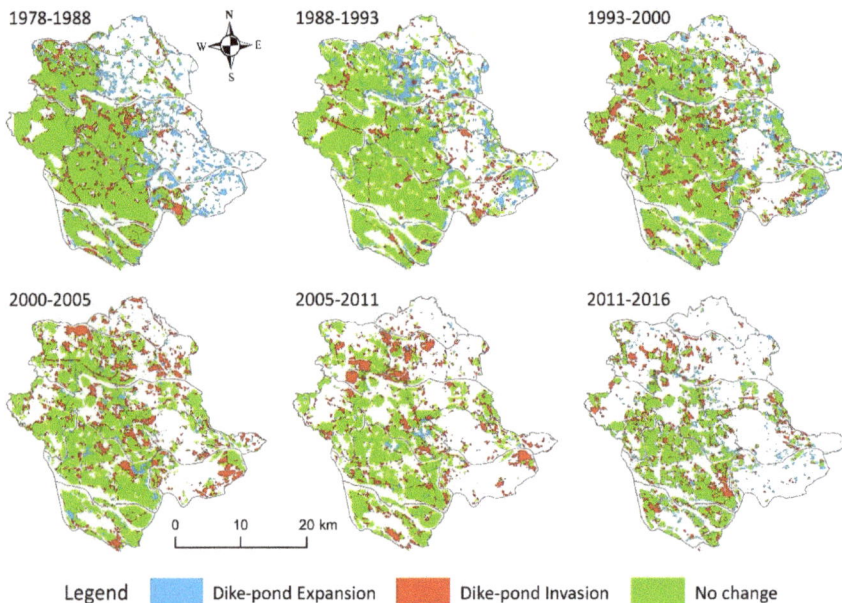

Figure 8. Dike-pond expansion/invasion during each period.

Table 6. Calculated AWDII (area weighted dike-pond invasion index) values.

Period Region	1978–1988	1988–1993	1993–2000	2000–2005	2005–2011	2011–2016
Beijiao	−1.000	−1.000	−0.986	−0.474	−0.232	−0.652
Chencun	−1.000	−0.891	−1.000	−0.200	−0.191	−0.956
Daliang	−1.000	−1.000	−1.000	−0.492	−0.409	−0.944
Junan	−0.847	−0.962	−1.000	−0.879	−0.665	−0.807
Lecong	−0.891	−0.980	−0.789	−0.582	−0.479	−0.481
Leliu	−0.908	−0.983	−0.759	−0.699	−0.726	−0.875
Longjiang	−0.890	−0.874	−0.610	−0.742	−0.673	−0.642
Lunjiao	−1.000	−1.000	−1.000	−0.698	−0.612	−0.827
Ronggui	−1.000	−1.000	−0.967	−0.395	−0.258	−1.000
Xingtan	−0.785	−0.953	−0.861	−0.889	−0.813	−0.750
Shunde	−1.000	−1.000	−0.871	−0.439	−0.456	−0.570

Between 1978 and 1993, the *AWDII* of Shunde District was −1, illustrating Shunde District was not invaded during this period. Among the towns, the *AWDII* values of Beijiao, Daliang, Lunjiao, and Ronggui remained at −1, and it could also be observed from Figure 8 that the area of these four towns kept increasing. Xingtan and Longjiang had the largest AWDIIs in 1978–1988 and 1988–1993, respectively, showing that they had experienced serious dike-pond invasion compared to the other towns.

During 1993–2005, the AWDII values of Shunde District rose from −0.871 to −0.439, mainly due to extensive reclamation of land and urban development. Dike-ponds began to be invaded by other land types, and the protection of dike-ponds was not as good as before. Between 1993 and 2000, Lecong and Leliu were occupied more densely than the other towns. After 2000, Chencun was the most seriously invaded, the AWDII of which reached to −0.2, followed by Ronggui (−0.395) and Daliang (−0.492).

The situation of expropriation of dike-ponds in 2005–2011 was slightly better than that in 2000–2005, but the invasion degree was larger. Among the ten towns, Chencun was occupied most heavily, the AWDII value of which was as high as −0.191, followed by Beijiao (−0.232) and Ronggui (−0.258).

Up to 2016, the ecological environment caused by the dike-pond invasion by other land types had attracted the public's eye, following the development of modernization. In recent years, the situation of dike-pond invasion has improved compared with previous periods. However, it can be clearly concluded that dike-ponds in Lecong, Longjiang and Beijiao were still invaded in spite of the control. There was obvious protection for Ronggui and Chencun on account of the low AWDII values. On the whole, the dike-ponds in Shunde District experienced steady development, rapid invasion and a gradual reduction of invasion in the period from 1978 to 2016.

3.4. Comparison of Typical Dike-Pond Trends

Two specific examples with different dike-pond evolutions were selected to describe their concrete changes. Figure 9 shows the changes in the dike-pond proportion of Xingtan and Longjiang (relative to the total area of the town) during the period from 1978 to 2016. As mentioned previously, in 1978, the proportions in Xingtan and Longjiang were 78.99% and 78.94%, respectively, which could be considered as them having the same starting point of their evolutionary trends. After approximately 40 years of development, the distribution scope in both towns has reduced. In 2016, the proportions in Xingtan and Longjiang decreased to 45.39% and slightly lower than 30%, respectively. The dike-pond change trend curves of the two towns are also displayed in Figure 9, which show that the decreasing trends of the two towns have been quite different since 1978, and especially after 1993, the dike-pond proportion of Longjiang dropped sharply. In contrast, the rate of decline of dike-ponds in Xingtan was relatively slow. Next, let us take 1993 as a boundary point, so that we further analyze two periods (i.e., before and after 1993). From 1978 to 1993, the decreasing trend of dike-pond proportion in both towns were similar, although the proportion in Xingtan was slightly lower than that in Longjiang in 1988. After 1993, the rate of decline in two towns began to diverge greatly. The decreasing rate of dike-pond proportion in Xingtan was scarcely different from that of the previous period, while Longjiang showed a drastic decline with a relatively larger rate from the previous period.

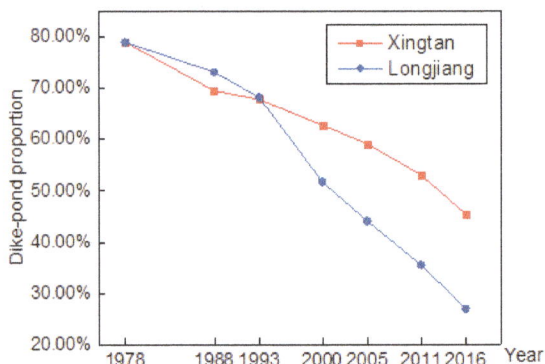

Figure 9. Comparison of dike-pond area proportions in Xingtan and Longjiang from 1978 to 2016.

As a result of a Southern tour of the chief architect of reform and opening up to Guangdong Province in 1992, the pace of reform in the Pearl River Delta was accelerated by the major historic event. It could be inferred that the development of dike-ponds in Shunde District was mainly affected by the expansion of urban land use. Therefore, we speculate that Longjiang pushed forward a period of construction and urbanization, which meant that more dike-ponds were occupied by built-up areas after 1993.

Based on the changes of dike-ponds and the other types of land use in Xingtan and Longjiang, the reasons for the differences in their dike-pond evolutions is discussed. Xingtan, a well-known water town in the Pearl River Delta region, is densely intertwined with river networks in the Southwest of Shunde. Dike-ponds used to be the main agricultural form in Xingtan because of their contained fertile land. Since the reform and opening up, Xingtan has paid attention to the adjustment and optimization of agricultural structures while developing industry, where the science and technology industry, ecological agriculture, and water country culture form the three major industrial strategy foci of the town [60]. Table 7 shows the land use changes quantitatively in Xingtan from 1978 to 2016. In 1978, the area of dike-ponds accounted for a large proportion, 79.06%, of the total area of the town, while about 35.60% of the town's dike-ponds were converted to built-up areas, where only slightly more than half of the dike-ponds were preserved. From 1978 to 2016, dike-ponds, cultivated lands, forests, and rivers were all converted to built-up areas to varying degrees, which increased from 2.01 to 38.69 km^2, and accounted for 31.82% of the town's total area in 2016. Despite the sharp change in dike-ponds and built-up areas, forests and rivers comprised only a little both in terms of area and proportion.

Longjiang is located in western Shunde. It has superior agricultural production conditions, and its river network is dense. When the dike-ponds in the Pearl River Delta emerged in the late Ming Dynasty [61], it also became a typical, and representative, type of argo-ecological model, which resulted in it being a major economic source. After the reform and opening up, Longjiang took manufacturing industry as the main body of economic development, the pillar industries of which include the manufacture of furniture, small household appliances, textiles, food, beverages, plastics, etc.

Table 8 shows the land use changes in Longjiang from 1978 to 2016. During this period, the proportion of dike-ponds in Longjiang decreased dramatically from 78.91% to 27.41% of the total town area. As of 2016, about 56.52% of the dike-pond area in Longjiang has been converted to built-up areas, and only about 33.29% of the dike-ponds have been preserved. While developing industry,

Longjiang focused on developing its agriculture in terms of producing high yields, high quality, and high economic efficiency. Therefore, the dike-ponds and livestock breeding have been maintained. Apart from dike-ponds, a large proportion of cultivated lands and forests were converted to built-up areas, which increased from 0.63 to 38.54 km^2, which account for a town proportion increase from 0.9% to 54.69%.

Table 7. Land use transfer matrix of Xingtan between 1978 and 2016 (unit: km^2).

2016 / 1978	Built-up Area	Cultivated Land	Dike-Pond	Forest	River	Total (1978)
Built-up area	**1.74** 86.52%	0.02 1.06%	0.25 12.42%			2.01 1.65%
Cultivated land	1.79 24.44%	**2.18** 29.82%	2.87 39.28%	0.10 1.38%	0.37 5.08%	7.32 6.02%
Dike-pond	34.22 35.60%	10.03 10.44%	**51.13** 53.19%	0.66 0.69%	0.08 0.08%	96.13 79.06%
Forest	0.15 6.66%	0.06 2.66%	0.20 8.79%	**1.84** 81.89%		2.25 1.85%
River	0.79 5.69%	0.40 2.89%	0.82 5.93%	0.23 1.68%	**11.64** 83.80%	13.88 11.42%
Total (2016)	38.69 31.82%	12.70 10.45%	55.28 45.46%	2.84 2.33%	12.09 9.94%	**121.59** 100.00%

Table 8. Land use transfer matrix of Longjiang between 1978 and 2016(unit: km^2).

2016 / 1978	Built-up Area	Cultivated Land	Dike-Pond	Forest	River	Total (1978)
Built-up area	**0.63** 99.54%					0.63 0.90%
Cultivated land	2.07 70.08%	**0.46** 15.68%	0.41 14.03%		0.01 0.21%	2.95 4.19%
Dike-pond	31.43 56.52%	5.18 9.31%	**18.51** 33.29%	0.44 0.79%	0.05 0.08%	55.61 78.91%
Forest	4.07 48.33%	1.42 16.92%	0.15 1.82%	**2.77** 32.93%		8.42 11.95%
River	0.34 11.89%	0.25 8.76%	0.24 8.23%		**2.03** 71.12%	2.86 4.06%
Total (2016)	38.54 54.69%	7.32 10.38%	19.32 27.41%	3.21 4.56%	2.09 2.96%	**70.47** 100.00%

To sum up, the dike-pond area proportions of Xingtan and Longjiang both showed decreasing trends from 1978 to 2016, but they declined at very different rates after 1993. The rate of dike-pond curtailment in Xingtan remained steady after 1993, while that in Longjiang performed saliently.

3.5. Relationship between River Network and Dike-Ponds

In the territory of Shunde, several major rivers run through the area, which are a primary means of communication among the ten towns. The influence of the main river on each town can be considered as consistent. Outside the main river, the rivers in Shunde District are densely distributed. They compose the river network which is one of the objects analyzed in this paper. The limitation of spatial resolution made it difficult to obtain information on rivers from the KH-9 image. Visually comparing the KH-4A image obtained in 1967 and the KH-9 remote sensing image in 1978, the river network distribution in Shunde District was not significantly different between these two periods.

Therefore, we used the high-resolution KH-4A image as a RS data source to obtain high-precision river network information.

The total length of the river network, the river network density (the ratio of the length of the river network to the total area of the town) and the dike-pond proportion in each town in Shunde District are shown in Table 9. Xingtan had the largest river length of 240.72 km, and its river network was also the largest, reaching 1.98 km per square kilometer. The length of the river network was only 76.67 km in Ronggui, which was the shortest, and its river network density was the smallest among the ten towns, which was only 0.96 km per square kilometer.

Table 9. Length and density of the river network in each town in Shunde in 1967.

Towns	Length (km)	Density (km/km^2)
Beijiao	165.06	1.80
Chencun	80.47	1.59
Daliang	92.71	1.16
Junan	105.72	1.36
Lecong	127.12	1.63
Leliu	141.79	1.56
Longjiang	134.31	1.82
Lunjiao	93.34	1.58
Ronggui	76.67	0.96
Xingtan	240.72	1.98

Comparing the dike-pond proportion in 1978 (Figure 5) with the river network information in 1967 (Table 9), in general, the trend of the river network density was in line with the trend of the dike-pond proportion (Figure 10). This means that the towns with the high density of river networks had a correspondingly larger proportion of dike-ponds. In particular, the density of river networks in Beijiao and Chencun were higher, but the proportion of dike-ponds did not correspondingly rise up. It is inferred that the relationship between the river network density and the proportion of dike-ponds in each town can be affected by other types of land use. To this end, a statistical analysis was performed on the land use classification information of 1978 in Shunde, which also considers the proportion of each type of land in each town (Table 10).

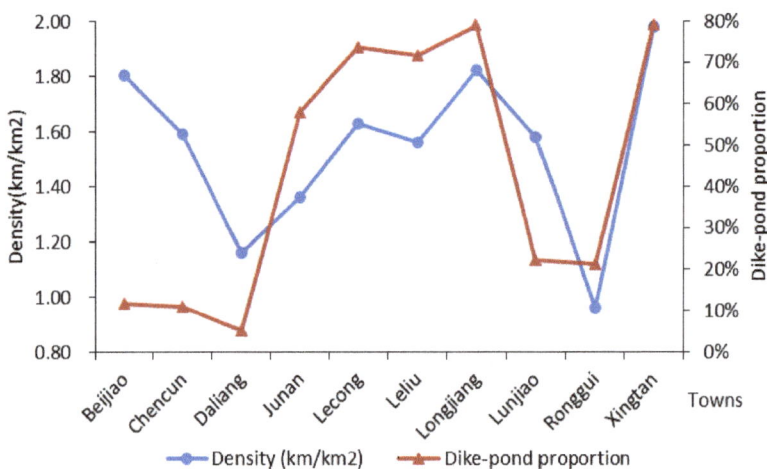

Figure 10. The relationship between river network density in 1967 and the dike-pond proportion in each town in 1978.

Table 10. The proportion of various types of land use in each town in 1978.

Towns	Built-Up Area	Cultivated Land	Dike-Pond	Forest	River
Beijiao	1.06%	80.93% *	11.60%	0.89%	5.53%
Chencun	1.66%	81.94% *	10.85%	2.30%	3.24%
Daliang	1.87%	79.88% *	5.12%	7.05%	6.07%
Junan	0.74%	16.41%	57.95% *	7.20%	17.69%
Lecong	2.64%	19.51%	73.54% *	0.00%	4.31%
Leliu	2.04%	17.58%	71.66% *	0.56%	8.16%
Longjiang	0.89%	4.18%	78.94% *	11.92%	4.06%
Lunjiao	1.03%	67.85% *	22.25%	0.00%	8.86%
Ronggui	2.99%	62.29% *	21.30%	2.08%	11.34%
Xingtan	1.65%	6.01%	78.99% *	1.87%	11.48%

Note: * means the largest proportion of area in the town.

It can be concluded from Table 10 that the dike-ponds in Junan, Lecong, Leliu, Longjiang, and Xingtan accounted for the largest proportion of the total area of the town. For the other towns, cultivated land was the main land use type. This was taken as a basis to divide the towns in Shunde District into two categories. Junan, Lecong, Leliu, Longjiang, and Xingtan were classified as Class C, and the others were Class D. This has led to a deeper understanding of the C and D-class towns, of which the dominant land use types were dike-ponds and cultivated land, respectively. In practice, the rivers in the Class D towns are a rich source of water for agricultural lands other than dike-ponds. As a result, the trends of the two curves were not identical, as shown in Figure 10, when considering the proportion of dike-ponds and river network density of the two classes of towns.

The curves of the river network density and the dike-pond proportion (to the total area of the corresponding town) are shown in Figure 11. For the towns of Class C, their river network density curves are consistent with the trends of the dike-pond proportion curve, illustrating that the higher network density is, the higher dike-pond proportion is. For towns of Class D, the towns with high river network densities generally had large dike-pond proportions. In particular, Ronggui had a large dike-pond area but a low river network density. This is because Ronggui is surrounded by a main river channel, and its dike-ponds are mainly distributed along the main river channel. In the river network analysis, only the length of the river stream was considered and the main channel information was ignored, thus inferring that the main river channel had a greater impact on the distribution of dike-ponds than the inner stream of the river. It can be concluded that the river network density had the same influence on dike-ponds of both Class C and D. With an increase of river network density, the proportion of dike-ponds increase accordingly. In general, the river network density of Class C towns, dominated by dike-ponds, was higher than that of Class D.

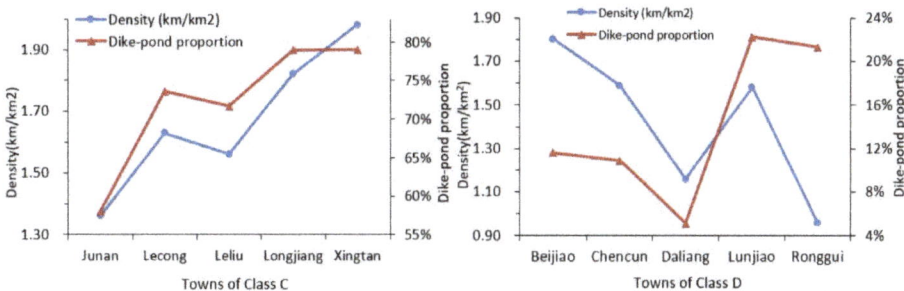

Figure 11. The relationship between river network density and dike-pond proportion.

4. Discussion

4.1. Impact of Government Policies on Dike-Pond Area Reduction

This study illustrated the government policies and river network both had significant influence on the evolution of dike-ponds in Shunde District. In order to discuss the decision-making activities of the government, especially the influence of local governments' decision in terms of the evolution of dike-ponds in Shunde, we selected the two sites as the research objects, which had the same starting point but discrepant end points. That is, the difference of the dike-pond proportion of both towns in 1978 was relatively small, but larger in 2016 (Figure 9). In 1978, China began to implement reforms and an opening-up policy, where the south took a leading role compared with other regions in the country. As the Pearl River Delta region is at the forefront of reform and opening-up, industrial restructuring, urban development and other measures became more compact. Over the past 38 years, the proportion of built-up areas in Shunde District has increased from 1.68% to 54.43%. Under the influence of built-up areas and other types of land use, the number and spatial distribution pattern of dike-ponds in Shunde District have changed, from which it could be inferred that the country's macroeconomic policies were inextricably linked to the evolution of dike-ponds in Shunde.

The dike-ponds in each town were analyzed, and it was shown that the total dike-pond area and its proportion (relative to the total area of the town) in different types of towns had different trends. Indeed, even in the same type of town, the evolution curve of dike-ponds was not exactly the same. The difference in the evolution of dike-ponds was closely related to the enthusiasm of local farmers for cultivation. The market demand, economic benefits, and social response brought about by farming products have a direct impact on the enthusiasm of farmers. Therefore, the policies of local government play an important, perhaps leading role, in the industrial structure and economic development of the whole region [62]. It was an important turning point in the process of China's reform and opening up in 1992 because the pace of development began to accelerate across the country, especially in the Pearl River Delta. After reform and opening up, Xingtan and Longjiang adopted different development strategies driven by government policies in two directions. After about four decades, the two towns, which had the same evolutionary starting point, have large differences in their conservation of dike-ponds. The great policy has promoted the expansion of the built-up areas, which in turn led to a reduction of dike-pond area. It can be concluded that the country's macroeconomic policies have had an indelible impact on dike-ponds.

4.2. Impact of River Network on Dike-Pond Distribution

Another important factor that affects dike-pond evolution is known as natural environment, more specifically, the river network. Some predecessors have conducted extensive research on the distribution area characteristics, ecological patterns, and system operation mechanisms of dike-ponds [63–65]. These investigations indicated that dike-ponds may be the most successful agro-ecological model for the transformation of low-lying waterlogged lands. In other words, the need for humans to transform low waterlogging land has promoted the formation and promotion of the dike-pond. From this, a main natural condition necessary for dike-pond construction must be a rich source of water.

After 1978, the dike-ponds in each town in Shunde District evolved with different trends under the influence of the national macro-policy and local government decision-making. Moreover, urbanization became a major factor in the evolution of dike-ponds after reform and opening up. Statistics showed that prior to reform and opening up, dike-ponds in Shunde District were dominant over other the types of land use in 1978. Therefore, in order to avoid the influence of other factors after the reform and opening up, the article took the dike-ponds in Shunde District that existed before the reform and explored the relationship between the distribution of the river network in Shunde District and the distribution of dike-ponds. More precisely, we explored the distribution of dike-ponds of the various towns in Shunde District in 1978 and the relationship with the river network distribution during this period. From what the study showed in 3.5, it can be inferred that the proportion of dike-ponds increase synchronously with a rise of river network density. Therefore, river network is another important factor besides government policies which is regarded as water resource of dike-ponds helping ecological cycle and large-scale development.

4.3. Impact of Rural Area Depopulation on Dike-Pond area Reduction

The migration of the rural population to the cities will affect the area of the dike-ponds from two aspects. First of all, the reduction of rural population is bound to decrease the effective development and management. The hysteretic sludge cleaning of dike-ponds will directly lead to insufficient water cycle. Additionally, the yield will not be able to keep up with the requirement due to the unsatisfied feeding, which will give rise to economic benefit reduction. In this case, a large amount of area of dike-ponds can be influenced in Shunde District. Secondly, the rural area depopulation can be a stimulus to the rise of urban population, thus increasing the urban burden to some extent and speeding up the process of urbanization. In need of more living space, more dike-ponds with relatively low economic output value are invaded by expanding urban construction area. As a result, the area of dike-ponds continue to decrease transferring to built-up area.

4.4. Changes of Dike-Pond Role in Water Management

Due to the long-term waterlogging in the low-lying areas of the Pearl River Delta in history, flood disasters were frequent in the past, thus bringing about original dike-ponds that helps drainage and disaster reduction. Since 1949, flood control in China has entered a new period of development [66]. The public concentrated on the construction and reinforcement of river embankments, opened new flood discharge channels, and completed water conservancy projects. Therefore, the flood control system of major rivers has been gradually enhanced and improved. The result of this is that the role of dike-ponds was gradually weakened in flood control and disaster mitigation.

In the 1970's and 1980's, the dike-ponds developed into an agricultural model with higher economic benefits [14]. Combining the terrestrial and aquatic ecosystems, the utilization efficiency of freshwater resources reached optimal. Moreover, considering its large amount of storage, the dike-ponds without industrial pollution are widely regarded as one of the main freshwater sources of domestic water in rural areas. As a result, the resource-based ecological service function has become the most essential part of water management.

With the acceleration of urbanization, the area of dike-ponds has been reduced, and the fragmentation has increased. While pipelines are generally installed in rural areas to supply water by reservoirs, the function of dike-ponds cannot be underestimated conserving water storage. Because the water level in dike-ponds is artificially controlled, the water is allocated to cultivated land for agricultural irrigation besides supplying the demand for its own aquatic products. Generating convenience and benefits, the major functions of dike-ponds has been changing with time and the current situation, but have always played a significant and irreplaceable role in water management.

4.5. Limitation

The development of dike-pond system can be influenced by many factors including natural environment and socio-economic aspects. Among them, the former contains regional climate, geological conditions, river basin locations, etc., while the latter includes population, regional agricultural structure, agricultural production efficiency, urbanization process, economic structure, macroeconomic policies, and so on. In the Results part of this paper, only government policies and river network were considered and discussed. In fact, some other factors such as irreversible urbanization process, also affect the evolution of dike-ponds. Thus, more research needs to be done on the contribution of different factors to dike-pond evolution.

In addition, the paper intends to analyze the relationship between dike-pond evolution and affecting factors from a quantitative point of view in the design of the experiment. However, the number of selected objects is relatively small that is not convincing enough. For example, only two representative towns were selected in the analysis of impact of government decisions on the dike-ponds. If more than a few can be chose for comparison, it will be more completely accepted.

In terms of the above discussion, a deeper exploration can be carried out from the following research. Firstly, the multi-temporal and high spatial resolution RS images of Shunde District is supposed to be acquired to explore the evolution characteristics of the internal structure in dike-ponds, comparing the differences between towns. What is more, combining with other non-RS data like statistical yearbooks and official reports, the factors affecting the dike-pond evolution can be analyzed more thoroughly and accurately, so as to provide a more comprehensive reference for the restoration and construction of dike-ponds. Lastly, we have already discussed the evolution and its factors regarding dike-ponds as target objects in the former part. However, in turn, considering that dike-pond is a typical wetland ecosystem, the area change and fragmentation of dike-ponds will also have a feedback effect on the environment, which is a significant field we can explore.

5. Conclusions

In this paper, DISP and Landsat images were used to derive land use information over the period from 1978 to 2016. Ultimately, seven land use maps were produced to extract the dike-pond distribution, which had a classification accuracy greater than 89% and a kappa coefficient of more than 0.86.

The results of comparison among the seven chosen periods indicate that the area of dike-ponds in 2016 was significantly reduced, and fragmentation had increased compared with that in 1978, after 38 years of development. It was seen that most of the dike-ponds that had disappeared by 2016 were due to their conversion into built-up areas, and to a lesser degree, cultivated land. Among the ten administrative regions in Shunde District, two main classes were constructed to describe the trend of dike-pond area proportion change of each town: Class A, where the proportion of dike-ponds continuously decreased, and Class B, where the proportion of dike-ponds increased first and then decreased. Combined with the calculated AWDII values, the dike-ponds in Shunde District experienced three main phases over the past 38 years: steady development, rapid invasion and a gradual reduction of invasion by other land use types. It indicated that AWDII are of great significance to explain other types of wetlands with fluctuated and complex trends. There are some main factors that may affect the evolution of dike-pond, namely government policies, the river network, and rural area depopulation, which all belong to socio-economic and the natural environmental factors.

To overcome the situation where dike-ponds showed a decreasing trend caused by continuous urbanization and intemperate cultivation, the harmony should be made a priority between economic development and ecological environment. Based on free satellite data and open source software, the processing workflow and eventual results can be used generally for spatio-temporal monitoring wetland evolution, more rational water resource management as well as supporting directives for the conversation of inland freshwater ecosystem.

Author Contributions: F.L., K.L. and H.T. designed the study, performed the experiments, analyzed the results and wrote the manuscript. L.L. contributed in designing the study, editing the manuscripts and providing funding to the project, and H.L. provided some useful suggestions for the paper's results and discussions.

Funding: This research is founded by the National Science Foundation of China (Grant No. 41001291, 41531178), the Natural Science Foundation of Guangdong (Grant No. 2016A030313261), the Science and Technology Planning Project of Guangdong Province (Grant No. 2017A020217003).

Acknowledgments: The authors would like to thank the anonymous referees and the editors for their valuable comments, and also appreciate International Science Editing (http://www.internationalscienceediting.com) for editing this manuscript.

Conflicts of Interest: The authors declare no conflict of interest.

Appendix A. Error Matrix of Classification Accuracies in Shunde District from 1978 to 2016

Table A1. Error matrix of classification in 1978.

Sample Number		Evaluated Image					
		River	Forest	Built-up Area	Cultivated Land	Dike-Pond	Sum
	River	57	1	3	1	2	64
	Forest	0	56	1	2	0	59
Reference	Built-up area	6	1	76	2	0	85
image	Cultivated land	0	3	2	64	2	71
	Dike-pond	2	0	0	3	83	88
	Sum	65	61	82	72	87	367
Producer's accuracy		94.00%	100.00%	96.60%	100.00%	100.00%	-
User's accuracy		100.00%	100.00%	96.60%	95.00%	100.00%	-
Overall accuracy: 91.55%; Overall kappa coefficient: 0.894							

Table A2. Error matrix of classification in 1988.

Sample Number		Evaluated Image					
		River	Forest	Built-up Area	Cultivated Land	Dike-Pond	Sum
	River	43	0	2	1	1	47
	Forest	0	45	1	2	0	48
Reference	Built-up area	2	0	71	5	0	78
image	Cultivated land	1	2	0	72	1	76
	Dike-pond	1	3	0	1	71	76
	Sum	47	50	74	81	73	325
Producer's accuracy		91.49%	93.75%	91.03%	94.74%	93.42%	-
User's accuracy		91.49%	90.00%	95.95%	88.89%	97.26%	-
Overall accuracy: 92.92%; Overall kappa coefficient: 0.911							

Table A3. Error matrix of classification in 1993.

Sample Number		Evaluated Image					
		River	Forest	Built-up Area	Cultivated Land	Dike-Pond	Sum
	River	61	1	2	1	1	66
	Forest	0	67	1	2	1	71
Reference	Built-up area	4	0	74	0	0	78
image	Cultivated land	1	2	2	83	1	89
	Dike-pond	2	3	2	1	91	99
	Sum	68	73	81	87	94	403
Producer's accuracy		92.42%	94.37%	94.87%	93.26%	91.92%	-
User's accuracy		89.71%	91.78%	91.36%	95.40%	96.81%	-
Overall accuracy: 93.30%; Overall kappa coefficient: 0.916							

Table A4. Error matrix of classification in 2000.

Sample Number		River	Forest	Built-up Area	Cultivated Land	Dike-Pond	Sum
				Evaluated Image			
Reference image	River	55	1	3	2	1	62
	Forest	0	55	3	0	4	62
	Built-up area	0	1	61	1	3	66
	Cultivated land	0	1	0	71	3	75
	Dike-pond	2	3	2	1	69	77
	Sum	57	61	69	75	80	342
Producer's accuracy		88.71%	88.71%	92.42%	94.67%	89.61%	-
User's accuracy		96.49%	90.16%	88.41%	94.67%	86.25%	-
Overall accuracy: 90.94%; Overall kappa coefficient: 0.886							

Table A5. Error matrix of classification in 2005.

Sample Number		River	Forest	Built-up Area	Cultivated Land	Dike-Pond	Sum
				Evaluated Image			
Reference image	River	38	2	0	5	1	46
	Forest	1	46	2	0	0	49
	Built-up area	2	1	65	1	3	72
	Cultivated land	1	1	1	51	3	57
	Dike-pond	0	3	1	3	64	71
	Sum	42	53	69	60	71	295
Producer's accuracy		82.61%	93.88%	90.28%	89.47%	90.14%	-
User's accuracy		90.48%	86.79%	94.20%	85.00%	90.14%	-
Overall accuracy: 89.49%; Overall kappa coefficient: 0.868							

Table A6. Error matrix of classification in 2011.

Sample Number		River	Forest	Built-up Area	Cultivated Land	Dike-Pond	Sum
				Evaluated Image			
Reference image	River	53	0	0	1	3	57
	Forest	0	45	1	1	0	47
	Built-up area	1	2	76	2	1	82
	Cultivated land	1	3	2	83	2	91
	Dike-pond	2	1	1	1	70	75
	Sum	57	51	80	88	76	352
Producer's accuracy		92.98%	95.74%	92.68%	91.21%	93.33%	-
User's accuracy		92.98%	88.24%	95.00%	94.32%	92.11%	-
Overall accuracy: 92.90%; Overall kappa coefficient: 0.910							

Table A7. Error matrix of classification in 2016.

Sample Number		River	Forest	Built-up Area	Cultivated Land	Dike-Pond	Sum
				Evaluated Image			
Reference image	River	51	0	1	1	2	55
	Forest	0	47	2	0	1	50
	Built-up area	1	2	64	3	2	72
	Cultivated land	1	3	2	73	1	80
	Dike-pond	3	1	1	2	69	76
	Sum	56	53	70	79	75	333
Producer's accuracy		92.73%	94.00%	88.89%	91.25%	90.79%	-
User's accuracy		91.07%	88.68%	91.43%	92.41%	92.00%	-
Overall accuracy: 91.29%; Overall kappa coefficient: 0.890							

References

1. Zhong, G. A Deeper Realization of Mulberry Dike-Fish Pond System. *Trop. Geogr.* **1984**, *4*, 129–135.
2. Zhong, G.; Cai, G. Eco-Economic Model of Basic (Cultivated Land) Pond System in China—A Case Study of the Pearl River Delta and the Yangtze River Delta. *Ecol. Econ.* **1987**, *53*, 15–20.
3. Liao, X. Theory and Technology Mode of Mulberry Based Fish Pond Ecosystem in Low-Lying Water Ponds. *Guangxi Seric* **1999**, *36*, 30–31.
4. Zhong, G. The Characteristics of Dike-Pond System and Practical Significance. *Sci. Geogr. Sin.* **1988**, *45*, 161–187.
5. Ding, J.; Wen, Y.; Shu, Q. Studies on Sustainable Development of Aquaculture in the Dike-Pond Ecosystem. *Chongqing Environ. Sci.* **2001**, *23*, 12–14.
6. Nie, C.; Luo, S.; Zhang, J.; Li, H.; Zhao, Y. The Dike-Pond System in the Pearl River Delta: Degradation Following Recent Land Use Alterations and Measures for Their Ecological Restoration. *Acta Ecol. Sin.* **2003**, *23*, 1851–1860.
7. Zhong, G. The Structural Characteristics and Effects of the Dyke-Pond System in China. *Outlook Agric.* **1989**, *18*, 119–123.
8. Zhong, G. The Types, Structure and Results of the Dike-Pond System in South China. *GeoJournal* **1990**, *21*, 83–89.
9. Zhong, G. Some Problems About the Mulberry-Dike-Fish-Pond Ecosystem on the Zhujiang Delta. *Chin. J. Ecol.* **1982**, 10–11. [CrossRef]
10. Zhong, G. Mulberry-Dike-Fish-Pond on the Zhujiang Delta—A Complete Artificial Ecosystem of Land-Water Interaction. *Acta Geogr. Sin.* **1980**, *35*, 200–209.
11. Korn, M. The Dike-Pond Concept: Sustainable Agriculture and Nutrient Recycling in China. *Ambio* **1993**, *25*, 6–13.
12. Lo, C. Environmental Impact on the Development of Agricultural Technology in China: The Case of the Dike-Pond ('Jitang') System of Integrated Agriculture-Aquaculture in the Zhujiang Delta of China. *Agric. Ecosyst. Environ.* **1996**, *60*, 183–195. [CrossRef]
13. Huang, Y.; Liu, T. Dike-Pond Ecosystes and Their Soils. *J. South China Agric. Univ.* **1995**, 102–107.
14. Li, H.; Luo, S. Eco-Economic Analysis on the Chinampa Systems in Maoming City. *Rural Eco-Environ.* **1992**, *8*, 41–45.
15. Yang, Y. Problems and Countermeasures of Sustainable Development of Agricultural Ecology in the Pearl River Delta. *Guangdong Agric. Sci.* **1995**, 14–16.
16. Zhang, J.; Zhong, G. Analysis of Harmonizing Human and Geographic Environment Relationship in the Dike-Pond Econ-Ecological System. *Ecol. Sci.* **1993**, 55–59.
17. Li, H.; Ling, W.; Yang, J.; Zhang, M. Biremediation of the Hchs Residue in Sediments of Dike-Pond System by Growing Ipomoea Aquatic and Using Different Fertilizer. *J. Agro-Environ. Sci.* **2007**, *26*, 2251–2256.
18. Li, H.; Luo, S.; Nie, C. Reconstruction and Control of Modern in Tensive Dike-Pond System in Shunde. *Chin. J. Ecol.* **2005**, *24*, 108–112.
19. Zhu, G. Ponderation Over the Flood Control Functions of the Artifical Landforms in Prd. *Trop. Geogr.* **2012**, *32*, 378–384.
20. Zhao, Y. The Effect of Scoeconomical Development on the Dike-Pond System in Zhujiang Delta. *J. Zhongkai Agrotech. Coll.* **2001**, *14*, 28–33.
21. Wu, J. Land-Pool Farming and Economic Restructuring in Shunde County during the Times of Ming, Qing Dynasties and the Republic of China. *Ancient Mod. Agric.* **2011**, *1*, 96–104.
22. Li, M.; Nie, C.; Long, X. Establishment of Indicator System for Assessing Eco-Environment Quality on Dike-Pond System. *J. Agro-Environ. Sci.* **2007**, *26*, 386–390.
23. Lu, H.; Peng, S.; Lan, S.; Chen, F. Energy Value Evaluation of Dike-Pond Agro-Ecological Engineering Modes. *Chin. J. Appl. Ecol.* **2003**, *14*, 1622–1626.
24. Han, X.; Yu, K.; Li, D.; Wang, S. Building the Landscape Security Pattern of Dike-Pond System with Urban Functions. *Area. Res. Dev.* **2008**, *27*, 107–110.
25. Lin, M.; Feng, R.; Ji, S. Analysis on Mode Change and Landscape Pattern of the Dike-Pond Agriculture in Zhongshan. *Guangdong Agric. Sci.* **2014**, *41*, 184–189.

26. Liu, K.; Wang, S.; Xie, L.; Zhuaang, J. Spatial Evolution Analysis of Dike-Pond Systems in Foshan City. *Trop. Geogr.* **2008**, *28*, 513–517.

27. Yang, D.; Ye, C. Analyses and Simulation on Landscape Fragmentation of Dike—Pond Based on Ca Model in Pearl River Delta. *Hubei Agric. Sci.* **2016**, *55*, 3932–3937.

28. Kennedy, R.E.; Townsend, P.A.; Gross, J.E.; Cohen, W.B.; Bolstad, P.; Wang, Y.Q.; Adams, P. Remote Sensing Change Detection Tools for Natural Resource Managers: Understanding Concepts and Tradeoffs in the Design of Landscape Monitoring Projects. *Remote Sens. Environ.* **2009**, *113*, 1382–1396. [CrossRef]

29. Schneider, A. Monitoring land Cover Change in Urban and Peri-Urban Areas Using Dense Time Stacks of Landsat Satellite Data and a Data Mining Approach. *Remote Sens. Environ.* **2012**, *124*, 689–704. [CrossRef]

30. Zurqani, H.A.; Post, C.J.; Mikhailova, E.A.; Schlautman, M.A.; Sharp, J.L. Geospatial Analysis of Land Use Change in the Savannah River Basin Using Google Earth Engine. *Int J. Appl Earth Obs.* **2018**, *69*, 175–185. [CrossRef]

31. Hauser, L.T.; Nguyen Vu, G.; Nguyen, B.A.; Dade, E.; Nguyen, H.M.; Nguyen, T.T.Q.; Le, T.Q.; Vu, L.H.; Tong, A.T.H.; Pham, H.V. Uncovering the Spatio-Temporal Dynamics of Land Cover Change and Fragmentation of Mangroves in the Ca Mau Peninsula, Vietnam Using Multi-Temporal Spot Satellite Imagery (2004–2013). *Appl. Geogr.* **2017**, *86*, 197–207. [CrossRef]

32. Ye, C. Change Characteristics and Spatial Types of Dike-Pond in the Pearl River Delta. *J. East China Inst. Technol.* **2013**, *36*, 315–322.

33. Wang, X.; Xia, L.; Deng, S.; Pan, Z. Spatial Temporal Changes in Dike-Pond Land in Nanhai District Based on RS and GIS. *Resour. Ind.* **2011**, *13*, 55–60.

34. Pang, Z. The Revelation of Contemporary American Photographic Reconnaissance Satellite. *China Ordnance Ind.* **2001**, 57–60.

35. Yang, X. An Overview of Foreign Reconnaissance Satellites. *For. Space Technol.* **1979**, 43–48.

36. Chen, B.; Chen, L.; Huang, B.; Michishita, R.; Xu, B. Dynamic Monitoring of the Poyang Lake Wetland by Integrating Landsat and Modis Observations. *ISPRS J. Photogramm.* **2018**, *139*, 75–87. [CrossRef]

37. Liu, Y.; Li, R.; Yang, L. Image Classification Method in Landuse Dynamic Detection Based on Multi-Source Remote Sensing Data—A Case Study in the Loess Plateau of Northern Shaanxi Province. *Bull. Soil Water Conserv.* **2006**, *26*, 63–66.

38. Ottinger, M.; Clauss, K.; Kuenzer, C. Large-Scale Assessment of Coastal Aquaculture Ponds with Sentinel-1 Time Series Data. *Remote Sens-Basel.* **2017**, *9*, 1–23. [CrossRef]

39. Zhai, T.; Jin, G.; Deng, X.; Li, Z.; Wang, R. Research of Wuhan City Land Use Classification Method Based on Multi-Source Remote Sensing Image Fusion. *Resour. Environ. Yangtze Basin* **2016**, *25*, 1594–1602.

40. He, Z.; Guan, L.; Kang, Q.; Wang, X.; Wang, C. Sustainable Development of Mulberry-Dike and Fish-Pond Farming in Pearl River Delta by Remote-Sensing Technology. *Acta Sci. Nat. Univ. Sunyatseni* **1998**, *37*, 68–73.

41. Wang, X.; Kang, Q. Application of tm Image in Investigating Changes of Dike-Pond Land on the Pearl River Delta—Samples as Shunde Nanhai Dike-Pond Region. *Remote Sens. Land Resour.* **1997**, *3*, 8–14.

42. Yee, A. New Developments in Integrated Dike-Pond Agriculture-Aquaculture in the Zhujiang Delta, China: Ecological Implications. *Ambio* **1999**, *28*, 529–533.

43. Turner, M.G. Landscape Ecology: The Effect of Pattern on Process. *Annu. Rev. Ecol. Syst.* **1989**, *20*, 171–197. [CrossRef]

44. Li, Y.; Zhou, L.; Cui, H. Pollen Indicators of Human Activity. *Chin. Sci. Bull.* **2008**, *53*, 1281–1293. [CrossRef]

45. Mercuri, A.M. Genesis and Evolution of the Cultural Landscape in Central Mediterranean: The 'Where, When and How' through the Palynological Approach. *Landsc. Ecol.* **2014**, *29*, 1799–1810. [CrossRef]

46. Xinwen, Z.; Chuanxiu, L.; Shuangxi, C.; Changsheng, H.; Zhuo, Z.; Haogang, D.; Min, Z.; Wen, C.; Fengmei, L.; Yiyong, L.; et al. Quantitative Reconstruction of Holocene Climate based on Pollen Data from Drill Hole qzk6 in Pearl River Delta. *Geol. Bull. China* **2014**, *33*, 1621–1628.

47. Yang, S.; Zheng, Z.; Huang, K.; Zong, Y.; Wang, J.; Xu, Q.; Rolett, B.V.; Li, J. Modern Pollen Assemblages from Cultivated Rice Fields and Rice Pollen Morphology: Application to a Study of Ancient Land Use and Agriculture in the Pearl River Delta, China. *Holocene* **2012**, *22*, 1393–1404. [CrossRef]

48. Almeida, C.M.D.; Monteiro, A.M.V.; Camara, G.; Soares-Filho, B.S.; Cerqueira, G.C.; Pennachin, C.L.; Batty, M. GIS and Remote Sensing as Tools for the Simulation of Urban Land-Use Change. *Int. J. Remote Sens.* **2005**, *26*, 759–774. [CrossRef]

49. Bhatta, B.; Saraswati, S.; Bandyopadhyay, D. Quantifying the Degree-of-Freedom, Degree-of-Sprawl, and Degree-of-Goodness of Urban Growth from Remote Sensing Data. *Appl. Geogr.* **2010**, *30*, 96–111. [CrossRef]

50. Kumar, J.A.V.; Pathan, S.K.; Bhanderi, R.J. Spatio-Temporal Analysis for Monitoring Urban Growth—A Case Study of Indore City. *J. Indian Soc. Remote Sens.* **2007**, *35*, 11–20. [CrossRef]

51. Liu, X.; Li, X.; Chen, Y.; Tan, Z.; Li, S.; Ai, B. A New Landscape Index for Quantifying Urban Expansion Using Multi-Temporal Remotely Sensed Data. *Landsc. Ecol.* **2010**, *25*, 671–682. [CrossRef]

52. Badjana, H.; Helmschrot, J.; Selsam, P.; Wala, K.; Flugel, W.; Afouda, A.; Akpagana, K. Land cover Changes Assessment Using Object-Based Image Analysis in the Binah River Watershed (Togo and Benin). *Earth Space Sci.* **2016**, *2*, 403–416. [CrossRef]

53. Zhang, L.; Nan, Z.; Xu, Y.; Li, S. Hydrological Impacts of Land Use Change and Climate Variability in the Headwater Region of the Heihe River Basin, Northwest China. *PLoS ONE* **2016**, *11*, 1–25. [CrossRef] [PubMed]

54. Liang, H. *Study on Lucc in Typical County of the Pearl River Delta*; Guangzhou University: Guangzhou, China, 2007.

55. Sinha, P.; Kumar, L.; Drielsma, M.; Barrett, T. Time-Series Effective Habitat Area (eha) Modeling Using Cost-Benefit Raster Based Technique. *Ecol. Inform.* **2014**, *19*, 16–25. [CrossRef]

56. Liu, X.; Li, X.; Chen, Y.; Qin, Y.; Li, S.; Chen, H. Landscape Expansion Index and Its Applications to Quantitative Analysis of Urban Expansion. *Acta Geopgr. Sin.* **2009**, *64*, 1430–1438.

57. Zeng, Y.; He, L.; Jin, W.; Wu, K.; Xu, Y.; Yu, F. Quantitative Analysis of the Urban Expansion Models in Changsha-Zhuzhou-Xiangtan Metroplan Areas. *Sci. Geogr. Sin.* **2012**, *32*, 544–549.

58. Zhou, X.; Chen, L.; Xiang, W. Quantitative Analysis of the Built-Up Area Expansion in Su-Xi-Chang Region, China. *Chin. J. Appl. Ecol.* **2014**, *25*, 1422–1430.

59. Wu, P.; Zhou, D.; Gong, H. A New Landscape Expansion Index Definition and Quantification. *Acta Ecol. Sin.* **2012**, *32*, 4270–4277.

60. Xingtan Government Online. Available online: http://www.shunde.gov.cn/xingtan/zjxt/?Sid=1 (accessed on 11 June 2018).

61. Manqiang, W. *An Analysis on the Status of Land-Use of Foshan Region in History*; Ji'nan University: Guangzhou, China, 2012.

62. Dai, W.; Si, S. Government Policies and Firms' Entrepreneurial Orientation: Strategic Choice and Institutional Perspectives. *J. Bus. Res.* **2018**, *93*, 23–36. [CrossRef]

63. Guo, C.X.; Xu, S.J. The New Visual Angle and Advance on Researches of the Dike-Pond System in China. *Wetland Sci.* **2011**, *9*, 75–81.

64. Li, Y.; Liu, K.; Liu, Y.; Zhu, Y. *The Dynamic of Dike-Pond System in the Pearl River Delta during 1964–2012*; Springer: Cham, Switzerland, 2017; pp. 47–59.

65. Ye, C.; Feng, Y. Ecological Risk Assessment for Pearl River Delta based on Land Use Change. *Trans. Chin. Soc. Agric. Eng.* **2013**, *29*, 224–232.

66. Wuying, S. *The Hubei Province and Flood Relief Study (1949–1956)*; Huazhong Normal University: Wuhan, China, 2013.

sustainability

MDPI

Article

Carbon Neutral by 2021: The Past and Present of Costa Rica's Unusual Political Tradition

Julia A. Flagg

Sociology Department and Environmental Studies Program, Connecticut College, New London, CT 06320, USA;
julia.flagg@conncoll.edu; Tel.: +1-860-439-2522

Received: 19 October 2017; Accepted: 15 January 2018; Published: 24 January 2018

Abstract: Costa Rica has pledged to become the first nation to become carbon neutral. This event raises the important question of how to understand this contemporary form of climate politics, given that Costa Rica has made an almost negligible contribution to the problem of global climate change. To understand this pledge, a case study spanning about 200 years situates the pledge within the country's unique historical profile. An analysis of interview data, archival research, and secondary data reveals that the pledge is the latest instance in Costa Rica's unusual political tradition. This political tradition dates back to the area's experience as a Spanish colony and as a newly independent nation. Several events, including the abolition of the army, the work on green development, and being awarded a Nobel Peace Prize were all foundational in forming Costa Rica's tradition as a place that leads by example and stands for peace and protection of nature. The carbon neutral pledge extends the political tradition that has been established through these earlier events. This case highlights the importance of understanding contemporary environmental politics through an analysis of long-term, historical data.

Keywords: Costa Rica; mitigation; carbon neutral; political tradition; environment; army; peace; case study

1. Introduction

In 2007, political elites in Costa Rica pledged to make the country the first in the world to become carbon neutral, or to emit net zero carbon emissions [1]. This decision seems somewhat unusual, given Costa Rica's nearly negligible contribution to the production of greenhouse gas emissions in recent years [2]. Insights from paleoenvironmental studies also indicate the country's history of abundant, and relatively stable, biodiversity and environmental conditions over recent millennia [3–5]. Through a historical case study, this paper situates Costa Rica's pledge within a longer historical narrative linking environment and society. To do so, this paper weaves together analyses of secondary data, interview data with key stakeholders in government, industry, science, and civil society, and historical documents. The analysis of these data reveals that Costa Rica's pledge represents another instance in the country's unusual "political tradition" [6] (p. 113). This paper describes the historical sources of this tradition, the more contemporary events that embody this tradition, and, through a narrative explanation, the ways in which the carbon neutral pledge represents yet another event in Costa Rica's political tradition of pursuing actions that highlight the country's democratic stability and environmental protection.

This paper is divided into several parts. The first section provides a historical background of the country, starting with its experience as a Spanish colony and tracing how elites came to provide generous public goods to its citizens. This historical experience set Costa Rica on a path toward making very different kinds of decisions about political and economic development from its neighbors. The second section explains the circumstances under which several events that have been foundational in the development of a contemporary political tradition in Costa Rica took place. These events are the abolition of the army, the work on "green" development, and receiving a Nobel Peace Prize. These events were influenced by Costa Rica's colonial experience and its experience as a newly independent nation. All of these events further established Costa Rica as a country that pursues seemingly iconic actions to promote peace and environmental protection without waiting for other countries to set a precedent. The third section explains the circumstances under which elites adopted the carbon neutral pledge and discusses the connections key stakeholders saw between the pledge and the country's earlier, seemingly iconic actions. The final section reviews the findings and discusses the broader relevance of the work. This case study spans nearly 200 years and reveals that the pledge can only be understood within the unique historical profile of the country. This research illustrates the centrality of environmental, political, and social history in an understanding of contemporary human-environment interactions.

2. Materials and Methods

The analytic approach of "process-tracing", as described by George and McKeown [7] is used to construct this narrative. This approach can help identify the intentional actors involved in a process, reconstruct how they arrived at particular junctures, identify and order the steps they took, and "develop a theory of action" [7] (p. 35). This research is an example of a "within-case analysis" [4] (p. 23), meaning that a single case of a carbon neutral pledging nation is studied over time. Though a single case study may have shortcomings in terms of generalizability, even studies of single cases have dynamism as scholars study these cases over time [8]. This current focus on Costa Rica stretches over 200 years, so concerns about the limited theoretical utility of the case should be minimized. The analysis traces events and processes over this 200-year period, and organizes each event as "a step in a causal chain" leading up to the outcome of interest—the carbon neutral pledge [9] (p. 177).

Interviews were conducted in August 2013 and July 2015 in San José, Costa Rica. Initial respondents were recruited from the Department of Climate Change in San José, Costa Rica, an extension of the Ministry of the Environment and Energy (MINAE). Each interviewee was asked to suggest other key stakeholders for interviews; thus, a snowball sample was assembled. Interviews were conducted in Spanish, English, or a combination of both languages. For the purposes of this paper, author-supplied translations of documents and interview data have been included. Archival research took place before, during, and after fieldwork, and consists of analyses of government reports, newspaper articles, and politicians' speeches.

A total of 22 interviews were conducted with 20 respondents. Two key informants were interviewed during the initial fieldwork in 2013 and again in 2015 to provide an update on work on the carbon neutrality pledge and to acquire contact information for additional interviewees. Interviewees come from the fields of government, science, industry, and civil society, though several had worked in more than one of these fields over the course of their careers. Despite not falling neatly into one of these four categories, on the recommendation of interviewees, some additional stakeholders were interviewed. These figures include a representative from AECID, the Spanish Agency for International Development Cooperation, which helped finance the carbon neutrality initiative, several academics (some of whom were involved in the initial work on carbon neutrality, and others who have written about the pledge), and a representative from INTECO (the Costa Rican Institute of Technical Norms). Interviews were not recorded because the sample includes political figures, some of whom were still in office at the time of the interview. The presence of a recorder may have led potential interviewees to decline the interview or to alter their comments. Thus, notes were hand-written by the author during each interview and were then typed up as field notes.

3. Results

3.1. Historical Context

This first section provides a brief historical overview of Costa Rica's experience as a Spanish colony and a newly independent nation. Costa Rica's experience during this time period provided the foundation for the development of the country's political tradition. Costa Rica was isolated and the "poorest" of the Spanish colonies [10] (p. 5). The territory's lack of both a large, organized indigenous society and valuable natural resources led the Spanish to largely avoid the area now recognized as Costa Rica [11,12]. Costa Rica's isolation endured even after national independence was won in 1821. In the immediate post-independence years, Costa Rica, El Salvador, Honduras, Guatemala, and Nicaragua formed the Federation of Central America [13]. Costa Rica endured as an isolated area of this regional political union, which worked to the country's advantage. According to Mahoney [12] (p. 193), " ... while all of the other provinces quickly became engulfed in warfare and political chaos, Costa Rica escaped such devastation and made tentative economic strides forward". Thus, Costa Rica's isolation during the Federation years enabled the country to avoid involvement in internecine war that came to consume the isthmus.

Costa Rica's isolation, and the associated benefits, endured after the collapse of the Federation in 1838. Again Mahoney [12] (p. 193) writes that Costa Rica's " ... autonomy enabled it to avoid the conservative resurgence that bolstered colonial heritages and blocked liberal progress among all the other Central American countries". The political period in the immediate post-Federation years instead brought about "state-centralization measures", including the creation of laws to protect rights and the rapid expansion of coffee exports [12] (p. 193). The military was also fundamental in strengthening the central state during the 19th century [14]. The coffee exports of this time and the associated wealth they created were enjoyed by a large class of small landholders in Costa Rica, a situation that was highly unusual for the Latin American context at this historical moment. Following independence, land ownership in nearby countries came to be centralized in the hands of a few private individuals [15] (p. 63). By contrast, the Costa Rican government divided communal land into private property and distributed it to small landholders [12] (p. 193). (In the early 1800s, before coffee, privatization of land was uncommon because land was perceived to be plentiful [16,17]. All members of society, including wealthy ones, grew crops on either communal lands or "unclaimed forest" [17] (p. 116).) In this way, the existence of open land created the opportunity for widespread economic opportunity and stalled the development of hierarchy, as has been the case in other frontier contexts [18]. The government's redistribution of open land had a leveling effect on the class structure, with Costa Rica having a large class of small landowners and small-scale coffee producers through to the 1950s [12,15,17,19]. These reforms would take shape in other Central American countries, but not until several decades later [12].

Historically, the elites that have existed in Costa Rica have made their wealth from processing coffee, not from the producing or land-owning, part of the coffee industry [15]. Research shows that the existence of a land-owning elite can present a formidable challenge to democracy [15]. Thus, the absence of a landed elite in Costa Rica has enabled the development of good governance and the continuation of a stable democracy.

This class structure fostered elites' ability to provide generous public goods for two reasons. First, elites were providing goods to people who tended to be similar to themselves. Second, elites anticipated that the goods they provided would improve the lives of most members of society, not just a portion of society [12] (p. 194). Again, elites' investment in the public good was expected to have a leveling effect on society. In the mid to late-1800s, the government created a commercial bank, a land registry, a system to provide credit to farmers, systems of justice, invested in infrastructure [12] (pp. 193–194) and established a comprehensive system of free public education [20] (p. 88). Consequently, Costa Rica became an early leader in measures of human development [12] (pp. 193–194). Costa Rica outpaced its neighbors in the proportion of resources it invested in education at this time, and elites used the 1921 centennial celebration to show off its achievements in the educational system [21]. A pattern of public good provision continued into the first half of the 20th century, with the creation of a social security system [20] (p. 93) and amendments made to the constitution in support of workers' rights [14,22].

The research reported so far has illuminated several historical factors and events that shaped the development of Costa Rica's political tradition. The country's geopolitical isolation both before and after independence helped it avoid the warfare and subsequent conservative pivot that consumed its neighbors, and allowed Costa Rica to instead pursue liberal reforms. In this context, elites divided communal land into private property, thereby creating an economy with a large group of small landholders and limiting the development of a powerful landed elite. Elites invested in the public good since these investments could "lift up" many members of society, not just a select few. This historical background set the stage for several events that have been foundational in the development of an unusual political tradition in Costa Rica. An analysis of the conditions under which these events took place is the focus of the next section.

3.2. Event Structure

This section describes the circumstances surrounding three events in Costa Rica that have been foundational in the development of the contemporary political tradition in Costa Rica. These events include the abolition of the army, the work on "green development", including the initial forestry law and subsequent efforts to stall deforestation, and receiving a Nobel Peace Prize. These events were influenced by the conditions described above (geopolitical isolation, relatively egalitarian agricultural economy, and elites' investment in public goods). All of these events established Costa Rica as a place that stands for peace, democracy, and nature. The 2007 carbon neutral pledge builds on this narrative for which Costa Rica has come to be known.

3.2.1. First Event: Abolishing the Army

One significant event that is part of the trajectory leading up to the carbon neutral pledge is the government's decision to abolish the army after a brief civil war. The 1948 presidential election featured a run off between experienced, Conservative politician Rafael Ángel Calderón and "Conservative newspaper publisher" Otilio Ulate [20] (p. 93). In response to the election results, which showed that Ulate had won, the "Calderonista majority in the Assembly voted to annul the election on grounds of fraud", [20] (p. 93–94). This upset the Social Democratic party, and one of its key members, José Figueres, who believed Ulate was the rightful victor of the election and leader of the country [14,20]. This provided the fodder for a civil war. When the Assembly tried to annul the election Figueres marched on the capital of San José with a 600-man army that he had trained at his rural home [14]. The fighting ended soon after, but not before causing between 1000–2000 casualties, mostly on the government army's side [14]. President-elect Ulate and Figueres subsequently signed a pact, which left Figueres in charge of the country for 18 months and made Ulate president after that time. As part of this pact, a commission to draft a new constitution was appointed. Continuing the 19th century pattern of public good provision, the new constitution included several generous provisions, many of them about workers' rights [14,22].

The new constitution also abolished the army. According to Hoivik and Aas [14] (p. 342), the decision to abolish the army originated with several young men affiliated with Figueres's Social Democratic Party, who saw the abolition of the army as " . . . an integral part of a modernized Costa Rican state." The abolition of a national army in any context is unusual, as the military represents one critical pillar of state power [23]. The Costa Rican army had played a critical role in the development of central state power during the 19th century [14]. However, an analysis of the Costa Rican decision reveals several forces that made the decision to abolish the army relatively uncontroversial.

First, the army was small and weak: it had just been defeated by an army that was developed at one man's home. Second, Costa Rica's pattern of relying on military assistance from the United States on prior occasions undermined the sovereignty of its own army [14]. In other words, maintaining an army was not worth it. Third, the abolition of the army builds on the political tradition of investing in public goods. Freeing up funds from military costs enabled the government to more heavily invest in education and natural resource protection [10] (pp. 27, 155) [24], a pattern that holds up across developing nations [25]. Edelman and Oviedo [24] (p. 23) explain that the Figueres-led junta "abolished the army in order to forestall a conservative restoration and to free up resources for social programs". They add that although "security-related expenditures" grew through the 1980s, Costa Rica maintained "one of the lowest ratios of military to social welfare spending of any country in the world" [24] (p. 30). The abolition of its army was foundational in establishing Costa Rica's place in the world as a peacekeeping nation that leads by example. According to Hoivik and Aas [14], there are several components of demilitarization, only one of which is the ideology of demilitarization. However, they conclude that Costa Rica scores very highly on this particular dimension, since the "abolition of the army is often mentioned as a sign of the essential peacefulness of the country" [14] (p. 350).

3.2.2. Second Event: Green Development

Costa Rica is also well known for its work in the area of environmental protection. The 1969 forestry law and the 1996 establishment of a Payments for Environmental Services (PES) program are the critical events in Costa Rica's work in the area of "green development". The pace of deforestation in Costa Rica increased dramatically post-1950 and by the early 1980s, Costa Rica had the highest deforestation rate in the world at about 4% of land deforested per year [10,26–30]. The production of beef for international markets fueled the rapid rise in Costa Rica's rate of deforestation at this time [26]. Government made several isolated attempts to slow deforestation, but it was not until 1969, that the government passed a forestry law [10]. The authors of the legislation looked to forestry laws of larger, dissimilar countries, including Venezuela, Mexico, and the United States, in drafting a law for Costa Rica [10] (p. 65). The law included protections for wildlife, promoted environmental education, set standards for the destruction of forests, and created the national parks department in the Ministry of Agriculture and Livestock [10] (pp. 66, 72). It aimed to curtail deforestation by making reforestation tax-deductible [31] (p. 1); however, its greatest area of success was in establishing national parks and biological reserves [10] (p. 71). In subsequent years, Costa Rica came to be known as a pioneer in the "biocultural" approach to land management [32].

By itself, the law did not dramatically curtail deforestation rates; in fact, significant deforestation has occurred since the passage of the law [10]. Despite this, the forestry law is a critical event in the development of Costa Rica's unusual political tradition for several reasons. First, the law symbolizes the Costa Rican government's move from a relatively hands-off land management approach and toward an "interventionist regime" of land management that characterized the 1970s and 1980s [27], see also [33] (p. 72). This interventionist approach included government's decisions in subsequent years to pay squatters to leave protected, public lands [10] (p. 77), [34] (p. 96). Second, the law illustrates that political leaders in Costa Rica did not wait for other, similar countries to set a precedent in land management; instead, leaders set their own precedent of state-led environmental protection. Again, authors of the plan consulted plans from dissimilar countries before drafting a plan for the Costa Rican context [10]. Third, the law provided a starting point for subsequent work on forestry, including the work from the 1990s on the Payments for Environmental Services program [29,35].

In the decades following the 1969 law, government used various means to establish financial benefits for reforestation [31]. However, by the mid-1990s, the government could no longer provide subsidies through certificates to land owners because of stipulations in the country's structural adjustment programs [29]. In this context, the government began a Payments for Environmental Services (PES) program in 1996 as part of its fourth national forestry law [29,31]. FONAFIFO, the National Forestry Financing Fund, oversaw and managed the program [29,31,35,36]. Under the PES program, the government compensates private landholders for the services that their lands provide. Four services are recognized: mitigation of greenhouse gas emissions, hydrological services, biodiversity conservation, or scenic beauty [29,31,35–37]. Initial funds to pay landowners came from a tax on fossil fuels and additional support came from the World Bank and the Global Environmental Facility [35,37,38].

The PES program did not, on its own, solve the problem of deforestation. By the time the PES program was established, Costa Rica already had about 25 percent of its land area in protected spaces [10,29]. By the mid-1980s, deforestation rates were dropping because of a decline in beef prices on international markets [10,28]. The decline in beef prices made the clearing of forests for the production of beef a less financially attractive option. However, there are three reasons why the creation of the PES program is a critical event in the development of an unusual political tradition in Costa Rica. First, Costa Rica developed this program early in comparison to other developing nations [37], and developed the most "elaborate" PES program in Latin America [39]. As with the forestry law, Costa Rican political elites developed and used the PES program without waiting for other countries to set a precedent. Second, the Costa Rican government has invested in the program at a level unmatched by other countries. As Daniels et al. [29] (p. 2118) explain, within the first ten years of the PES's existence, the country invested 0.43% of its annual budget in the program. Costa Rica's immense financial commitment to the program illustrates elites' level of commitment to it. The authors add, "Nominally influencing land use management through positive incentives, as opposed to regulatory land use restrictions, on such a large fraction of privately-held land *may be unparalleled in the world*" (emphasis added) [29] (p. 2118). Third, the PES program symbolizes political efficacy in having "dealt" with the problem of deforestation. It illustrates that Costa Rica was "out front" in responding to land use change. Costa Rica is often heralded as a success story for PES [31] and the country's program has served as a model for other countries' programs [40,41].

The establishment of the PES program is similar to the abolition of the army. As the earlier section demonstrated, the army was abolished (in part, at least) because the army was small and weak and because Costa Rica received consistent military support from the US. The confluence of these factors enabled the country to abolish the army, a feat that would be difficult in almost any other context. Rather than being an isolated event, the PES program is part of a trajectory of financial incentives to remedy land use problems in the country [29,35]. By the time the PES program started, the worst of the deforestation problem was over. The PES program, like the abolition of the army, illustrates to an international audience Costa Rica's priorities of protecting nature and investing in the public good.

3.2.3. Third Event: Nobel Peace Prize

President Óscar Arias Sánchez's 1987 Nobel Peace Prize, like the abolition of the army and the creation of the PES program, further established Costa Rica as a country with a history of taking on exceptional political actions that appeal to an international audience. Arias, a descendant of the country's coffee elite, ran on a platform of peace in the 1986 presidential election and won the election by a narrow margin [15,42]. Subsequently, Arias became involved in the Central American peace process. In February 1987, Arias proposed what became known as the "Arias Peace Plan", which recognized all governments of Central America and called for the abatement of support to rebel groups, including the contras in Nicaragua [15] (pp. 42–43). Following the 1979 overthrow of the Somoza dictatorship by the Sandinistas, Nicaragua became a "revolutionary socialist state" [15] (p. 5). The Contra-Sandinista conflict raged both before and after the signing of the peace treaty in Nicaragua [43]. El Salvador experienced a civil war and several military dictatorships. On 7 August 1987 the presidents of Costa Rica, Nicaragua, Honduras, Guatemala, and El Salvador signed the agreement and aimed to secure peace throughout the region [44,45].

According to Wehr and Lederach [43], Arias and the presidents of Guatemala (Vinicio Cerezo) and Nicaragua (Daniel Ortega) helped make the peace process a reality. Arias oversaw the event and mediated the countries' involvement, Cerezo agreed to host the treaty-signing event and lobbied for Nicaragua's inclusion, and Ortega made "important concessions" in the process [43] (p. 90). Despite five presidents' involvement and the key leadership of three of them, Arias was the only one to be awarded the 1987 Nobel Peace Prize. Arias was no doubt "central" to the process [43] (p. 90), but the extent to which he was seen as the only deserving candidate of the prize is less clear.

Arias's Nobel Peace Prize, like the abolition of the army, original forestry law, and creation of the PES program, is another critical event in the development of Costa Rica's unusual political tradition. There are several reasons for this. First, Arias was the first Central American to be awarded a Nobel Peace Prize [46]. This further established Costa Rica's position as a beacon of democracy and stability in Central America, with authors subsequently calling Costa Rica the "Switzerland of Central America" [47]. Second, and relatedly, the Nobel Peace Prize helped usher in a dramatic rise in tourism. Beginning in the late 1980s, international tourism rates to Costa Rica increased dramatically. According to Honey, Vargas, and Durham [48] (p. 16), "Between 1989 and 1994, international tourism more than doubled, from 376,000 to 761,000." The majority of this growth was seen in a particular style of tourism, ecotourism, which attracted people who were interested in hiking and other outdoor experiences. By 1993, tourism surpassed traditional exports, such as coffee and bananas, to become the industry that contributed the most to the economy [49]. This dramatic rise in tourism was due to a confluence of events. The country's proximity to the United States, democratic tradition, leadership in human development indicators, and reasonable infrastructure made Costa Rica an attractive location [48] (p. 16). Once Arias was awarded the Nobel Prize, tourists noticed the country as a safe and stable place in which they could visit rainforests.

What is exceptional about this rise in international tourism is that this was a time when international tourism in Central America was unpopular. According to data from 1995, Costa Rica far surpassed its Central American neighbors and some larger countries in South America in annual international tourist visits (Costa Rica: 785,000; Belize: 131,000; Ecuador: 440,000; El Salvador: 235,000; Guatemala: no data; Honduras: 271,000; Nicaragua: 281,000; Panama: 345,000; Peru: 479,000) [50]. Amid the civil wars and crises of the late 1980s in Central America, Costa Rica stood out as a symbol of democratic stability and peace.

The abolition of the army, the nation's work on green development, and being awarded a Nobel Peace Prize are critical events in the development of Costa Rica's unusual political tradition. All of these events helped define the kind of place Costa Rica is.

3.3. Carbon Neutrality

This section describes the conditions under which Costa Rica's pledge was adopted and how interviewees saw the carbon neutral pledge as a further development in the country's unusual political tradition. Óscar Arias Sánchez began his second presidential term in May 2006 and by the end of the year, his government had started to work on the idea of creating a "Peace with Nature" political coalition. According to the leader of this coalition, the idea of establishing peace with nature originated with the late Alvaro Ugalde, former head of the country's national parks system. Arias and Ugalde had known each other since the late 1970s when they served on a strategic committee about the national parks. According to this respondent, in 2006, Ugalde met Arias on a plane and said to him, "you know, you brokered peace in Central America" during your first presidency, "why not broker peace with nature during your second?" (Author's interview field notes 16 July 2015).

In early December 2006, the president declared, in the public interest, the initiative "Peace with Nature" [51]. In addition to calling the broad, new government-initiated agenda "Peace with Nature", Arias and his advisors also established a 30-person committee called "Peace with Nature", to execute the carbon neutral pledge (Author's interview field notes 16 July 2015) [52]. Several officials associated with this coalition, along with others, began to write a national plan on climate change in 2007 and finished it in 2009. This "National Strategy on Climate Change" describes the country's dual focus on national and international agendas on climate change, as well as its plan for carbon neutrality [53]. The national agenda includes six primary areas of focus: (1) mitigation, (2) adaptation, (3) metrics, (4) development of capacities and technological transfers, (5) public sensitivity, education, and culture, and (6) finances. This indicates that while mitigation is a clear area of focus in the national strategy, it is not the only national climate priority. Within the area of mitigation, three sub-goals are articulated: (1) the reduction of greenhouse gas emissions by source, (2) carbon capture and storage, and (3) the development of a national carbon market. Within the first sub-goal (the reduction of greenhouse gas emissions by source), eight key sectors for emission reductions are identified: (1) energy, (2) transport, (3) agriculture and livestock, (4) industrial, (5) solid wastes, (6) tourism, (7) water, and (8) land use change [53] (p. 48).

There are some important omissions from this section. The primary driver of Costa Rica's economy is international tourism, yet, the greenhouse gas emissions generated from tourists' air travel are not mentioned in the sectors on transport or tourism [53] (p. 49–51). In addition, the section on energy generation prioritizes almost exclusively the development of renewable energy. Costa Rica's energy portfolio was already drawn almost exclusively from renewable sources. Thus, while achieving carbon neutrality is more feasible in such a context, it is interesting to note that the national plan prioritizes the development of even more renewable energy rather than taking up questions about the overall use of energy.

Although even prior to its work on carbon neutrality, Costa Rica was known for its dependence on renewable energy and its weak reliance on fossil fuels, recent news articles suggest great interest in further developing Costa Rica's highway transportation systems [54,55]. These are likely to be very fossil fuel intensive projects. Overall, an analysis of this national strategy reveals that the plan is both very broad, and perhaps by consequence, somewhat vague. By trying to capture so many different climate goals across so many different sectors, specific, and major, contributors to emissions may be excluded.

Although carbon neutrality is a national goal, government has encouraged individual institutions at sub-national scales to become carbon neutral (Author's interview field notes 6 July 2015) [53]. The original target for becoming a carbon neutral nation was set at 2021, the same year as the country's bicentennial celebration of independence [53]. Interviewees explained that the 2021 target was selected for celebratory and symbolic reasons, rather than scientific ones (Author's interview field notes 8 July 2015; 14 July 2015; 15 July 2015). However, other research has illustrated that while Costa Rica's pledge began largely as a symbolic commitment, over time the pledge has taken on substance that may lead to emission reductions [56]. Following norm INTE.12.016, businesses follow three steps to become carbon neutral. First, a business develops an inventory of its emissions. Second, a business takes steps to reduce emissions. Third, a business pays FONAFIFO to offset the emissions that it continues to emit, even after taking steps to reduce emissions (Author's interview field notes 15 July 2015; 22 July 2015).

Insights from Interview Data

Interviewees often mentioned how the historical profile of Costa Rica (as discussed in Section 3.1) catalyzed the work on carbon neutrality. A university official involved with carbon neutrality verification processes said that Costa Rica is more "open" to something like a carbon neutrality pledge because of the country's historical legacy of democracy and the population's high level of education. He added that democracy is a political style that "permits sensibility with respect to environmental issues," (Author's interview field notes 10 July 2015). A different respondent, a biologist, government consultant, and civil society member, expressed similar views about the importance of Costa Rica's past with respect to its carbon neutral pledge. When asked why he thought Costa Rica was the kind of place that would make a carbon neutrality pledge, he started by explaining the kinds of places that would not make a pledge. It would be impossible, he said, to make a carbon neutrality pledge in countries such as Chile or Nicaragua. It would be impossible in countries with armies or in countries where hunger is the main issue, such as countries in Africa. As he put it, it is never going to happen there. He went on to say you do not need to be a developed country, because "we're not", but you've got to have certain things, such as human rights and water, to make a carbon neutral pledge. For emphasis, he added, "it is not going to happen in Haiti". Finally, he said that things happen in Costa Rica that do not happen elsewhere, like the elimination of the army (Author's interview field notes 13 July 2015). His words make clear that he, like others, attributed these unusual events in Costa Rica's past to its historical profile that developed in the nineteenth and early twentieth centuries.

In addition, some interview and archival data reveal the relevance that Costa Rica's actions throughout the twentieth century, including the abolition of the army, the work on green development, and being awarded a Nobel Peace Prize, had to the carbon neutral pledge. For example, several days after making a national announcement about the carbon neutrality pledge in San José, former President Arias published an article in Costa Rica's major national newspaper, La Nación. In the article, he makes an explicit comparison between achieving carbon neutrality and both the abolition of the army and his own Nobel Peace Prize. He even goes as far as to use the very same word (abolish/abolir) to draw a connection between achieving carbon neutrality and eliminating the army. He claims, "Abolishing net carbon will be, for us, the equivalent of the abolishment of the army that Don Pepe [Figueres] did and the pacification of Central America, which we achieved during my first presidency" [57]. His comments here position the carbon neutral pledge as the next event in Costa Rica's unusual political tradition, building on the abolition of the army and the 1987 Nobel Peace Prize.

Two interviewees who were involved at the highest political levels with the development of the carbon neutral idea and writing the carbon neutral national strategy saw the pledge as a way to bolster the nation's efforts to encourage reforestation and slow deforestation. As discussed above, in the early 1990s the government put a tax on gasoline and the funds went to FONAFIFO. Then FONAFIFO used this money to pay landowners if their land provided any one of four possible services. Several interviewees said that FONAFIFO "worked" and that by the early 2000s, the country had recuperated forest. Thus, there was interest in the continuing existence of the program. However, by 2006–2007, interviewees said that the initial budget in FONAFIFO (that came from the tax on fossil fuels) was exhausted and the country had to pursue other sources of funding (Author's interview field notes 8 July 2015; 13 July 2015). Other sources confirm that FONAFIFO's funds have been insufficient to pay landowners over time [41,58]. Since part of the process of a business becoming carbon neutral involves paying FONAFIFO to compensate for remaining emissions (described above), the carbon neutral norm was a way to boost funding for FONAFIFO and thus bolster preexisting national forest protection efforts.

Several other interviewees described how events in Costa Rica's history of green development led to the work on carbon neutrality. Though not specifically tied to the 1969 forestry law or the development of the payments for environmental services program, one interviewee claimed that the carbon neutrality goal was necessary because other available mechanisms for mitigation, including the Clean Development Mechanism (CDM) from the Kyoto Protocol were not useful in Costa Rica. A government official involved at the highest levels with the national work on carbon neutrality explained that the CDM had "limitations" for making progress on mitigation in Costa Rica. This is because (according to him) the two factors that the CDM can address are the generation of clean electricity and deforestation. In both of these themes, Costa Rica "has advanced" beyond other developing countries because by the time the CDM was created to help establish clean energy projects and slow deforestation in developing countries, Costa Rica had already tackled these issues. He added, "business as usual" in Costa Rica already involved a lot of renewable energy and reforestation. He added that because we couldn't get much out of the CDM, we did carbon neutrality: "We needed our own plan," (Author's interview field notes 6 July 2015). A university professor with prior experience working in government articulated how Costa Rica's prior successes in green development made the unusual carbon neutrality pledge a necessity. The country's success with reforestation efforts and the development of renewable energy portfolio (about 98%), left Costa Rica in a position where it "needed something to push the agenda", since the functions of Kyoto did not work in the country (Author's interview field notes 8 July 2015).

4. Discussion

The carbon neutral pledge is best understood as the latest in a long line of seemingly exceptional actions that Costa Rican elites have undertaken. Thus, the carbon neutral pledge builds on and extends the unusual Costa Rican political tradition. First, the pledge builds on themes established with the elimination of the army. It is worth recalling that some of the rationales given for the elimination of the army is that the army was small and weak [14]. There was thus little opposition to the idea of abolishing this depleted institution. In this sense, there was little to lose by eliminating the army. There are parallels between the ease with which the army was eliminated and the ease with which the carbon neutral pledge was made. Because Costa Rica's pledge is a voluntary national commitment that was not overseen by a governing agency in the years immediately following the commitment, "going" carbon neutral was a very low cost political commitment. Costa Rica really cannot lose by going carbon neutral.

The carbon neutral pledge also builds on some of the central tenets established with the 1969 forestry law. Again, passage of this law signaled the government's move into the "interventionist period" of the 1970s and 1980s [27], when government became more directly involved in forestry protection and environmental oversight. The passage of this law stands out because it showed that Costa Rica did not wait for other countries to set a precedent in forestry laws. In addition, the forestry law provided the legal basis for subsequent work. The carbon neutral pledge also displayed an unwillingness to wait for other countries to set a precedent for carbon mitigation. As the world's first country to make a carbon neutral pledge, the country set a precedent for others to follow. Several other nations have since made identical pledges [1].

In addition, the carbon neutral pledge provided a starting point for later work on climate change. All countries were called upon to make Intended Nationally Determined Contributions (INDCs) ahead of the 2015 Conference of the Parties (COP) meeting in Paris. In its INDC, Costa Rica reaffirmed its commitment to become a carbon neutral country, albeit by a later date (2085) [59]. While the pushback of the target end date might be a sign of declining national ambition in the country around the theme of carbon neutrality, as of 2015, there was still enthusiasm for the government's continuing work on carbon neutrality (Author's interview field notes 6 July 2015). It is possible that the new end date reflects a more realistic, more scientifically based target end date for achieving carbon neutrality. As several interviewees described, the initial target of 2021 was chosen for its cultural and political importance, not because research showed that the country could achieve the goal by this date. The carbon neutral pledge served as a starting point for the future national climate pledge in a way that is similar to how the initial forestry law was part of a trajectory for subsequent work on forestry and environmental services.

The carbon neutral pledge also shares some common characteristics with the PES program. Again, Costa Rica developed its PES program early [37] and elaborately [39] for the Latin American context. Importantly, the PES program also signals a political efficacy in having "dealt" with the problem of deforestation. There are parallels between the PES program and the carbon neutral pledge, with Costa Rica again standing out as a regional leader in climate mitigation issues. As of September 2017, Climate Action Tracker ranked Costa Rica as the only Latin American country that has national policies in line with keeping the globe to a 2-degree Celsius increase [60]. In addition, in 2014, Costa Rica was the most highly ranked Central American country in its preparedness for climate adaptation [61]. The carbon neutral pledge also illustrates a common theme in Costa Rican politics—a tendency to set a precedent, rather than to wait for one to be established, and to take pride in having done so. Several interviewees took pride in knowing that Costa Rica was the first country to pledge to become carbon neutral (Author's interview field notes 8 July 2015; 14 July 2015), though some were happy knowing that Costa Rica was one of the first (Author's interview field notes 10 July 2015). This was even true at the level of individual businesses, with several pointing out that they were one of the first industries in the country to become carbon neutral (Author's interview field notes 1 August 2013; 21 July 2015), or the first company in a particular city to become carbon neutral (Author's interview field notes 21 July 2015).

The pledge, like the Nobel Peace Prize, also has the potential to reinvigorate the national tourism industry. With airlines, hotels, and rental car companies all pledging to become carbon neutral [62], the pledge has created opportunities for the continued growth of ecotourism that is purported to have a small carbon footprint.

The research reported on here has shown that the carbon neutral pledge is the latest in a long line of seemingly exceptional actions undertaken by the Costa Rican state. These actions, including the abolition of the army, the work on "green development", and being awarded a Nobel Peace Prize, formed an image for an international audience about the kind of place Costa Rica is. The carbon neutral pledge reinforces the specific political narrative that has come to characterize Costa Rican history, strengthens the nation's largest industry, and bolsters the country's status as a place that is "out front" on issues that speak to the values of peace and protection of nature.

Though this work has focused on the specific case of Costa Rica, the implications are much broader. While some research in the social sciences has explored how nations' positions on climate change are influenced by the ways in which they were incorporated into the global economy [63], less research has focused on one single nation and provided a detailed historical account of how responses to global climate change build on (or fail to build on) a nation's prior achievements or experiences. Future research on nations' environmental actions can seek to embed nations' recent decisions about climate change within long-term, political narratives. This kind of work is possible, as nearly all nations have made individual commitments to address climate change. It may be possible to better understand how a country approaches the future by first understanding how it has acted in the past. In addition, political narratives can be merged with existing paleoenvironmental research i.e., [3–5] to understand nations' human-environment interactions over long periods of time. In short, future research could benefit from integrating insights from across diverse fields to better understand how humans and landscapes influence each other over time, as some researchers have started to do [64,65].

This case study illustrates the importance of using long-term historical data to understand environmental issues. Writing specifically about the use of paleoecology in the field of conservation, Davies and Bunting [66] claim that scientists could benefit from using a broader range of types of evidence as we enter a period of extreme climate variability. This period of unknown climate variability is coupled with rapid policy and social changes. More specifically, ideas about which nations ought to respond to climate change has shifted dramatically from the Kyoto Protocol era, in which the responsibility for action fell to wealthy nations, to the Paris Agreement, in which all nations were called upon to act. At the same time, the world is undergoing rapid social changes due to globalization, demographic shifts, and migration. This confluence of unknown climate variability, extensive political changes, and rapid social shifts suggests that research on contemporary climate issues would benefit from the use of more, not fewer, types of data to understand social responses to climate change. One way to do this is to situate developments in climate decisions within longer historical narratives.

Acknowledgments: This research was supported by funding from Rutgers University Departments of Human Ecology and Sociology, the Rutgers University Graduate School, and Columbia University's Hertog Global Strategy Initiative. Start-up research funding from Connecticut College support the costs to publish in open access.

Conflicts of Interest: The author declares no conflict of interest.

References

1. Flagg, J.A. Aiming for Zero: What Makes Nations Adopt Carbon Neutral Pledges? *Environ. Sociol.* **2015**, *1*, 202–212. [CrossRef]
2. Brechin, S.R. Climate Change Mitigation and the Collective Action Problem: Exploring Country Differences in Greenhouse Gas Contributions. *Sociol. Forum.* **2016**, *31*, 846–861. [CrossRef]
3. Horn, S.P. Prehistoric fires in the highlands of Costa Rica: Sedimentary charcoal evidence. *Rev. Biol. Trop.* **1989**, *37*, 139–148.
4. Horn, S.P. Postglacial vegetation and fire history in the Chirripó Páramo of Costa Rica. *Quat. Res.* **1993**, *40*, 107–116. [CrossRef]

5. Hooghiemstra, H.; Cleef, A.M.; Noldus, G.W.; Kappelle, M. Upper Quaternary vegetation dynamics and palaeoclimatology of the La Chonta bog area, Cordillera de Talamanca, Costa Rica. *J. Quat. Sci.* **1992**, *7*, 205–225. [CrossRef]
6. Booth, J.A.; Seligson, M.A. Paths to Democracy and the Political Culture of Costa Rica, Mexico, and Nicaragua. In *Political Culture and Democracy in Developing Countries*; Diamond, L.J., Ed.; Lynne Rienner Publishers: Boulder, CO, USA, 1993; pp. 107–138, ISBN 1555875157.
7. George, A.; McKeown, T. Case Studies and Theories of Organizational Decision-Making. In *Advances in Information Processing in Organizations*, 2nd ed.; Coulam, R.F., Smith, R.A., Jai, P., Eds.; Jai Press: Greenwich, CT, USA, 1985; pp. 21–58, ISBN 0892324252.
8. Gerring, J. What Is a Case Study and What Is It Good for? *Am. Polit. Sci. Rev.* **2004**, *98*, 341–354. [CrossRef]
9. George, A.L.; Bennett, A. *Case Studies and Theory Development in the Social Sciences*; MIT Press: Cambridge, MA, USA, 2004.
10. Evans, S. *The Green Republic: A Conservation History of Costa Rica*; University of Texas Press: Austin, TX, USA, 1999.
11. Fitch, J.S. *The Armed Forces and Democracy in Latin America*; The Johns Hopkins University Press: Baltimore, MD, USA, 1998; ISBN 0-8018-5917-4.
12. Mahoney, J. *Colonialism and Postcolonial Development: Spanish America in Comparative Perspective*; Cambridge University Press: New York, NY, USA, 2010.
13. A guide to the United States' History of Recognition, Diplomatic, and Consular Relations, by Country, Since 1776: Central American Federation. Available online: https://history.state.gov/countries/central-american-federation (accessed on 22 October 2015).
14. Hoivik, T.; Aas, S. Demilitarization in Costa Rica: A Farewell to Arms? *J. Peace Res.* **1981**, *18*, 333–351. [CrossRef]
15. Paige, J.M. *Coffee and Power: Revolution and the Rise of Democracy in Central America*; Harvard University Press: Cambridge, MA, USA, 1997.
16. Gudmundson, L. Costa Rica before Coffee: Occupational Distribution, Wealth Inequality, and Elite Society in the Village Economy of the 1840s. *J. Lat. Am. Stud.* **1983**, *15*, 427–452. [CrossRef]
17. Gudmundson, L. Peasant, Farmer, Proletarian: Class Formation in a Smallholder Coffee Economy, 1850–1950. In *Coffee, Society, and Power in Latin America*; Roseberry, W., Gudmundson, L., Kutschbach, M.S., Eds.; The Johns Hopkins University Press: Baltimore, MD, USA, 1995; pp. 112–150.
18. The Significance of the Frontier in American History. Available online: https://www.learner.org/workshops/primarysources/corporations/docs/turner2.html (accessed on 16 June 2016).
19. Lehoucq, F.E. The Institutional Foundations of Democratic Cooperation in Costa Rica. *J. Lat. Am. Stud.* **1996**, *28*, 329–355. [CrossRef]
20. Peeler, J.A. Elite settlements and democratic consolidation: Colombia, Costa Rica, and Venezuela. In *Elites and Democratic Consolidation in Latin America and Southern Europe*; Higley, J., Gunther, R., Eds.; Cambridge University Press: New York, NY, USA, 1991; pp. 81–112.
21. Fumero-Vargas, A.P. National Identities in Central America in a Comparative Perspective: The Modern Public Sphere and the Celebration of Centennial of Central American Independence September 15, 1921. Ph.D. Thesis, University of Kansas, Lawrence, KS, USA, 2005.
22. Wilson, B.M. When social democrats choose neoliberal economic policies: The case of Costa Rica. *Comp. Polit.* **1994**, *26*, 149–168. [CrossRef]
23. Collins, R. Mann's Transformations of the Classic Traditions. In *An Anatomy of Power: The Social Theory of Michael Mann*; Hall, J.A., Schroeder, R., Eds.; Cambridge University Press: New York, NY, USA, 2006; pp. 19–32.
24. Edelman, M.; Oviedo, R.M. Costa Rica: The non-market roots of market success. *Rep. Am.* **1993**, *26*, 22–44.
25. Adeola, F.O. Military expenditures, health, and education: Bedfellows or Antagonists in third world development? *Arm. Forces Soc.* **1996**, *22*, 441–467. [CrossRef]
26. Augelli, J.P. Modernization of Costa Rica's Beef Cattle Economy: 1950–1985. *J. Cult. Geogr.* **1989**, *9*, 77–90. [CrossRef]
27. Brockett, C.D.; Gottfried, R.R. State policies and the preservation of forest cover: Lessons from contrasting public-policy regimes in Costa Rica. *Lat. Am. Res. Rev.* **2002**, *37*, 7–40.

28. Daniels, A.E. Forest Expansion in Northwest Costa Rica: Conjuncture of the Global Market, Land-Use Intensification, and Forest Protection. In *Reforesting Landscapes: Linking Pattern and Process*; Nagendra, H., Southworth, J., Eds.; Springer: Dordrecht, The Netherlands, 2010; pp. 227–252.

29. Daniels, A.E.; Bagstad, K.; Esposito, V.; Moulaert, A.; Rodriguez, C.M. Understanding the impacts of Costa Rica's PES: Are we asking the right questions? *Ecol. Econ.* **2010**, *69*, 2116–2126. [CrossRef]

30. Nygren, A. Development Discourses and Peasant-Forest Relations: Natural Resource Utilization as Social Process. *Dev. Chang.* **2000**, *31*, 11–34. [CrossRef]

31. Payments for Ecosystem Services in Costa Rica and Forest Law No. 7575: Key Lessons for Legislators. Available online: http://www.agora-parl.org/sites/default/files/090422_e-parliament_forests_initiative.pdf (accessed on 26 October 2015).

32. Janzen, D.H. Guanacaste National Park: Tropical Ecological and Biocultural Restoration. In *Rehabilitating Damaged Ecosystems: Volume II*; Cairns, J., Jr., Ed.; CRC Press: Boca Raton, FL, USA, 1988; pp. 143–192.

33. Miranda, M.; Dieperink, C.; Glasbergen, P. The Social Meaning of Carbon Dioxide Emission Trading. *Environ. Dev. Sustain.* **2002**, *4*, 69–86. [CrossRef]

34. Horton, L.R. Buying up nature: Economic and social impacts of Costa Rica's ecotourism boom. *Lat. Am. Perspect.* **2009**, *36*, 93–107. [CrossRef]

35. Miranda, M.; Dieperink, C.; Glasbergen, P. Costa Rican Environmental Service Payments: The Use of a Financial Instrument in Participatory Forest Management. *Environ. Manag.* **2006**, *38*, 562–571. [CrossRef] [PubMed]

36. Fletcher, R.; Breitling, J. Market mechanism or subsidy in disguise? Governing payment for environmental services in Costa Rica. *Geoforum* **2012**, *43*, 402–411. [CrossRef]

37. Pagiola, S. Payments for Environmental Services in Costa Rica. *World Bank*, 2006. Available online: http://www.ncsu.edu/project/amazonia/for414/Readings/Pagiola_PSA2006.pdf (accessed on 9 October 2015).

38. Zúñiga, J.M.R. Paying for forest environmental services: The Costa Rican experience. *Unasylva* **2003**, *54*, 31–33.

39. Pagiola, S.; Arcenas, A.; Platais, G. Can Payments for Environmental Services Help Reduce Poverty? An Exploration of the Issues and the Evidence to Date from Latin America. *World Dev.* **2005**, *33*, 237–253. [CrossRef]

40. Fletcher, R. Making 'Peace with Nature': Costa Rica's Campaign for Climate Neutrality. In *Climate Change Governance in the Developing World*; Roger, C., Heldpp, D., Nag, E., Eds.; Polity Press: London, UK, 2013; pp. 155–173.

41. Pagiola, S. Payments for Environmental Services in Costa Rica. *Ecol. Econ.* **2008**, *65*, 712–724. [CrossRef]

42. Dunkerley, J. *Power in the Isthmus: A Political History of Modern Central America*; Verso: New York, NY, USA, 1988.

43. Wehr, P.; Lederach, J.P. Mediating Conflict in Central America. *J. Peace Res.* **1991**, *28*, 85–98. [CrossRef]

44. Oscar Arias Sanchez–Facts. Available online: https://www.nobelprize.org/nobel_prizes/peace/laureates/1987/arias-facts.html (accessed on 15 June 2016).

45. Americas, Central America: Efforts toward Peace. Available online: http://www.un.org/en/sc/repertoire/89-92/Chapter%208/AMERICA/item%2009_Central%20America_.pdf (accessed on 19 April 2016).

46. All Nobel Peace Prizes. Available online: https://www.nobelprize.org/nobel_prizes/peace/laureates/ (accessed on 15 April 2016).

47. Weinberg, B. *War on the Land: Ecology and Politics in Central America*; Zed Books Ltd.: Atlantic Highlands, NJ, USA, 1991.

48. Impact of Tourism Related Development on the Pacific Coast of Costa Rica: Summary Report. Available online: http://www.responsibletravel.org/resources/documents/coastal-tourism-documents/summary%20report/summary_report_-_impact_tourism_related_development_pacific_coast_costa_rica.pdf (accessed on 6 November 2015).

49. LePree, J.G. Certifying sustainability: the efficacy of Costa Rica's certificate for sustainable tourism. *Flor. Atl. Comp. Stud. J.* **2009**, *11*, 57–78.

50. The World Bank DataBank. Available online: http://databank.worldbank.org/data/home.aspx (accessed on 15 October 2015).

51. The President of the Republic Declares the Public Interest Initiative "Peace with Nature". Available online: http://www2.eie.ucr.ac.cr/~jromero/sitio-TCU-oficial/normativa/archivos/leyes_nac/Declara_de_Interes_Publico_la_Iniciativa_Paz_con_la_Naturaleza.pdf (accessed on 4 September 2015).

52. Ponchner, D.; Vargas, A. Government will Launch the Initiative "Peace with Nature". *La Nacion*, 2007. Available online: http://wvw.nacion.com/ln_ee/2007/julio/04/aldea1155221.html (accessed on 25 September 2015).

53. National Strategy on Climate Change. Ministry of Environment, Energy, and Telecommunications, 2009. Available online: http://cambioclimaticocr.com/2012-05-22-19-42-06/estrategia-nacional-de-cambio-climatico (accessed on 6 May 2013).

54. Arias, L. Government pledges to speed up construction of San Carlos highway. *The Tico Times*. Available online: http://www.ticotimes.net/2015/12/10/government-pledges-speed-construction-san-carlos-highway (accessed on 14 December 2017).

55. Arias, L. Construction to begin soon to curb heavy traffic on Circunvalacion beltway. *The Tico Times*. Available online: http://www.ticotimes.net/2016/09/08/costa-rica-heavy-traffic (accessed on 14 December 2017).

56. Flagg, J.A. From Symbol to Substance: Costa Rica's Carbon Neutral Pledge. Under review.

57. Sanchez, O.A. Peace with Nature. *La Nacion*, 2007. Available online: http://www.nacion.com/opinion/Paz-naturaleza_0_915508509.html (accessed on 6 November 2015).

58. Blackman, A.; Woodward, R.T. User financing in a national payments for environmental services program: Costa Rican hydropower. *Ecol. Econ.* **2010**, *69*, 1626–1638. [CrossRef]

59. Government of Costa Rica: Ministry of Environment and Energy, Costa Rica's Intended Nationally Determined Contribution. Available online: http://www4.unfccc.int/submissions/INDC/Published%20Documents/Costa%20Rica/1/INDC%20Costa%20Rica%20Version%202%200%20final%20ENG.pdf (accessed on 6 January 2016).

60. Climate Action Tracker: Costa Rica. Available online: http://climateactiontracker.org/countries/costarica.html (accessed on 25 September 2017).

61. Luxner, L. Costa Rica tops Central American ranking in latest climate change adaptation index. *The Tico Times*, 2014. Available online: http://www.ticotimes.net/2014/11/10/costa-rica-tops-central-american-ranking-in-latest-climate-change-adaptation-index (accessed on 26 September 2017).

62. Flagg, J.A. Aiming for Zero: What Makes a Nation Adopt a Carbon Neutral Pledge? Ph.D. Thesis, Rutgers University, New Brunswick, NJ, USA, 2016.

63. Roberts, J.T.; Parks, B.C. A Climate of Injustice: Global Inequality, North-South Politics, and Climate Policy. MIT Press: Cambridge, MA, USA, 2007.

64. Izdebski, A.; Holmgren, K.; Weiberg, E.; Stocker, S.R.; Buntgen, U.; Florenzano, A.; Gogou, A.; Leroy, S.A.G.; Luterbacher, J.; Martrat, B.; et al. Realising consilience: How better communication between archaeologists, historians and natural scientists can transform the study of past climate change in the Mediterranean. *Quat. Sci. Rev.* **2015**, *136*, 5–22. [CrossRef]

65. Marignani, M.; Chiarucci, A.; Sadori, L.; Mercuri, A.M. Natural and human impact in Mediterranean landscapes: An intriguing puzzle or only a question of time? *Plant Biosyst.* **2017**, *151*, 900–905. [CrossRef]

66. Davies, A.L.; Bunting, M.J. Applications of Paleoecology in Conservation. *Open Ecol. J.* **2010**, *3*, 54–67.

sustainability

MDPI

Article

The Tragedy of Forestland Sustainability in Postcolonial Africa: Land Development, Cocoa, and Politics in Côte d'Ivoire

Symphorien Ongolo [1,*], Sylvestre Kouamé Kouassi [2], Sadia Chérif [3] and Lukas Giessen [4]

[1]   Chair Group of Forest and Nature Conservation Policy, Georg-August University, 37077 Göttingen, Germany
[2]   Department of Geography, University Alassane Ouattara, BP V 18 01 Bouaké, Côte d'Ivoire;
      kouamsylvestre@yahoo.fr
[3]   Department of Sociology, University Alassane Ouattara, BP V 18 01 Bouaké, Côte d'Ivoire; scherif@gmx.fr
[4]   European Forest Institute, 53113 Bonn, Germany; Lukas.Giessen@efi.int
*   Correspondence: songolo@uni-goettingen.de; Tel.: +49-551-39-33410

Received: 8 October 2018; Accepted: 23 November 2018; Published: 5 December 2018

Abstract: Tropical countries are often blamed for not managing their natural resources sustainably. But what if overexploitation is inherent in political structures and policies—rooted in foreign colonial order—and is consistently detrimental in the contemporary use of forestlands? This article argues that post-colonial land development policies and related political interests seriously impede the sustainability of forest ecosystems in Côte d'Ivoire. Methodologically, the study builds on a historic contextualisation of forestland use policies in Sub-Saharan Africa, with Côte d'Ivoire serving as a case study. The results indicate that the increasing development of so-called rent crops clearly follows the historical dynamics of 'land grabbing' and a post-colonial agrarian model. This situation benefits agribusiness entrepreneurs and, more recently, sustainability standards. The study discusses the findings based on recent literature and empirical evidence. In conclusion, the post-colonial heritage and the manipulation of the related patterns by elites and policy-makers largely explains the present-day unsustainable forestland conversions in Côte d'Ivoire.

Keywords: land politics; forestland governance; African politics; development; Côte d'Ivoire; Deforestation

1. Introduction

In tropical zones, the dynamics of economic prosperity and development are closely connected to forestland use. They are often guided by the logic of expanding or intensifying agricultural models of a type applied to perennial shrub-land crops [1–4]. In short, the notion of forestland refers to an area of land covered by trees or forest ecosystem and that can be used for other specific purposes such as farming, natural resource extraction, hunting zone, ecotourism, or building.

This paper analyses why and how the stakes of ecological sustainability have often been marginalised in 'capitalisation' policies or forestland use strategies in tropical Africa. In most cases in tropical Africa, the plantation economy, [5] with its postcolonial dynamics, is usually supported by a cycle of massive conversion of forestlands into croplands given to more or less long occupancy (oil palms, cocoa, rubber, etc.). This hasty replacement of forest ecosystems by crops in the tropical areas is also part of a relatively recent process called 'land grabbing' [6]. As for the relation with the market economy, the dynamics of the plantation economy precede, or in some cases, follow the increased appeal of the agricultural sector for private investment, and business companies' transnationalisation of the agricultural production circuits [1,7].

Besides pressure from the business world, connected to the capitalistic development of the plantation economy [5,8,9], the large-scale transformation of tropical African forestlands is also bolstered by several political lines of logic for ecological adjustment. Coupling the consolidation of the plantation economy guided by the logic of capitalism on the one hand and the politisation of forestland use in tropical Africa on the other, forms the foundation of what can be called: "the political capitalisation of forestlands".

In most tropical regions, intact forestlands and biodiversity are declining. An estimated 80% of the new agricultural lands (including plantations) are taken from the forestlands [10]. This situation has major social and environmental consequences for the local people who depend on the forestlands for their daily survival [11], and at the global level in terms reduction of greenhouse gas emissions. In Sub-Saharan Africa (SSA), pressure from agriculture and changes in the vocation of the forestlands accelerate deforestation [12]. Côte d'Ivoire is a significant example of a country where land use policies based on large-scale agrobusinesses (in this case cocoa) almost wipe out the majority of the national forestlands.

From a paleo-ecological perspective to contemporary studies, the African continent has been a very informative case study to understand the complex interactions between plants and societies. To scrutinize how cultural developments affect environmental transformations in the long-term such in-depth understanding requires interdisciplinary approach including historical, social perspectives as well as agrarian and ecosystem diversity studies [13]. To better understand the diachronic evolution of land development practices and strategies in SSA since the end of the 19th century, this article has adopted an analytical approach based on historical sociology and agrarian studies. This approach requires scrutinisation, identification, and rigorous selection of lines of logic that have been documented in literature related to the topic addressed. The aim is to better understand contemporary realities and the dynamics of the recent past and the present in SSA land use policies. We are interested in the question of the history of economic thought on agricultural rent, the colonial heritage of agricultural dynamics in SSA as of the end of the 19th century [14]; The impact of the "State privatisation" processes [15–18] in the 1990s; The extraversion capacities [19] of the postcolonial state and the capacity of their state bureaucracies to pretend [20] when confronted with external demands for 'good governance' of forestlands.

2. Methodological Approach: Analysing the Present Using the Logics of History

This article is rooted in the complexity, pluralism and apparent political disorder that characterize the postcolonial societies. More specifically, post-colonialism here refers to "a given historical trajectory that of societies recently emerging from the experience of colonization and the violence which the colonial relationship, par excellence, involves" [21].

Our focus on forest sustainability in the postcolonial African context emerged from the recurrent debates and disputes between domestic actors who viewed forests as resources and potential farmlands on the one hand, and the increasing global actors' view of forests as service providers [4]. Second, many independent postcolonial countries in tropical Africa have often claimed their right to use forestland as a sovereign resource that should be exploited for national development policy purposes [5,22,23]. Since the 2000s, this discourse has been strongly supported in tropical African countries by proponents of the goal for these countries to reach the emerging economic nation status in 10, 20 or 30 years. In our two cases, Côte d'Ivoire is a good example of how state bureaucracies can promote deforestation as a policy for economic development.

This article uses a rationale based on the historic contextualisation of forestland use policies in Sub-Saharan Africa. We combine history and political sociology, and treat them as complementary disciplines. This combination, historical sociology, entails the analysis of the temporality of social facts and transformations of the present time, using carefully identified and contextualised variables that are contingent on their past. Despite the similarities between earlier situations in reference and the situations we are focusing on, we are not trying to position the present in a simple chronological series based on a type of path dependency [24]. The brief is the opposite. We need to find a historical logic that can better explain the complexity of what is observed at a given moment in time. The purpose of this approach, for us, is to construct a meticulous analysis that includes historicity, in other words, the importance of temporalities in the development of land exploitation policies in the countries of the former French colonial empire in Africa.

From an empirical viewpoint, our reflection is based on the expansion of export crops, e.g., cocoa in Côte d'Ivoire. This country was also selected mainly because the authors have spent many years researching the question of land politics, and agricultural and forest policies there. Our choice of case study and the field observation is also justified by the symbolic, pioneering role that Côte d'Ivoire have played in the way land development policies are imagined in postcolonial context. The degree of politicisation of the stakes and challenges of sustainability in the forestlands of the former French colonial empire in tropical Africa was also one of our motivation to focus on this country.

In our analysis, we use a twofold approach that combines a comprehensive, critical analysis of the qualitative data together with a descriptive statistical analysis of the quantitative data. Our field observations include consulting experiences and first-hand data in Côte d'Ivoire. Other data come from scientific empirical research including doctoral studies of two of the authors carried out in Côte d'Ivoire between 2008 and 2016. The most recent field observations and the interviews were carried out in the Nawa region, (last major cocoa production basin in Côte d'Ivoire) in May 2016 and have contributed greatly to this article.

In qualitative terms, we reviewed the genealogy of the logic behind land "optimisation" in the history of the political economics of agricultural dynamics on the one hand and the African development policies on the other. Besides this historiographic analysis that covers the end of the 19th century to the contemporary forestland transformation, we reviewed the scientific literature on the uses and exploitation of these spaces in tropical Africa in general and especially in Côte d'Ivoire.

In quantitative terms, our statistical analyses (selection of databases, graphic representations, and interpretation) are supported by long series of data on the evolution of the forest covers and the deforestation rate in Côte d'Ivoire [25,26]. The official monitoring data of the Food and Agricultural Organization [27] were used to study the historical path of cocoa production in Côte d'Ivoire from 1960 to 2014. Although the quantitative data we use is official, we recognise their potential limits [28,29].

In the case of the FAO, the academics have debated the question of the validity and reliability of statistics on monitoring the evolution of global forest resources [30,31]. Since there are no uncontested equivalent data, it is almost impossible, even in social science, to avoid using FAO quantitative data, especially to support empirical observations and evidence by comparing them to measurable long-term dynamics in the agricultural and forestry sectors [3,32]. With regard to 20th c. deforestation rates in West Africa, [33] pointed out the potential for exaggeration since global quantitative analyses have a limited capacity to assess local realities. In this article, where we use a method that combines qualitative and quantitative approaches, we constantly cross data and our angles of interpretation. To a certain extent, this enables us to substantially reduce the risk of introducing a bias to our analyses and the related results.

3. Theoretical and Conceptual Frameworks

Depending on the context and the subject of discussion, the notion of "optimisation" can refer to a process of creation or consolidation, or a process of relative or absolute increase in the value of tangible or intangible goods or entities. In this article, *land optimisation* refers to a colonial doctrine that equates the intrinsic value of the land to the value of its uses, the utilisation method and the nature of the investments granted by a legitimate or illegitimate land user.

3.1. The Colonial Roots of 'Land Enhancement' in Tropical Africa

The French colonial empire in tropical Africa, that created Côte d'Ivoire in its present political system, nurtured the "capitalisation" principle for its colonies based on a production and development process for the territories that had been conquered in the 1870s. This does imply the total absence of pre-colonial forestland structures, e.g., the example [34] of the local shea butter production systems in northern Ghana. This paper, however, focuses on the impact of colonial patterns on the postcolonial dynamics of forestland use. In his book on the development of the French colonies, Albert Saraut [35], the then Minister of the Colonies, made a distinction between two components of the French colonial government's land enhancement policy. The first component focused on developing 'human skills', in other words, the capacity to use and draw maximum benefits from the 'indigenous workers', be they works supervisors, soldiers, or simple labourers. The second component concerned the capacity to benefit from the natural wealth of the land, in other words the capacity to (over)exploit the soil and subsoil resources in the colonies to enrich, the 'motherland', mainland France.

We are mainly interested in the second component and what it bequeathed to the postcolonial agricultural policies in the former French colonies, using Côte d'Ivoire as an example. There are very few pioneer studies devoted to plantation economies in Sub-Saharan Africa [8,36] and more specifically in Côte d'Ivoire [14,22,37–40], in addition to contributing in a comprehensive understanding of this issue, this article seeks to analyse the sustainability stakes and challenges relating to the forestland enhancement processes in Côte d'Ivoire.

3.2. Land Value and Political Economy Theory

With reference to the "good" use of land, the most widespread and widely supported idea for the last two centuries has been that the best way to capitalise land assets is to assign them multiple uses that can generate a maximum of revenue as pointed out by Johann von Thünen [41]. From the conceptual vantage point, the question of land enhancement was an important issue until the beginning of the 20th century because of the central importance of agriculture and land transactions as part of economic activities of that era [42]. Economic thought first considered the subject of land value at the beginning of the 19th century. The works of David Ricardo on the theory of land rent, then Karl Marx on the links between the agrarian question and capital-labour were actually precursors to the problem of forestland enhancement under pressure from agriculture, the question that concerns us in this article. For Ricardo [43], the question of rent for the property owners can be summarised by these latters' quest for payment for the use of their land and for the scarcity factor of the fertile land they had to offer.

Since the quantity of fertile lands was limited, their intrinsic value, —as basic capital for agriculture, —rose, especially since the demand for agricultural products had to keep abreast of growing population figures. This situation led to constant emphasis on the need to start by making the fertile lands profitable, and then turn to the less fertile, more readily available lands [44,45]. Marx partly inspired by Ricardo, explains how the tension connected to controlling and using the lands is expressed through various transactions. The landowners, organised as an oligarchy of as an oligopoly, offer their lands to capitalistic agro-entrepreneurs in exchange for a rent [18,46,47]. Regardless of differences in context and era, this short summary of the earliest economic thought on the theory of rent is especially enlightening for our reflection on forestland enhancement policies in Sub-Saharan Africa. In the Ivorian example, the rent logic applied to land use for agriculture is explicit or underlying, depending on the case. Deforestation has been long accepted as a legitimate way to draw profit from 'forest rent' and to increase the productivity of the cocoa crops. Cocoa being considered as a 'cash crop', similar to other export crops like coffee and rubber, was produced mainly for foreign industrial and commercial markets.

3.3. On the Dominant Logics of African Land-Use Policies

The creation, control and clearing of forests have often been at the heart of land access strategies in tropical zones [48,49]. In tropical Africa, plantations of slow-growing trees and the production of perennial crops that take many years to develop and bear their fruits are examples of strategies that implicitly suggest land ownership. In some cases, e.g., the wood industry and the control of forest areas are closely tied to the stakes of land control that must last at least long enough to guarantee a return on investment. Similarly, clearing and deforestation of wooded lands are ways to materialise controls and privatise lands. The stakeholders use this method to show their exclusive user rights and, hence, the right to exclude any third party from the land concerned.

The logics of soil and sub-soil exploitation outweighing forest ecosystem sustainability is the basis of one of the popular and influential slogans about land development in Africa: "the land belongs to those who work it" [14]. This slogan has almost become synonymous with the name of Houphouët-Boigny the first president (1960–1993) of Côte d'Ivoire who was also considered to be the pioneer of postcolonial plantation economies in Africa. The principle of these economies was to consolidate the status of territories given to natural resources extraction from 'quarries' and to agricultural production for the benefit of industrial and commercial circuits belonging to the former colonial powers. Although the magnitude, form and degree of dependency of these plantation economies has changed, the status of the territories has not; enterprises belonging to the former colonial powers continue to wield their influence and produce (or obtain) raw materials. One of the changes is clearly shown through global land grabbing practices. Some scholars [50,51] provided mode details on this subject in their works on the domination of the capitalist logic in land transactions in developing countries.

3.4. On Sustainable Use of Forestlands in Tropical Africa

In most cases, the land control use or access strategies in tropical Africa are carried out at the expense of resource sustainability. Despite the differences between the dry tropical zones (forest resources very scarce) and the humid tropical zones (much of the forestland relatively undisturbed) disputes and conflicts are often connected to the management of the available resources. But the implementation of these reforms has often brought out the severe politisation of the stakes of sustainability in forestland governance in topical Africa.

To understand the stakes and evolutionary process of forestland use sustainability in tropical Africa requires both observation of the recent dynamics as seen in Côte d'Ivoire, and a detailed analysis of the historical narrative of earlier dynamics of the precolonial, colonial and postcolonial ages in this country. In this domain, the aim is to critically examine the evolution of public policies for the agricultural and land sectors and to understand their effects on forestland sustainability. The originality of this research lies in its conceptual contribution based on extensive literature review and empirical evidence about the links between sustainability and what we call 'the political capitalisation of forestlands' in tropical Africa from the colonial to the contemporary period. One of the paradoxes of postcolonial land development policies in Sub-Saharan Africa is that they emerged from and still co-exist through a set of conflicting forestland uses including logging, hunting, agro-industrial concessions and protected areas for biodiversity conservation. These complex dynamics have major environmental, social and socio-political effects on forest ecosystems, domestic political systems and the livelihoods of forest-dependent people.

Hardin [47] pointed out that concessionary regimes in tropical Africa were characterized by a mix of brutality, privilege, cronyism and patronage. In the case of the Congo basin region, for example, intense rivalry between German and French colonial forces led to overexploitation of forestlands, captive workers of logging companies, coercive expropriation of local lands for the creation of protected areas mainly dedicated to leisure-time activities of European settlers such as sport-hunting [47,52,53]. This model of forest overexploitation combined with privileged use of forests by dominant local and western elites for their enjoyment have strongly influenced the politics of forestland use in postcolonial tropical Africa since the 1960s [54,55].

The hypothesis put forward in this article is that the postcolonial land development policies are one of the main impediments to the sustainability of forest ecosystems in tropical Africa. Empirically, this paper draws on a case study of the relations between cocoa development policies and deforestation in Côte d'Ivoire. The main reason for selecting Côte d'Ivoire from the former French colonial empire was its long experience in forest clearing as a strategy to legitimate access to and control over the forestland for agricultural development.

4. Why Forest Sustainability Does Not Matter in Postcolonial Africa

Beginning in the early 1960s many postcolonial states in Africa including Côte d'Ivoire adopted the principles of land in the 'public domain', in other words, the so-called vacant land (*terres vacantes*) or lands that are not officially registered automatically belong to the State. This situation provided the basis for a system called 'land grabbing' in the former French colonies after they became independent states [56]. One of the effects of this land grabbing system has been the establishment of vast state-owned or semi-state-owned agro-industrial plantations in the forest areas, or else the peasants' hastily clearing the parts of their ancestral lands that are wooded to incontrovertibly establish their family property or to develop export crops production.

As a reaction to the anthropic and commercial pressure on natural resources globalisation of ecological stakes has become more intense and has given rise to an urgent demand for greater sustainability, especially with regard to the forest ecosystems in tropical Africa. In Côte d'Ivoire where most of the forestland biodiversity was destroyed by agriculture and especially cocoa plantations, the hope of saving the last intact forest refuges will probably depend on at least two important policy measures: the capacity and political interest of state bureaucracies to reinforce the rule of the law (or a positive use of coercion) for a sustainable use of forestlands and the generalisation of long-term and significant incentives to maintain biodiversity in the agricultural systems. Incentives for the environmental services include sustainable development and certification standards especially designed to bring about changes in agrarian practices, encouragement for tree plantations and/or the development of niche markets for the cocoa products labelled 'fair trade and sustainable'.

4.1. The Expansion of 'Rent Crops' as the Post-Colonial Agrarian Model

The notion of 'rent crops' is commonly used in the former French colonies to refer to agricultural export products whose production system was keyed to providing mainland France with products from the colonies. In the land capitalisation doctrine, one of the levers of rent crop expansion (cocoa, cotton, rubber, oil palm, sugar palm, sugarcane, etc.) was promises that farmers and agricultural entrepreneurs who developed their lands would be given secure access and right of use. This promise was also supported by the implicit principle of "the land belongs to those who cultivate it". This historical background is discussed in earlier sections of this paper.

Colonial archives reveal that Côte d'Ivoire was one of the first postcolonial African states to test the above-mentioned "the land belongs to those who cultivate it" doctrine. If we take this doctrine a step forward, the fact that the forestland is assimilated to 'vacant' land that is not exploited, i.e., converted to an assigned use, or has not been registered in the land registry office automatically puts it in the State's 'private domain'. Since the 1960s, the rise of the cocoa plantation economy has contributed to the accelerated decline of forest ecosystems and the collapse of the biodiversity. The development of cocoa cropping stems largely from the 'forest rent' principle, in other words, the influx of agricultural entrepreneurs looking for newly cleared forestlands whose soil fertility was still intact. This seriously contributed to reducing the Ivorian forests from about 12 million hectares in the 1960s to around 1 million in 2014.

4.2. Cocoa in Côte d'Ivoire: From Economic Success to Ecological Disaster

By choosing to build economic prosperity on the cocoa sector, at the expense of the forest ecosystems, Côte d'Ivoire played the hand of the agricultural entrepreneurs. This started in August 1960, the country's Independence Day [57,58]. The postcolonial government condoned deforestation as a tool for good land management. Emphasis was officially given to the accelerated conversion of forestlands by the then President, Houphouët-Boigny in a speech to the Ivorian National Assembly on January 1962.

To heed the President's political will the National Assembly (dominated by the PDCI, the State's party) voted a law on 20 March 1963. At the 6th Congress of the Côte d'Ivoire Democratic Party (PDCI), on 30 October 1970, Houphouët-Boigny confirmed his determination and introduced the slogan on land development:

> *"Côte d'Ivoire is three-fifth the size of France and has a population of 5 million people, including our foreign brothers. There is enough cultivatable land for all of us; what we lack is manpower. The Government and the Party have thus decided, in the interest of the country, to grant to all Ivorian nationals by birth or adoption, the right of enjoyment to a piece of land he has developed, regardless of size. This right is final and transmissible to the person's heirs"* [59].

There were several interpretations of this decisive speech which was summed up as *the land belongs to those who cultivate it!* Since the 1970s this slogan has become a powerful political mantra that affects the evolution and/or the status quo of land tenure regimes and forestland use policies in tropical Africa. Many exaggerated interpretations have had detrimental effects on speeds of forestland conversion and biodiversity loss in Côte d'Ivoire.

The pace of deforestation and forestland conversion into farmlands and cocoa plantations has soared. The Ivorian forestlands, that were estimated at about 15.8 million hectares in 1880 were down to 11.8 million in 1956, in other words, a loss of 4 million hectares in 76 years. In 1986 the forest cover in Côte d'Ivoire amounted to a mere 2.9 million hectares, which meant a loss of close to 8.9 million hectares of forestland in 30 years. This acceleration in the pace of deforestation during the second period, —annual figures are more than five times higher than during the first period, —coincides with the first boom in cocoa production in Côte d'Ivoire (Figure 1).

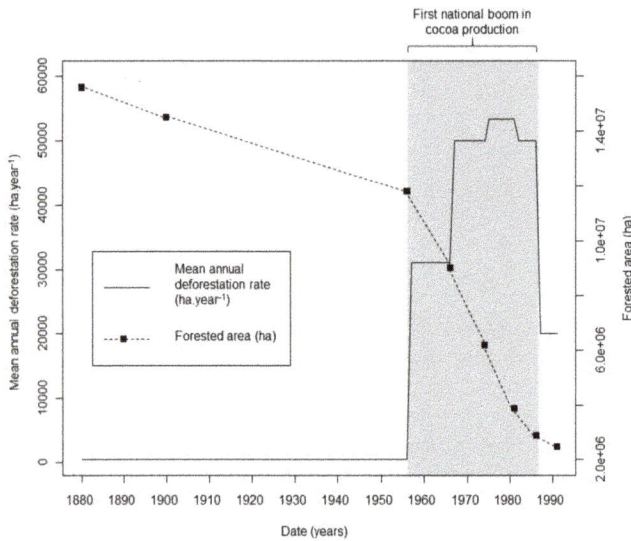

Figure 1. Evolution of deforestation and forestland cover in Côte d'Ivoire from 1880 to 1991. Source: authors (based on data from [25,27]).

These free land access policies were boosted by the encouragement given to the neighbouring countries, especially Burkina Faso, to send thousands of peasant workers to the cocoa plantations. These migratory flows were tragic for the Ivorian forests [58,60,61]. Deforestation dynamics in the Marahoué National Park (MNP) is a good example MNP, created in 1968, is a protected area of about 101,000 hectares located in an area approximatively 300 km northeast of Abidjan. A decade prior to the creation of this protected area, the *Gouro*, the major indigenous populations of the Marahoué region, were forcibly expropriated from their ancestral forestlands by colonial and early postcolonial administrations to impose MNP as a nature conservation policy instrument. A few years later, while the Gouro continued to claim their lands, the Ivorian postcolonial state bureaucracies and elites encouraged the installation of irregular camps of migrants in the MNP. By supporting and even facilitating the establishment of the Baoulé (internal migrants from the centre region of Côte d'Ivoire) and most importantly Mossi (from Burkina Faso), the postcolonial administration was implementing its doctrine of the land belongs to those who cultivate it without implementing its postcolonial land development policy.

In 1989 when the government decided to evaluate the situation, notably because of the increasing pressure on global environmental sustainability, the forest bureaucracy reported that 1397 migrants had developed cocoa plantations on the MNP land mainly Baoulé (49.6%), Gouro (29.6%), Mossi (17.8%) and about 3% of other local minorities. In 1999 this number was estimated at 2,635 migrant-farmers. Because of a 10-year politico-military crisis that struck Côte d'Ivoire in 2002, about 30,000 migrant-farmers and their families were established in the MNP in 2016. This population included 35% Baoulé, 32% Mossi, while the Gouro native populations decreased by 16% [62].

As a result of these active forestland conversion to agriculture and 'free land access' policies, implemented by massive flows of internal and external migrant-farmers, the country of Houphouët-Boigny has become the world leader in cocoa production. The national cocoa production rose from 85,000 tons/year in 1961 to 180,700 tons/year in 1969. By 1977 Côte d'Ivoire had become the world leader with an annual production of 303,621 tons. From then on, the cocoa production rate rose exponentially and exceeded 500,000 tons per year as of 1984. At the death of Félix Houphouët-Boigny in December 1993, the annual cocoa production in Côte d'Ivoire was 803,799 tons/year. In 2000, Côte d'Ivoire produced 1.4 million tons (Figure 2). The Ivorian miracle, in other words, the terrific economic prosperity obtained from the great productivity of one export crop became, more than ever before, the economic model for the other countries in tropical Africa.

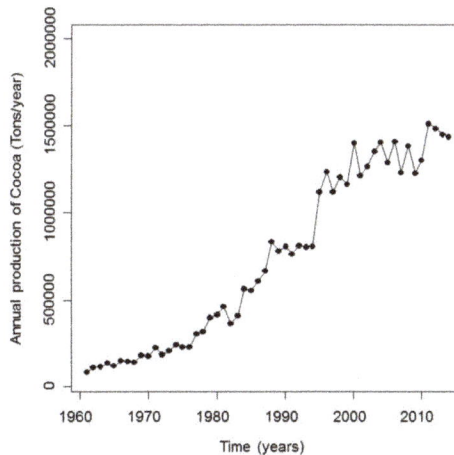

Figure 2. Evolution of cocoa production in Côte d'Ivoire from 1961 to 2010. Source: Authors (From [27] data).

Besides consolidating its position as leading producer of cocoa worldwide, Côte d'Ivoire raked in other economic trophies for several export crops that strongly violated the forestlands. The country of Houphouët became one of the main coffee producing countries, the top African bananas, pineapples, and oil palm producer and was in second place for rubber in the 1980s and1990s [63].

More than half a century after the beginning of the 'Ivorian miracle', in 2016, cocoa still occupied close to 2,000,000 hectares, managed by some 1,000,000 planters and provided about 40% of the world offer. Cocoa, thus, is a driver of economic prosperity in Côte d'Ivoire, providing more than half the export earnings and more than 10% of the national GDP [64,65].

Moreover, the attractiveness of the 'Ivorian model' and its influence over countries with large forest cover such as the countries of Central Africa portends an accelerated pace of forestland conversion in these countries. In most of them, the governments praise medium term 'emergence', which amounts to a quest for economic prosperity based on faster exploitation of natural resources and the conversion/development of 'available' forestlands.

Setting aside the short-term effects of price fluctuations on the world cocoa market, deforestation was encouraged by the gradual return of political stability in Côte d'Ivoire, the macro-economic situation that was affected by the structural adjustment programmes. This can be notably explained by the increase in the number of agricultural entrepreneurs and a 'return to the land' tendency stemming from the fact that the wages of the former civil servants had been halved and many had lost their jobs. As a result of this fast pace of forestland conversion, Côte d'Ivoire has been one of the countries with the highest deforestation rates in tropical Africa.

4.3. State Fragility and Ecological 'Adjustment'

President Houphouët-Boigny had reigned for 33 years. When he died, in December 1993, Côte d'Ivoire lapsed into a period of political and economic crisis and insecurity that lasted until 2012, when the first signs of peace and emergence reappeared. Between 1980 and 2000 the introduction of 'good governance' of people (human rights), institutions (democratisation), and goods (accountable natural resources management) seriously shook the African countries. This 'good governance' doctrine was based on a series of suppositions including the widely held postulate that breaking down State hegemony by making the bureaucracies weaker would lead to optimal co-management of public goods in the countries receiving official development aid.

In Côte d'Ivoire, external pressure from the Structural Adjustment Programmes (austerity policies) affected the pace of forestland conversion. The tendency to use soil and sub-soil resources for utilitarian purposes, encouraged during the colonial period and perpetuated by the postcolonial State bureaucracies, became especially common as of the 1970s. Forestlands were more readily and rapidly given to agriculture and the ecosystems were more readily, rapidly and formally exploited. The purpose of this accelerated conversion rate was either to increase export growth and earnings for governments facing reimbursement obligations under the structural adjustment scheme during the 1980–2000 period, or else as an alternative for civil servants who lost their jobs thus as part of a government streamlining plan. In other cases, this enthusiasm for forestland conversion or exploitation reflected the peasants' internal organisation strategies in a sector where the withdrawal of the State had been especially harmful. The farmers reacted by cultivating the newly cleared forestlands where the primary fertility compensated for the lack of farm inputs that were no longer subsidised for crops such as cocoa.

During the structural adjustment period, the international financial institutions including the World Bank and the International Monetary Fund were very anxious for the African governments, the 'clients' of their debt contraction or debt repayment programmes, to become creditworthy again. As Jarret et Mathieu [66] pointed out with regard to Côte d'Ivoire, by the end of the 1980s, measures to boost the export crops led to "surplus production of coffee and cocoa [in Côte d'Ivoire] which, together with surplus production in other African (. . .) and Asian countries that used the same model, led the market to collapse".

In the early 1990s, the World Bank reacted to the boost in cocoa and coffee production especially in Côte d'Ivoire that had prevented a return to growth by adopting a series of measures that were the opposite of the decisions that had led to the collapse of the market prices. The order was to do everything possible to lower the cocoa and coffee production levels, in particular through drastic decreases in the price support for the planters. In the 1990s, the World Bank forced Côte d'Ivoire to abandon its price stabilisation fund system by making this a prerequisite to the Ivorian debt relief negotiations.

From the ecological angle, besides the growing deforestation of the 1970s (see Figure 1), the upheavals in the cocoa and coffee sectors were equally harmful to the life of the forest ecosystems. Unable to lower the speed of deforestation in a country where the so-called 'cocoa loop' had already consumed large parts of the forestlands, the macro-economic pressure on the cocoa sector supported, or even reinforced the forest and biodiversity destruction process in the name of the 'land capitalisation' doctrine.

Actually, the prosperity of the cocoa crops and the related tragedy of forest ecosystem sustainability in Côte d'Ivoire can also be traced to a type of land development whose main goal is to benefit from the 'forest rent', in other word, rapid and excessive use of the primary fertility reserves from the recently cleared forestlands to boost growth in the young cocoa plantations. This approach to forestland development to serve the cocoa plantation economy contributed to the destruction of the Ivorian forests. Schematically, the estimated direct impact was: one ton of cocoa produced for one hectare of forestland destroyed. For more information on the connection between 'forest rent' and cocoa plantation economy in Côte d'Ivoire, consult the [57] study and Francois Ruf's works. After more than half a century, the relentless search for more 'forest rent' finally got the better of the Ivorian forests. This includes the forests with relatively restricted use ('classified forests') and the protected areas. Forest cover losses in the protected areas were also recorded during the post-structural adjustment period; between 2001 and 2004, the protected areas apparently lost about 38,000 ha/year in Côte d'Ivoire (Figure 3).

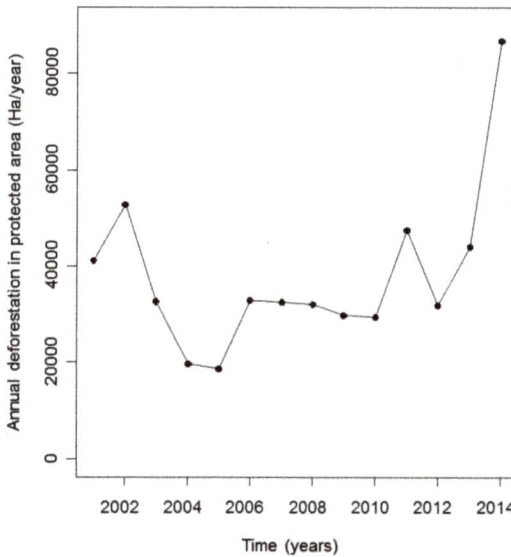

Figure 3. Recent trend of forest cover loss in protected areas in Côte d'Ivoire (From [26] data).

4.4. The Nawa Region, Rupture and Continuity

The paradox of the Nawa region is that it is home to both the last major intact block of primary rainforest in West Africa—the Taï national park (a protected area that contains the major last intact primary forest ecosystem in Western Africa. This park consists of about 454,000 hectares.)—, and the country's region that most recently developed major cocoa production. As such, the Nawa region is a very interesting case study to understand recent developments and future perspectives of forestland policies in Côte d'Ivoire. Despite the difficulty in obtaining reliable long-term quantitative data on the monitoring of forest cover and agricultural activities for the Nawa region, a recent study, co-authored by the lead author of this paper, provided an informative overview of the situation of land use in this region (Table 1).

Table 1. Mapping of land use cover/change in the Nawa region of Côte d'Ivoire in 2015 (From [67] data).

		Area (ha)	Ratio
Non-croplands	Dense forest	122,636	12.5%
	Degraded forest	47,121	4.8%
	Habitat	9107	0.9%
	Watercourse	35,640	3.6%
Sub-total of non-croplands		214,505	21.9%
Cropland and fallow	Cocoa	296,353	30.3%
	Coffee	105,417	10.8%
	Rubber	149,040	15.2%
	Oil palm	48,615	5.0%
	Food crops	75,972	7.8%
	Fallow land	88,668	9.1%
Sub-total of croplands		764,065	78.1%
Total (ha)		978,569	100.0%

Between 2003 and 2015 the total area of croplands was increased from 302,789 to 675,397 hectares, in other words, about 70% of the total forestlands of this region (apart from the Taï national park) were turned into farmland by 2015. Approximately 30% (296,353 hectares) was used for cocoa production making the Nawa region the largest major cocoa production region in Côte d'Ivoire. The remaining potential of degraded forests represented approximatively 5%, which means the prospects to continue expanding cocoa production by drawing on forestlands has largely dwindled away. As a result, the future of the last intact forest ecosystems, especially the Taï national park, in the Nawa region is threatened more than even by the cocoa pioneers (Figure 4).

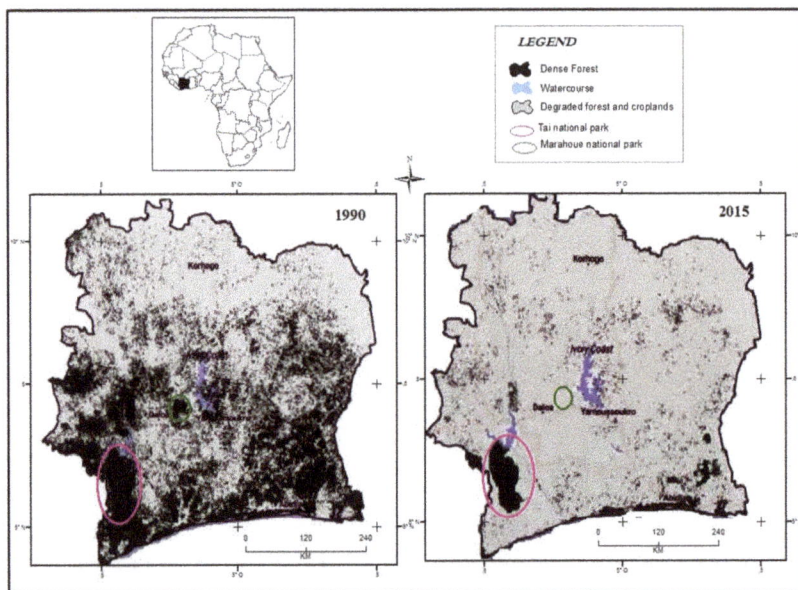

Figure 4. Visualisation of the deforestation dynamics in Côte d'Ivoire from 1990 to 2015 (source: adapted from [68]).

The dynamics of forestland conversion in the Nawa region and other similar contexts, a tragedy to forest sustainability, is likely to continue for at least three reasons. First, given the status of the main cocoa production region in Côte d'Ivoire, local and migrants farmers will try to continue benefiting from deforestation to maintain high cocoa production levels. As such, the tragic option to stay the course of this extensive production model will increase pressure to convert Taï national park. Our field observations indicated that farmers, elites and bureaucratic authorities considered the forestlands of this protected area as a good reserve of 'vacant land' potentially available for regional development projects. Second, the tragedy of forest sustainability in the Nawa region can be traced to state bureaucracies' claim to ownership of most of the lands but without a sufficient capacity and political interest to ensure regulated access or sustainable use of these lands. The Marahoué national park is an example of how such a situation can produce social dispute while destroying forest biodiversity. The exceptionally strict conservation of the Taï national park has been mainly due to massive external aid and the geographical isolation of this park 'thanks' to poor access roads. Third, population growth and migration flow in the Nawa region are a decisive variable that will contribute to reversing or reinforcing the business-as-usual ecological tragedy of cocoa production and the related social and political risks in the region. The deforestation rate in tropical regions rises most sharply when the population density exceeds $8.5/km^2$ [12]. In the Nawa region, the population in 2015 was about one million people, in other words, about $102/km^2$, with an estimated population growth rate of 3%/year.

4.5. On the Promise of Sustainability Through Environmental Labelling

Environmental certification standards first appeared in 1993 with the creation of the Forest Stewardships Council (FSC). In most cases, the principle of these environmental standards or labels is based on a voluntary decision to try to improve business practices whose production lines may affect the social systems equity and ecosystem durability [69]. The proliferation of certification standards often introduced by transnational conservation NGOs can be explained by the lack of public regulations to genuinely solve the problem of damage done by resource exploitation to both the society and the environment. The purpose of these standards is to produce customer guarantees that the products offered by the producers, who aspire to the labels, meet a set of predetermined ethical, social and ecological norms that are verified independently by a certification body.

In Côte d'Ivoire, certification standards for the cocoa industry include social and environmental measures such as respect for workers' rights and biodiversity preservation in the production systems [70,71]. Cocoa certification initially concerned micro-sectors such as responsible production and consumption connected to fair trade. Since 2000 however, this certification also applies to the 'classical' production sector in Côte d'Ivoire. Responding to pressure from conservation NGOs like Rainforest Alliance and the chocolate agro-food industries that increasingly look to buy raw materials from sustainable sources, a series of socio-ecological criteria were adopted. The main elements are ethics, improving the working conditions of the workers, and being more mindful about minors working on the Ivorian plantations. From the ecological angle besides the drastic reduction in the use of pesticides, producers applying for the sustainability labels have to improve their cropping techniques so as to reduce their impact on deforestation. On this last point, the planters are requested to make their production methods respect agro-forestry type systems, in other words, while producing cocoa, leave a reasonable number of forest trees standing so that in time the forest can reconstitute itself after the cocoa production cycles have ended.

But at this stage, it is difficult to say that the certification system in Côte d'Ivoire assures greater respect for the forest through major changes in agricultural practices for the cocoa sector. The system needs to pay more attention to some of the most critical issues affecting the sustainable use of forestlands such as land tenure. It is not clear, and conflict over the legal provisions still encourage strategies advocating conquest and land appropriation through deforestation, in some cases mounting tension between migrant-farmers and indigenous populations. In addition to the critical issue of land tenure system, the majority of the cocoa producers in Côte d'Ivoire do not belong to cooperative, and continue to operate without certification. Some of the labelling criteria for private standards give great importance to the involvement of the planters in the local bodies that organise sectors with cooperatives.

Despite these limits, there is a tendency to support the development of the sustainability mechanisms in the Ivorian cocoa sector. Besides the incentive measures (bonuses to planters based on kilogrammes of certified cocoa, access to niche fair-trade markets, etc.), the public authorities seem to be supporting the sustainable and certified cocoa production process in Côte d'Ivoire while other African governments are sceptical about these private governance mechanisms advocated by western conservation NGOs.

5. Discussion and Conclusions

This article argues that post-colonial land development policies and related political interests seriously impede the sustainability of forest ecosystems in Sub-Saharan Africa. The argument is conceptualised using notions from political economy, post-colonial studies, political anthropology and sustainable forestland use theories. Much of the reason for the lack of sufficient attention to forest ecosystems sustainability in SSA can be traced to the land use policies introduced during the colonial period and perpetuated by the postcolonial authorities. In countries of the former French colonial empire in Africa, the formulation and implementation of public actions related to agricultural policies were often influenced by external/foreign stakeholders. Krott's works [72,73] pointed out the constellation of actor groups with substantial power resources (coercion, incentive/disincentive, dominant information) who often benefit the most from the exploitation of forest resources. This may include private companies, specific categories of workers and citizens, associations and political parties as well as state bureaucracies. In the case of this research, our work revealed that the ecological tragedy of cocoa production in Côte d'Ivoire mainly benefits agribusiness actors, farmer-migrants, political elites. More recently, environmental entrepreneurs with their agri-environmental and sustainability standards have become the new players of the game. In some cases, the impact of these external interventions has been detrimental to forest sustainability. We find examples in the rush of the new rural residents (including foreign migrants) to overexploit the forestlands even inside protected areas, e.g., Marahoué national park.

But putting the demands of sustainability on a back burner and favouring a development model based on overexploitation of forestlands for large-scale consumerism is not specific to Sub-Saharan Africa, including Côte d'Ivoire of course. In some Southeast Asian regions with high potential tropical forests, agribusinesses have already led to the destruction of the biodiversity and the degradation of forestlands. And the race to first place on the world soya and beef production and sales markets is accelerating the destruction of the forest ecosystems in the countries of the Amazon basin.

Along the same line, literature abounds with examples of the impact of external—especially World Bank—interference in forestland use policies in developing countries. There are many examples of this type of postcolonial domination in domestic politics in western and other tropical countries. For example, the World Bank played a similar role in the natural resources exploitation policies of post-Soviet Armenia [74], and strongly influenced public development aid agencies in forest and biodiversity governance in Bangladesh [75]. While together, the WB and IMF exercised pressure on the formulation of forestland use policies in Indonesia [76].

In the case of Côte d'Ivoire, most of the recent deforestation is also closely connected to the post-electoral and security situation during the last decade. According to a United Nations Environment Programme report [77], the 2002–2010 'political-military' crises had a strong impact on forest ecosystem sustainability in Côte d'Ivoire. The number of classified forestlands ('forêts classées') that were transformed into agricultural lands doubled during the crisis period, according to the UNEP estimations from 575,300 ha in 2002 to close to 1,300,000 ha in 2012. Populations figures for these forests more than doubled during this same period (from 90,600 to 229,500 persons). Changing forestlands into farmlands had a considerable impact on wood removal as firewood that the resident populations needed more than ever for their survival during this period of great uncertainty and insecurity, connected to the war. This example of the post-conflict situation in Côte d'Ivoire reflects several other cases of more or less intense conflicts that in most cases added pressure to ecosystem sustainability and the dynamics of natural resources offtake in Africa.

However, the relative unsustainability of forest ecosystems management in Sub-Saharan Africa cannot be attributed solely to factors connected to its colonial heritage or to some form of path dependency although they have a considerable effect on the more-or-less contemporary use and misuse of forestlands in postcolonial Africa, as we showed in this article. We also need to consider several forms of logic based on private interests and cunning governments [78] that contributed to the major delay in the adoption of legislation to ensure a certain balance between the exploitation and preservation of forestlands in tropical Africa. This line of logic, which is based on private interests and the manipulation of rules related to forestland use policies, in some situations could be equated to 'gecko politics' [79] intentionally supported by the state bureaucracies. The gecko politics, in such a context, refers to tactics based on cunning, taking advantage of the fragility of domestic institutions or adopting a 'laisser-faire' position that banks on erratic policy coordination. This is similar to what happens in the land use policies in African countries that must deal with external pressure pushing for policy reforms in the forest governance domain. In sum, the challenges to forest ecosystems sustainability in Africa tend to become deeply marginalised when institutional 'disorder' in land use policies, undermined State authority, and other forms of deregulation, (whether endured, orchestrated or manipulated) are able to generate short-term individual benefits to the people in power.

In this situation, the arrival on the scene of other forms of 'hybrid governance arrangements' [80], such as those proposed by the certification standards could contribute to ensuring the inclusion of sustainability in land use policies in tropical Africa, on the condition that the standards' system of self-management provides autonomy and avoids the risk of a certain cronyism connected to their business model, i.e., the principle that allows the client being certified to pay the certifier. Since most tropical forestlands are owned by the State, it is essential to rally the State bureaucracies to the sustainability principle. In tropical countries with highly centralised political systems like those in tropical Africa, this is a precondition to achieving a resources management system that is socially more equitable and primarily concerned with preserving the countries' forestland potential and long-term value of their above- and below-ground resources.

Author Contributions: Conceptualization, S.O.; Data curation, S.K.K. and S.S.; Formal analysis, S.O.; S.S. and L.G.; Funding acquisition, S.O.; Investigation, S.K.K.; Methodology, S.O.; Writing—original draft, S.O. and S.K.K.; Writing—review and editing, S.O.

Funding: This research was funded by The Alexander von Humboldt Foundation, grant number: 3.4-CMR-1189288-GF-P), Bonn—Germany. The APC was funded by Georg-August-Universität Göttingen—Germany.

Acknowledgments: The authors gratefully acknowledge the participants to the international symposium on development studies entitled "Agriculture, rurality and development" held from 22 to 24 May 2017 at the Sociology Institute of the University of Brussels for their valuable comments on the first version of this article. We would especially like to thank the proof-readers of the early versions of this manuscript whose suggestions and corrections greatly improved the text. We would also like to thank Jean-Pierre Chauveau from the French research institute for development (IRD) for his valuable advice and very helpful suggestions. We also thank the anonymous referees for helpful criticism, comments and suggestions. Opinions and deficiencies in this paper remain our own.

Conflicts of Interest: The authors declare no conflict of interest.

References

1. Lambin, E.F.; Meyfroidt, P. Global land use change, economic globalization, and the looming land scarcity. *Proc. Natl. Acad. Sci. USA* **2011**, *108*, 3465–3472. [CrossRef] [PubMed]
2. Agrawal, A.; Chhatre, A.; Hardin, R. Changing Governance of the World's Forests. *Science* **2008**, *320*, 1460–1462. [CrossRef] [PubMed]
3. Rudel, T.K.; Schneider, L.; Uriarte, M.; Turner, B.L.; DeFries, R.; Lawrence, D.; Geoghegan, J.; Hecht, S.; Ickowitz, A.; Lambin, E.F.; et al. Agricultural intensification and changes in cultivated areas, 1970–2005. *Proc. Natl. Acad. Sci. USA* **2009**, *106*, 20675–20680. [CrossRef] [PubMed]
4. Lambin, E.F.; Turner, B.L.; Geist, H.J.; Agbola, S.B.; Angelsen, A.; Bruce, J.W.; Coomes, O.T.; Dirzo, R.; Fischer, G.; Folke, C.; et al. The causes of land-use and land-cover change: Moving beyond the myths. *Global Environ. Change* **2001**, *11*, 261–269. [CrossRef]

5. Amin, S. *Le Développement du Capitalisme en Côte d'Ivoire*; Editions de Minuit: Paris, France, 1973.

6. Borras, S.M., Jr.; Franco, J.C.; Wang, C. The challenge of global governance of land grabbing: Changing international agricultural context and competing political views and strategies. *Globalizations* **2013**, *10*, 161–179. [CrossRef]

7. Koh, L.P.; Wilcove, D.S. Is oil palm agriculture really destroying tropical biodiversity? *Conserv. Lett.* **2008**, *1*, 60–64. [CrossRef]

8. Amin, S. Underdevelopment and dependence in black Africa: Origins and contemporary forms. *J. Mod. Afr. Stud.* **1972**, *10*, 503–524. [CrossRef]

9. Bates, R.H. *Markets and States in Tropical Africa: The Political Basis of Agricultural Policies*; University of California Press: Berkeley, CA, USA, 2005.

10. Foley, J.A.; Ramankutty, N.; Brauman, K.A.; Cassidy, E.S.; Gerber, J.S.; Johnston, M.; Balzer, C. Solutions for a cultivated planet. *Nature* **2011**, *478*, 337–342. [CrossRef] [PubMed]

11. Assembe-Mvondo, S. Local communities' and indigenous peoples' rights to forests in Central Africa: From hope to challenges. *Africa Spectr.* **2013**, *48*, 25–47.

12. Mayaux, P.; Jean-Franois, P.; Baudouin, D.; Franois, D.; Andrea, L.; Frédéric, A.; Marco, C.; Catherine, B.; Andreas, B.; Robert, N.; et al. State and evolution of the African rainforests between 1990 and 2010. *Philos. Trans. R. Soc. Ser. B* **2013**. [CrossRef]

13. Mercuri, A.M.; D'Andrea, A.C.; Fornaciari, R.; Höhn, A. (Eds.) *Plants and People in the African Past: Progress in African Archaeobotany*; Springer: Berlin/Heidelberg, Germany, 2018.

14. Chauveau, J.-P. Mise en valeur coloniale et développement. In *Paysans, Experts et Chercheurs en Afrique Noire. Sciences Sociales et Développement Rural*; Boiral, P., Lanteri, J.-F., Olivier-de-Sardan, J.-P., Eds.; Karthala: Paris, France, 1985; pp. 143–166.

15. Chabal, P.; Daloz, J.-P. *Africa Works: Disorder as Political Instrument*; International African Institute in Association with James Currey: Oxford, UK, 1999.

16. Hibou, B. Retrait ou redéploiement de l'Etat? *Critique Internationnale* **1998**, *1*, 152–168. [CrossRef]

17. Mbembe, A. Du gouvernement privé indirect Politique Africaine. *Politique Afr.* **1999**, *73*, 103–121. [CrossRef]

18. Mkandawire, T. Neopatrimonialism and the political economy of economic performance in Africa: Critical reflections. *World Politics* **2015**, *67*, 563–612. [CrossRef]

19. Bayart, J.-F.; Ellis, S. Africa in the world: A history of extraversion. *Afr. Aff.* **2000**, *99*, 217–267. [CrossRef]

20. Mbembe, A. *On the Postcolony*; University of California Press: Berkeley, CA, USA, 2001.

21. Mbembe, A. Provisional notes on the postcolony. *Africa* **1992**, *62*, 3–37. [CrossRef]

22. Chauveau, J.-P.; Richard, J. "Une "périphérie recentrée": À propos d'un système local d'économie de plantation en Côte-d'Ivoire. *Cahiers D'études Africaines* **1977**, *17*, 485–523. [CrossRef]

23. Hecht, R.M. The Ivory Coast economic 'miracle': What benefits for peasant farmers? *J. Mod. Afr. Stud.* **1983**, *21*, 25–53. [CrossRef]

24. Kay, A. A critique of the use of path dependency in policy studies. *Public Adm.* **2005**, *83*, 553–571. [CrossRef]

25. Lauginie, F. *Conservation de la Nature et Aires Protégées en Côte d'Ivoire*; CEDA/NEI Hachette et Afrique Nature: Abidjan, Côte d'Ivoire, 2007; p. 668.

26. GFW. Global Forest Watch Climate. 2017. Available online: http://climate.globalforestwatch.org/ (accessed on 15 August 2017).

27. FAO. Food and Agriculture Data. 2017. Available online: http://www.fao.org/faostat/en/#home (accessed on 15 August 2017).

28. Hindess, B. *The Use of Official Statistics in Sociology: A Critique of Positivism and Ethnomethodology*; Macmillan: London, UK, 1973.

29. Ackroyd, S. Review Article: The Quality of Qualitative Methods: Qualitative or Quality Methodology for Organization Studies? *Organization* **1996**, *3*, 439–451. [CrossRef]

30. MacDicken, K.G. Global forest resources assessment 2015: What, why and how? *For. Ecol. Manag.* **2015**, *352*, 3–8. [CrossRef]

31. Keenan, R.J.; Reams, G.A.; Achard, F.; de Freitas, J.V.; Grainger, A.; Lindquist, E. Dynamics of global forest area: Results from the FAO Global Forest Resources Assessment 2015. *For. Ecol. Manag.* **2015**, *352*, 9–20. [CrossRef]

32. Rudel, T.K. The national determinants of deforestation in sub-Saharan Africa. *Philos. Trans. R. Soc. B* **2013**, *368*, 20120405. [CrossRef] [PubMed]

33. Fairhead, J.; Leach, M. *Reframing Deforestation: Global Analyses and Local Realities: Studies in West Africa*; Routledge: Abingdon-on-Thames, UK, 2003.
34. Wardell, A.; Fold, N. Globalisations in a nutshell: Historical perspectives on the changing governance of the shea commodity chain in northern Ghana. *Int. J. Commons* **2013**, *7*, 367–405. [CrossRef]
35. Saraut, A. *La Mise en Valeur des Colonies Francaises*; Payot: Paris, France, 1923.
36. Cooper, F. Africa and the World Economy. *Afr. Stud. Rev.* **1981**, *24*, 1–86. [CrossRef]
37. Ruf, F. Stratification Sociale en Économie de Plantation Ivoirienne. Ph.D. Thesis, Université de Paris X Nanterre, Paris, France, 1988; p. 700.
38. Balac, R. Gens de Terres, Gens de Réseaux: Mécanismes de Production et Lien Social: Pour Une Nouvelle Mise en Perspective de L'économie de Plantation en Côte d'Ivoire. Ph.D. Thesis, Institut d'Etudes Politiques de Paris, Paris, France, 1998.
39. Colin, J.-P.; Ruf, F. Une économie de plantation en devenir. L'essor des contrats de planter-partager comme innovation institutionelle dans les rapports entre autochtones et étrangers en Côte d'Ivoire. *Rev. Tiers Monde* **2011**, *207*, 169–187. [CrossRef]
40. Colin, J.-P. Securing Rural Land Transactions in Africa. An Ivorian Perspective. *Land Use Policy* **2013**, *31*, 430–440. [CrossRef]
41. von-Thünen, J. *Isolated State*; Pergamon Press: New York, NY, USA, 1966.
42. Angelsen, A. *Forest Cover Change in Space and Time: Combining the Von Thünen and Forest Transition Theories*; World Bank Publications: Washington, DC, USA, 2007.
43. Ricardo, D. *Principes of Political Economy and Taxation*; Everyman Library: London, UK, 1911.
44. Dorfman, R. Thomas Robert Malthus and David Ricardo. *J. Econ. Persp.* **1989**, *3*, 153–164. [CrossRef]
45. Bernstein, H. 'Changing Before Our Very Eyes': Agrarian Questions and the Politics of Land in Capitalism Today. *J. Agrar. Change* **2004**, *4*, 190–225. [CrossRef]
46. Fine, B. On Marx's theory of agricultural rent. *Econ. Soc.* **1979**, *8*, 241–278. [CrossRef]
47. Hardin, R.; Bahuchet, S. Concessionary politics: property, patronage, and political rivalry in central African forest management. *Curr. Anthropol.* **2011**, *52*, s113–s125. [CrossRef]
48. Bernstein, H.; Woodhouse, P. Telling Environmental Change Like It Is? Reflections on a Study in Sub-Saharan Africa. *J. Agrar. Change* **2001**, *1*, 283–324. [CrossRef]
49. Ribot, J.C.; Peluso, N.L. A theory of access. *Rural Sociol.* **2003**, *68*, 153–181. [CrossRef]
50. Robertson, B.; Pinstrup-Andersen, P. Global land acquisition: Neo-colonialism or development opportunity? *Food Policy* **2010**, *2*, 271–283. [CrossRef]
51. Borras, S.M.; Hall, R.; Scoones, I.; White, B.; Wolford, W. Towards a better understanding of global land grabbing: An editorial introduction. *J. Peasant Stud.* **2011**, *38*, 209–216. [CrossRef]
52. Coquery-Vidrovitch, C. *Le Congo au Temps des Grandes Compagnies Concessionnaires 1898–1930*; Mouton: Paris, France, 1972.
53. Auzel, P.; Hardin, R. Colonial history, concessionary politics, and collaborative management of Equatorial African rain forests. In *Hunting and Bushmeat Utilization in the African Rain Forest*; Conservation International: Washington, DC, USA, 2000; pp. 21–38.
54. West, P.; Igoe, J.; Brockington, D. Parks and peoples: The social impact of protected areas. *Annu. Rev. Anthropol.* **2006**, *35*, 251–277. [CrossRef]
55. Joiris, D.V.; Logo, P.B. *Gestion Participative des Forêts d'Afrique Centrale*; Editions Quae: Versailles CEDEX, France, 2010.
56. Dumont, R. *L'Afrique Noire est Mal Partie*; Editions du Seuil: Paris, France, 1962.
57. Léonard, E.; Ibo, J.G. Appropriation et gestion de la rente forestière en Côte-d'Ivoire. *Politique Afr.* **1994**, *53*, 25–36.
58. Woods, D. The tragedy of the cocoa pod: Rent-seeking, land and ethnic conflict in Ivory Coast. *J. Mod. Afr. Stud.* **2003**, *41*, 641–655. [CrossRef]
59. Ibo, J.G. Phénomène d'acquisition massive des terres et dynamiques socio-foncières en milieu rural ivoirien: Enjeux socio-économiques et culturels. In *Conférence-Débat sur les Acquisitions Massives des Terres Agricoles en Afrique et les Droits des Communautés Rurales, Abidjan*; INADES-Formation International: Abidhan, Côte d'Ivoire, 2012.
60. Colin, J.-P. *La Mutation D'une Économie de Plantation en Basse Côte d'Ivoire*; Ostom: Paris, France, 1990.
61. Ruf, F. Tree crops as deforestation and reforestation agents: The case of cocoa in Côte d'Ivoire and Sulawesi. In *Agricultural Technologies and Tropical Deforestation*; Angelsen, A., Kaimowitz, D., Eds.; CABI Publishing: Wallingford, SA, USA, 2001; pp. 291–315.

62. Kouassi, S.K. Analyse prospective des aspects conflictuels de la dynamique migratoire dans le parc national de la Marahoué en Côte d'Ivoire, in le Journal des Sciences Sociales. *Revue scientifique du Groupement Interdisciplinaire en Sciences Sociales (GIDIS)* **2014**, *11*, 139–155.

63. Assi, A.; Boni, D. *Développement Agricole et Protection de la Forêt: Quel Avenir Pour la Forêt Ivoirienne?* Compte-Rendu de la XIIe réunion plénière de l'AETFAT: Hamburg, Germany, 1990.

64. Tano, A.M. Crise Cacaoyère et Stratégies des Producteurs de la Sous-Préfecture de Méadji au Sud-Ouest Ivoirien. Ph.D. Thesis, Université Toulouse le Mirail, Toulouse, France, 2012; p. 261.

65. Kouassi, K.E. Introduction des Innovations en Milieu Paysan Ivoirien: Impacts Techniques et Socio-Économiques des Projets de Cacaoculture Durable Financés par le Conseil café-Cacao à Travers le FIRCA. Ph.D. Thesis, Université Alassane Ouattara, Bouaké, Côte d'Ivoire, 2015; p. 352.

66. Jarret, M.-F.; Mahieu, F.R. Ajustement structurel, croissance et répartition: L'exemple de la Côte-d'Ivoire. *Revue Tiers Monde* **1991**, *32*, 39–62. [CrossRef]

67. Biotope. *Recommandations pour l'intégration du mécanisme REDD+ dans le Schema Regional d'Aménagement du Territoire (SRADT) de la Nawa*; Ministère du Plan et du Développement: Abidjan, Côte d'Ivoire, 2016; p. 116.

68. Higonnet, E.; Bellantonio, M.; Hurowitz, G. *Chocolate's Dark Secret: How the Cocoa Industry Destroys National Parks*; Mighty: Washington, DC, USA, 2018; p. 24.

69. Bernstein, S.; Cashore, B. Complex global governance and domestic policies: Four pathways of influence. *Int. Aff.* **2012**, *88*, 585–604. [CrossRef]

70. Ingram, V.; Waarts, Y.; Ge, L.; van Vugt, S.; Wegner, L.; Puister-Jansen, L.; Tanoh, F.R.R. *Impact of UTZ Certification of Cocoa in Ivory Coast. Assessment Framework and Baseline*; LEI Wageningen UR (University & Research Centre): Wageningen, The Netherland, 2014.

71. Vaast, P.; Somarriba, E. Trade-offs between crop intensification and ecosystem services: The role of agroforestry in cocoa cultivation. *Agrofor. Syst.* **2014**, *88*, 947–956. [CrossRef]

72. Krott, M. *For. Policy Anal.*; Springer: Dordrecht, The Netherlands, 2005.

73. Krott, M.; Bader, A.; Schusser, C.; Devkota, R.; Maryudi, A.; Giessen, L.; Aurenhammer, H. Actor-centred power: The driving force in decentralised community-based forest governance. *For. Policy Econ.* **2014**, *49*, 34–42. [CrossRef]

74. Burns, S.L.; Krott, M.; Sayadyan, H.; Giessen, L. The World Bank Improving Environmental and Natural Resource Policies: Power, Deregulation, and Privatization in (Post-Soviet) Armenia. *World Dev.* **2017**, *92*, 215–224. [CrossRef]

75. Rahman, M.S.; Sarker, P.K.; Giessen, L. Power players in biodiversity policy: Insights from international and domestic forest biodiversity initiatives in Bangladesh from 1992 to 2013. *Land Use Policy* **2016**, *59*, 386–401. [CrossRef]

76. Sahide, M.A.K.; Nurrochmat, D.R.; Giessen, L. The regime complex for tropical rainforest transformation: Analysing the relevance of multiple global and regional land use regimes in Indonesia. *Land Use Policy* **2015**, *47*, 408–425. [CrossRef]

77. UNEP. *Côte d'Ivoire Post-Conflict Environmental Assessment*; United Nations Environment Programme (UNEP): Nairobi, Kenya, 2015; p. 169.

78. Ongolo, S.; Karsenty, A. The politics of forestland use in a cunning government: Lessons for contemporary forest governance reforms. *Int. For. Rev.* **2015**, *17*, 195–209. [CrossRef]

79. Ongolo, S. On the banality of forest governance fragmentation: Exploring 'gecko politics' as a bureaucratic behaviour in limited statehood. *For. Policy Econ.* **2015**, *53*, 12–30. [CrossRef]

80. Glin, L.C.; Oosterveer, P.; Mol, A.P. Governing the Organic Cocoa Network from Ghana: Towards Hybrid Governance Arrangements? *J. Agrar. Change* **2015**, *15*, 43–64. [CrossRef]

MDPI

St. Alban-Anlage 66

4052 Basel

Switzerland

Tel. +41 61 683 77 34

Fax +41 61 302 89 18

www.mdpi.com

Sustainability Editorial Office

E-mail: sustainability@mdpi.com

www.mdpi.com/journal/sustainability

www.ingramcontent.com/pod-product-compliance
Lightning Source LLC
Chambersburg PA
CBHW051726210326
41597CB00032B/5624